Advances in Spatial Science

Springer
Berlin
Heidelberg
New York
Barcelona
Budapest
Hong Kong
London
Milan
Paris
Singapore
Tokyo

Titles in the Series

C. S. Bertuglia,
M. M. Fischer and *G. Preto* (Eds.)
Technological Change,
Economic Development and Space
XVI, 354 pages. 1995. ISBN 3-540-59288-1

H. Coccossis and *P. Nijkamp* (Eds.)
Overcoming Isolation
VIII, 272 pages. 1995. ISBN 3-540-59423-X

L. Anselin and *R. J.G.M. Florax* (Eds.)
New Directions in Spatial
Econometrics
XIX, 420 pages. 1995. ISBN 3-540-60020-5

H. Eskelinen and *F. Snickars* (Eds.)
Competitive European Peripheries
VIII, 271 pages. 1995. ISBN 3-540-60211-9

J. C.J.M. van den Bergh, P. Nijkamp
and *P. Rietveld* (Eds.)
Recent Advances in Spatial
Equilibrium Modelling
VIII, 392 pages. 1996. ISBN 3-540-60708-0

P. Nijkamp, G. Pepping
and *D. Banister* (Eds.)
Telematics and Transport Behaviour
XII, 227 pages. 1996. ISBN 3-540-60919-9

D. F. Batten and
C. Karlsson (Eds.)
Infrastructure and the
Complexity of Economic Development
VIII, 298 pages. 1996. ISBN 3-540-61333-1

T. Puu
Mathematical Location and
Land Use Theory
IX, 294 pages. 1997. ISBN 3-540-61819-8

Y. Leung
Intelligent Spatial Decision
Support Systems
XV, 470 pages. 1997. ISBN 3-540-62518-6

C. S. Bertuglia, S. Lombardo
and *P. Nijkamp* (Eds.)
Innovative Behaviour in Space
and Time
X, 437 pages. 1997. ISBN 3-540-62542-9

A. Nagurney and *S. Siokos*
Financial Networks
XVI, 492 pages. 1997. ISBN 3-540-63116-X

M. M. Fischer and *A. Getis* (Eds.)
Recent Developments in Spatial Analysis
X, 434 pages. 1997. ISBN 3-540-63180-1

R.H.M. Emmerink
Information and Pricing
in Road Transportation
XVI, 294 pages. 1998. ISBN 3-540-64088-6

P. Rietveld and *F. Bruinsma*
Is Transport Infrastructure Effective?
XIV, 384 pages. 1998. ISBN 3-540-64542-x

P. McCann
The Economics of Industrial Location
XII, 228 pages. 1998. ISBN 3-540-64586-1

Lars Lundqvist · Lars-Göran Mattsson
Tschangho John Kim (Eds.)

Network Infrastructure and the Urban Environment

Advances in Spatial Systems Modelling

With 53 Figures
and 52 Tables

Springer

Professor Lars Lundqvist
Professor Lars-Göran Mattsson

Royal Institute of Technology
Department of Infrastructure and Planning
Division of Transport and Location Analysis
S-100 44 Stockholm, Sweden

Professor Tschangho John Kim
University of Illinois at Urbana-Champaign
Department of Urban and Regional Planning
111 Temple Buell Hall
611 E. Lorado Taft Dr.
Champaign, IL 61820, USA

ISBN-13: 978-3-642-72244-8 e-ISBN-13: 978-3-642-72242-4
DOI: 10.1007/978-3-642-72242-4

Cataloging-in-Data applied for
Die Deutsche Bibliothek - CIP-Einheitsaufnahme
Network infrastructure and the urban environment : advances inspatial systems modelling / Lars Lundqvist ... (ed.). -
Berlin; Heidelberg; New York; Barcelona; Budapest; Hong Kong; London; Milan; Paris; Singapore; Tokyo : Springer, 1998
(Advances in spatial science)

Softcover reprint of the hardcover 1st edition 1998

Hardcoverdesign: Erich Kirchner, Heidelberg
SPIN 10655475 42/2202-5 4 3 2 1 0 - Printed on acid-free paper

Preface

This volume is the result of an international collaboration, which started with a conference at Smådalarö Gård in Sweden. The workshop was supported by the National Science Foundation of the USA (INT-9215114) and by the Swedish National Road Administration, the Swedish Council for Building Research, the Swedish Transport and Communications Research Board and the Swedish Council for Planning and Coordination of Research. This support is gratefully acknowledged. The collaboration started as a bilateral U.S.-Swedish endeavour but was soon widened to other scholars in Europe, Asia, Australia and South-America.

Network Infrastructure and the Urban Environment is a policy area of growing importance. Sustainable cities and sustainable transport systems are necessary for attaining a sustainable development. The research and policy field, represented in this volume, comprises a number of challenging contrasts:

- the contrast between infrastructure investments, mobility and environmental sustainability;
- the contrast between policy contexts, modelling traditions and available decision support systems in various parts of the world;
- the contrast between available best practice methods and the majority of models applied in planning;
- the contrast between static models of cross-sectionary equilibria and dynamic models of disequilibrium adjustments; and
- the contrast between state-of-the-art operational land-use/transport models and new demands for land-use/transport/environment models due to changing policy contexts.

Bridging some of these gaps constitutes important research tasks, that are discussed in the twenty-two chapters of this book. A number of emerging research directions are identified in the introduction and summary chapter.

In addition to cross-refereeing by the authors of this volume, many colleagues have contributed to the improvement of its content by careful reviews of earlier versions of the chapters: Alex Anas, Roland Artle, Joseph Berechman, Erik Anders Eriksson, Stephen Fournier, Stewart Fotheringham, Marc Gaudry, Yoshitsugu Hayashi, James Hicks, Roger Mackett, Norbert Oppenheim, Michael Patriksson, Poulicos Prastacos, David Simmonds, Kenneth Small, Tony Smith and Roger Vickerman. Their efforts are gratefully acknowledged. David Batten was instrumental in organising the workshop and in the planning of this publication.

The editing work has been done within the Department of Infrastructure and Planning, Royal Institute of Technology, Stockholm. Sixteen of the chapters have been edited in Word or WordPerfect by Ingrid Johansson and six of the chapters have been edited in LaTeX by Anders Karlström. We thank both of them for efficient and conscientious work. Steve Scott-Robson has assisted in improving the language of non-English speaking authors. The work has been coordinated by Lars Lundqvist and Lars-Göran Mattsson, earlier within the Division of Regional Planning and currently within the recently established Division of Transport and Location Analysis.

Stockholm in April, 1998.

Lars Lundqvist Lars-Göran Mattsson Tschangho John Kim

Contents

Preface v

1 **Network Infrastructure and the Urban Environment: Introduction and Summary** 1
Lars Lundqvist, Lars-Göran Mattsson and Tschangho John Kim

Part I ISSUES

2 **Infrastructure and Economic Milieu: Swedish Contributions 1960–1995** 19
Börje Johansson

3 **The Changing Context of Transportation Modeling: Implications of the New Economy, Intermodalism and the Drive for Environmental Quality** 53
T.R. Lakshmanan

4 **Excess Commuting in U.S. Metropolitan Areas** 72
Edwin S. Mills

5 **Household Commuting: Implications of the Behavior of Two-Worker Households for Land-Use/Transportation Models** 84
Peter Gordon, Yu-chun Liao and Harry Richardson

Part II DYNAMICS AND EQUILIBRIA IN NETWORK MODELLING: NEW THEORETICAL AND METHODOLOGICAL DEVELOPMENTS

6 **Disequilibrium Network Design: A New Paradigm for Transportation Planning and Control** 99
Terry L. Friesz, Samir Shah and David Bernstein

7 **Infinite Dimensional Formulations of Some Dynamic Traffic Assignment Models** 112
David Bernstein and Terry L. Friesz

8 **Introduction to Projected Dynamical Systems for Traffic Network Equilibrium Problems** 125
Anna Nagurney and Ding Zhang

9 **Worker and Workplace Heterogeneity and Residential Location: A Historical Perspective on Stockholm** 157
Björn Hårsman and John M. Quigley

10 **Parameter Estimation for Combined Travel Choice Models** 177
David E. Boyce and Yu-Fang Zhang

11 **Discrete Spatial Price Equilibrium** 194
Sven Erlander and Jan T. Lundgren

12 **Integration of Freight Network and Computable General Equilibrium Models** 212
Terry L. Friesz, Zhong-Gui Suo and Lars Westin

13 **A Spatial Computable General Equilibrium Approach for Measuring Multiregional Impacts of Large Scale Transportation Projects** 224
Toshihiko Miyagi

Part III INTEGRATED ANALYSIS OF ACTIVITY LOCATION AND TRANSPORTATION IN URBAN AND REGIONAL SYSTEMS

14 **Applied Models of Urban Land Use, Transport and Environment: State of the Art and Future Developments** 245
Michael Wegener

15 **Results from Implementation of Integrated Transportation and Land Use Models in Metropolitan Regions** 268
Stephen H. Putman

16 **Improved Logit Formulations for Integrated Land Use, Transport and Environmental Models** 288
Tomás de la Barra

17 **Modelling Land-Use and Transport Interaction: Policy Analyses Using the IMREL Model** 308
Christer Anderstig and Lars-Göran Mattsson

18 **A Combined Model for Analysing Network Infrastructure and Land-Use/Transportation Interactions** 329
Lars Lundqvist

19 **Development of a Compact Urban Simulation Model** 344
John R. Roy, Leorey O. Marquez, Michael A.P. Taylor and Takayuki Ueda

20 **An Interactive Computer System for Land-Use Transport Analysis** 361
Geoffrey G. Roy and Folke Snickars

21 **A Combined Economic Activity and Transportation Model: A Solution Procedure and Application to Multi-Regional Planning** 381
Paul F. Hanley and Tschangho John Kim

22 **Road Infrastructure and Corridor Development** 395
Sytze A. Rienstra, Piet Rietveld, Maarten T.H. Hilferink and Frank R. Bruinsma

1 Network Infrastructure and the Urban Environment: Introduction and Summary

Lars Lundqvist
Lars-Göran Mattsson
Department of Infrastructure and Planning
Division of Transport and Location Analysis
Royal Institute of Technology
S-100 44 Stockholm, Sweden

Tschangho John Kim
Department of Urban and Regional Planning
University of Illinois at Urbana-Champaign
111 Temple Buell Hall
611 E. Lorado Taft Dr.
Champaign, IL 61820, USA

1.1 Challenging Issues

Transportation and communications networks together with the built environment constitute a major share of a society's infrastructure. The amount of investment in this area in any country, particularly in developed countries, is immense. The impact of infrastructure on economic efficiency, productivity, and competitiveness, as well as on people's welfare is also tremendous. But there is a negative side to this as well, especially to the expansion of transportation networks and the accompanying increased motorised mobility. Air emissions, the use of non-renewable resources, accidents and congestion are a few examples of issues that have been raised in the growing public debate. Sustainability, apart from the fact that it is not easily defined, has become a key concept in transport policy forming. A proper understanding of the role of infrastructure for welfare – including positive and negative externalities – and economic growth will be a decisive factor for a successful urban and regional policy that can balance different objectives in an appropriate way.

This very broad picture leads to a number of challenging research issues.

Even if it has been possible to connect infrastructure provision with economic growth, much remains to be done before the exact nature of this relationship has been fully understood. Improved transportation infrastructure will reduce travel times and transportation costs. But are these the only effects? Some approaches look at how infrastructure influences the development of a regional economic milieu and what factors can explain location decisions. What makes certain regions or cities more prosperous? Contact opportunities, interaction patterns, regional attributes, accessibility and agglomeration economies are important concepts in this context.

Reductions in travel time and transportation cost are typically evaluated by cost-benefit analysis. This requires, among other things, that all relevant effects of individuals and firms on transportation and location behaviour are incorporated in the underlying demand models. Both theory and data place restrictions on what can be included in a demand model. There is then a risk that certain effects are simply excluded by assumption. It is desirable to improve the methodology to cover impacts of infrastructure investments such as induced traffic, relocation and productivity changes in a more precise way.

Freight transportation models are less developed than passenger travel models. This can be related both to the lack of accurate databases and the inherently more complex decisions for freight transport. One promising avenue of research is to explore the possibilities of spatial computable general equilibrium models, ideally by integrating such models and network models.

The growing environmental awareness of the negative impacts of transport raises the demands upon the contents of the models. If it were necessary to reduce mobility, or rather its consequences, to what extent could this be accomplished by measures such as higher residential densities, adjustment of the urban form to public transport provision and various pricing and regulating schemes? A proper evaluation of such measures requires that the interaction between transport and land-use can be treated in a consistent way, which in turn requires suitable land-use/transport models.

One particular question is the way transport infrastructure should be priced. Pricing proposals have been based on different, and partly conflicting, principles, such as increasing the internal efficiency of the transport system (e.g. congestion pricing), raising necessary financial funds for investment (e.g. cordon tolls) or reducing environmental impacts by influencing individual travel decisions (e.g. petrol taxes). Transport planning practice has clearly indicated that a road pricing proposal must be carefully designed and launched, should it have any chance of gaining political acceptance.

The emerging Information Technology (IT) society will profoundly change the conditions for interaction and communication. This will raise a number of challenging questions concerning the design, assessment, optimisation and proper integration of different subsystems within the all-embracing transportation system. IT will also change the cost of communication in a broader sense. This will have impacts on the spatial structure of urban areas and on the interplay among regions. We still do not know to what extent IT communication and physical mobility are substitutes or complements. It can also be expected that IT will deeply affect the economic organisation within and among firms with location implications that are not well understood.

This brief exposition points to a potential conflict between economic growth stimulated by transport infrastructure investment and negative environmental impacts from the on-going increased mobility. One of the most challenging issues for urban policy making is to resolve this conflict, as far as possible, by finding a strategy for sustainable development.

The purpose of this book is to put together a number of important contributions concerning modelling network infrastructure and the urban environment, contributions that elaborate in depth most of these challenging issues. The spectrum ranges from discussions of pertinent issues and behavioural tendencies, overviews of model development, classification and assessment of available models, innovative model applications for policy analysis to theoretical advances in spatial systems modelling. This introductory chapter attempts to give a background and summary of the chapters to provide the readers with a better understanding of the links between the contributions.

1.2 Network Infrastructure

Infrastructure is a concept that despite its frequent use has no commonly accepted definition even if public availability of durable capital in time and space can be regarded as a basic dimension (for an overview, see Rietveld, 1989).

Johansson (Chap. 2) recognises physical infrastructure as properties of the built environment in a region and points out three characteristics: *temporal scale, spatial collectivity* and *function*. Temporal scale refers to the fact that infrastructure is a kind of capital that changes slowly, with respect both to its capacity and spatial distribution, in comparison with social and economic activities. By spatial collectivity is meant "the services from infrastructure are supplied to a collective of users, the spatial extension is limited, and the benefits reduce as accessibility declines." Function requires infrastructure to have at least one of the properties: "(i) polyvalence, (ii) generality over time, and (iii) a systems or network function that generates accessibility."

The main focus in *Johansson's* (Chap. 2) thorough exposition and survey is on the relationship between infrastructure and economic milieu. The basis is the observation that productivity and economic growth have been unevenly distributed over space at any level of spatial resolution and that these differences can be associated with variations in location attributes or the regional economic milieu. He identifies a particular Swedish research tradition, or an implicit and informal *research programme* to use the terminology of Lakatos, concentrated on (i) infrastructure, (ii) location attributes and (iii) accessibility. This research programme includes both theoretical and empirical studies by economists, geographers and applied mathematicians, and has given special emphasis to spatial networks. Local interaction conditions including face-to-face contacts are considered important for the economic performance of a region. These conditions are often expressed in form of accessibility indicators that are viewed as multidimensional. Transportation networks and travel conditions affect how the supply of attributes are spatially discounted in these indicators.

The analysis of *Rienstra et al.* (Chap. 22) fits well into the general framework of *Johansson*. The distinction is that they pay special attention to network infrastructure in a corridor perspective. For this to be interesting there needs to be a significant number of entry and exit points to the network within the corridor. For that reason such a perspective is mainly relevant for road infrastructure. They distinguish between short-term and long-term effects on the demand and supply side. Their focus is on programme effects of transportation infrastructure investment, i.e. long-term effects on the supply side concerning productivity and location of activities.

The environment in which transport infrastructure planning takes place has changed dramatically. *Lakshmanan* (Chap. 3) relates this transformation in a U.S. perspective to (i) a structural change in industrial production and market order, (ii) the public policy debate concerning the context and scope of transportation planning leading to increasing deregulation and expansion of market incentives in the transport sectors and a broadening of the planning objectives, and (iii) a change in institutional context as a consequence of the Intermodal Surface Transportation Efficiency Act of 1991 and the Clean Air Act as amended in 1990. The demand for the knowledge base, including the modelling tools, can be expected to increase as the transportation policy tool kit is being expanded. The integrated land-use and transportation models and the models of regional development that are presented and discussed in several of the contributions to this volume represent the state-of-the-art in this respect, and also the point of departure for many of the improved tools that can be expected to be operational in the near future.

As a result of major advances in sensing technology, and computers and communication technology, opportunities exist for the development of various intelligent transportation systems (ITS) such as advanced traveller information and guidance systems, advanced vehicle control and traffic management systems, automated vehicle debiting systems (which allow the introduction of sophisticated schemes of economic incentives), and intelligent transit systems. Such new technologies have a potential of improving the efficiency of the transportation system by smoothing traffic flow (reducing congestion), and effectively expanding the road capacity, possibly reducing the risk of accidents, and optimising the coordination of the public transport system. Research issues include dynamic route guidance technology and methods as well as real-time data collection and control. This will further increase the demand for new knowledge.

The theoretical contributions to this volume on dynamic network models and spatial equilibrium analysis are in many cases designed to address precisely this kind of issue.

Many network infrastructure planning problems, such as strategic investment decisions, can be formulated as network design problems. In the equilibrium case, the problem is to find an optimal network design, when the network flow pattern is constrained to be an equilibrium. *Friesz et al.* (Chap. 6) discuss the limitations with such an approach and the potential occurrence of a temporal version of Braess' paradox. To address these limitations they introduce a dynamic framework for urban networks, on which they base a disequilibrium network design paradigm. This new paradigm takes into account both the underlying time varying nature of the network and the disequilibrating effects that capacity enhancements to the network may produce.

Traffic assignment models are fundamental to the analysis of transportation network usage. Their practical importance is related to the role of congestion in road networks. The simple idea that is considered is that the route choices of the travellers both influence and are influenced by the prevailing level of congestion in the network. Those traffic assignment models that are used in practice for modelling actual behaviour in metropolitan networks are of the Wardropian user equilibrium type – all used routes that connect an origin and a destination have equal (and then minimum) travel time (cost) and no unused route has a lower travel time (cost). The behavioural interpretation is that if this holds for the whole network, then no single traveller can reduce the travel time (cost) by unilaterally changing to another route.

As is discussed in *Bernstein and Friesz* (Chap. 7) this framework involves a number of simplifications. Usually the models of congestion are very simple; the average travel time (cost) on a link during a period is an increasing function of the average number of vehicles using that link during the period. This means that the dynamic nature of congestion is ignored. Also other short-term adjustments of route and departure-time choices in response to, say, congestion, information, guidance and dynamic road pricing are ignored. If traffic assignment models really are to be useful for many of the ITS applications mentioned above, it will be necessary to incorporate the dynamic nature of traffic. This is the background to the new paradigm of dynamic traffic assignment models that is introduced by *Bernstein and Friesz.*

Dynamic traffic network problems are also in focus of the contribution by *Nagurney and Zhang* (Chap. 8). For more than a decade now, it has been well known that network equilibrium problems can be formulated as variational inequality problems. This theory does not by itself allow any study of the dynamics of the system, but a natural dynamic system, termed the *projected dynamical system*, can be associated with a variational inequality problem. Using this powerful idea, it is possible to develop new algorithms for the computation of not only traffic network equilibria, but also the dynamic behaviour of the travellers.

In comparison with passenger travelling the advances in understanding and modelling freight transportation have been more moderate. *Friesz et al.* (Chap. 12) discuss some reasons for this situation. In this volume, however, several contributions address theoretical and application-oriented issues related to freight transportation.

Erlander and Lundgren (Chap. 11) investigate the question of spatial price equilibrium within a discrete framework. This enables them to derive the equilibrium conditions in a new way, and, even more interesting, allows them to interpret in economic terms the objective of the optimisation problem, to which the equilibrium conditions are the solution.

To achieve consistency between the predictions of regional economic activities and prices on the one hand and the freight flows on the other it is necessary to integrate or link freight network models with spatial computable general equilibrium models. This is the topic of both *Friesz et al.* (Chap. 12) and *Miyagi* (Chap. 13). *Lundqvists* and *Hanley and Kim's* (Chap. 21) contributions can also be said to have similar ambitions.

1.3 The Urban Environment

Present urban and regional systems are complex. Spatial analysis models cannot produce useful recommendations for decision makers unless they are founded on an understanding of the continuously changing nature of structures and behaviour. What do we understand about the functions of modern urban areas that are distinctively different from rural areas? Some of the major urban characteristics may be summarised as follows:

- Land is used intensively in urban areas. The intensive use of land connotes a higher density of population and employment, which in turn causes land and transportation congestion. It also allows for specialisation and agglomerations economies.
- Both private and public decision makers affect shaping of urban functions, forms, and structures. Both private and public sectors produce goods and

services, however, the public sector imposes regulations on almost all private activities in modern cities through zoning, building codes, and other land use regulations and pricing.

- Inputs to urban industries include not only people, but also primary and intermediate goods, making urban economic activities complex and diverse. Transportation systems in urban areas need to provide services not only for the movement of people, but also for moving raw materials, intermediate goods, and finished products.
- Most urban areas have been built over decades, if not centuries. The current urban form and structure of a given urban area is the result of incremental additions, alterations, and demolition by both man and nature.

The increasing female labour force participation rate, the increasing fraction of multi-worker households, the declining birth-rate, the changing household size, the changing economic status of particular social and ethnic groups, and continuing decentralisation of economic activity from the traditional CBD are a few examples of changes in lifestyles that would have a profound impact on the future form, function, and structure of urban and regional systems. Undoubtedly, these changes, along with changes in demographics and social patterns, will affect locational behaviour and travel demand in urban areas.

Gordon et al. (Chap. 5) provide a number of empirical findings concerning household commuting in the U.S. In particular they study the differences between single-worker and multi-worker households. To understand location behaviour this distinction is very important and has to a large extent been neglected. Continuous urban economics models as well as integrated land-use and transport models based on discrete choice theory assume, almost exclusively, single-worker households. This distinction also has implications for the discussion about "wasteful" or "excess" commuting (see *Mills*, Chap. 4). If one accepts the over-simplification that proximity to work is the single location criterion for a household, commuting distances for a two-worker household will on the average be longer than for a single-worker household.

Another reason for "excess" commuting is the heterogeneity of education and experience of the labour force and of the job requirements. "Employers are far from indifferent among workers and workers are far from indifferent among jobs, employers or dwellings" to cite *Mills* (Chap. 4). Using Stockholm data for a 20-year period *Hårsman and Quigley* (Chap. 9) investigate trends in the spatial pattern (segregation) of employment and residential location and relate these trends to sex, education, occupation, and industry. The study illustrates the usefulness of sophisticated social indicators to evaluate the historical development in a policy area of primary interest for regional planning.

Mills (Chap. 4) criticises the way "excess" commuting has been operationalised in the literature and suggests that it should be considered from the angle of welfare economics. Excess commuting then occurs if the price of commuting is below its welfare optimum. In metropolitan areas unpriced congestion could be a reason for this. *Mills* argues that an increase in fuel taxes would be the easiest way to correct the price. According to his calculations for the U.S., the cost of commuting should be raised by about 25 percent. This would correspond to more than a tenfold increase in the fuel tax. Commuting is estimated to be reduced by about 10–15 percent, partly by relocation, or more, if the tax increase were accompanied by attempts to improve public transit.

The growth of new IT facilities is another factor that will transform the function, form, and structure of urban areas, and also have impacts on the travel patterns. Easy access to telecommunication facilities, including microcomputers, facsimile machines, and picture-phones are some examples. Some travel may be replaced by using one or more of those telecommunication means. People, who do not need to be at a particular working site because of equipment or access to face-to-face contacts, may choose a location based on personal criteria such as schools, amenities, access to green areas, low criminality, ethnicity, etc. In fact, two decades of rapid advances in information and computer technology have led some analysts to claim that mankind is now entering an IT society. This will also affect the location opportunities for production activities. With the growing use of telecommunication systems, it is becoming possible for activities such as manufacturing, distribution, management, accounting, and inventory control to be located at different places and yet operated efficiently as a single enterprise.

As we already have touched upon, the scope and policy context of infrastructure planning have changed (*Lakshmanan*, Chap. 3). Strategic decision making in the future will go far beyond being simply a question of capacity expansion. It will be necessary to adopt a systems view, in which claims on mobility, economic efficiency, natural resource use, safety, and environmental quality have to be balanced. The arsenal of modelling tools has to be enlarged to cope with demand for spatially detailed environmental assessment data, to be capable of estimating the likely behavioural responses to transport control measures (economic incentives, regulations, and physical changes in the networks), to transport demand management measures (intended to affect e.g. departure time, destination choice, mode of travel), to social and structural changes in the emerging IT society (changing working conditions, household structure, lifestyles, and activity and location patterns), and to the introduction of new ITS facilities.

One of the strategic long-term issues is the interaction between transport and land-use, to which many of the contributions to this volume are specially devoted. Many operational models and real-life applications were presented and discussed in the anthology edited by Hayashi and Roy (1996). Recently Wilson (1998), one of the leading scholars in the area, provided an excellent review of the historical development and future prospects of land-use/transport models. A few years ago a special issue of the *Journal of the American Planning Association* was dedicated to the same theme. That issue included also the much-cited article by Wegener (1994), in which he surveys and classifies operational urban models. In the present book *Wegener* (Chap. 14) follows up that article by discussing how urban land-use/transport models might be extended to urban land-use/transport/environment models, and to what extent this has already been achieved.

It is interesting to note that the growing awareness of the environmental problems associated with transport has actually contributed to the renewed interest in these models. This is manifested in the new U.S. legislation (*Lakshmanan*, Chap. 3), which specifically requires that transport planning must consider the interaction between transport and land-use. In Europe there are intense discussions about sustainable development, including the conditions for a sustainable transportation system. Considerable research efforts can be expected to be devoted to this issue in future. This will also raise the interest in environmental impacts analysis of land-use and transport policies using modern advanced methods.

Wegener (Chap. 14) reinforces the previous argument that the new policy context will require more versatile models and also new generations of models such as activity-based travel demand models. The most challenging task in the future will be to link or integrate models of the urban environment to existing or renewed land-use/transport models to establish a full-fledged land-use/transport/ environment model. The requirements on dynamics and spatial resolution may be different and more demanding. Models of air dispersion, noise propagation, surface and ground water flows, and microclimate analysis typically need highly disaggregate spatial data. Geographic information systems (GIS) may have a great potential here, not only for handling and visualising large amounts of geographic and environmental data, but also for the possibilities of building activity-based microsimulation models.

Recent extensions and innovative applications of several of the operational models that *Wegener* reviews are treated in the remaining contributions to this volume. Some of them like *Boyce and Zhang* (Chap. 10) and *de la Barra* (Chap. 16) elaborate on methodological issues such as model estimation techniques and the design of suitable evaluation indicators of land-use and transport policies. Others report on their experience on integrating land-use and network equilibrium-based travel demand models either through iteration (*Putman*, Chap. 15; *Anderstig and Mattsson,* Chap. 17; *Roy et al.*, Chap. 19; *Hanley and Kim*, Chap. 21) or by building an integrated framework (*Lundqvist*, Chap. 18). They also discuss actual or potential model applications concerning environmental impacts of transport and land-use policies.

One reason why advanced urban models have not been used more widely, is the demand they place on practising planners for an understanding of their complex anatomy. The lack of transparency, or black box syndrome, associated with many models was discussed by Lee (1973) in his seminal critique of large-scale models. One way of counteracting such tendencies, and in the long run to lay the table for more advanced models, is to provide the users with tools that are easy to set up, understand and use. This is the aim behind *Roy and Snickars'* (Chap. 20) development of the MacNetwork package for transport and land-use analysis.

1.4 Advances in Spatial Systems Modelling

The core of this volume is devoted to advances in spatial systems modelling. Innovative approaches to dynamic modelling of infrastructure networks are reported in Section 1.4.1. Theoretical and empirical analyses of equilibria in urban and regional systems are summarised in section 1.4.2. State-of-the-art models of urban land-use/transport interaction are discussed in Section 1.4.3 and some examples of operational tools for urban analysis are presented. Finally, two examples of efforts to analyse impacts of infrastructure on regional development are provided in Section 1.4.4.

1.4.1 Dynamic Network Modelling

Network modelling is an important field in its own right. In analyses of infrastructure and the urban environment, network properties are fundamental. This is obvious in the case of transportation and other technical infrastructure. Also social and economic interactions are often discussed in network terms. Information networks are rapidly changing the conditions for production, transportation (goods, persons) and the development of urban and regional systems. While most network modelling has relied on equilibrium assumptions, the need for studies of the dynamics of traffic networks is becoming more and more apparent stimulated by the introduction of various ITS facilities.

Three contributions to this volume report on advances in dynamic modelling of transportation networks. *Friesz et al.* (Chap. 6) treat the network design problem with traffic flows being generated by dynamic disequilibrium models. As a starting-point a tatonnement model is formulated to describe the disequilibrium dynamics of highway passenger traffic. This process relates the change of path flows and perceived origin-destination costs to excess travel costs and excess travel demands. It parallels the tatonnement adjustments of microeconomic theory. The disequilibrium trajectories of path flows and costs tend to be entirely within the feasible region (i.e. in the simple case of only non-negativity constraints on path flows and costs) and this has spurred the term *global projective dynamics*. *Friesz et al.* refer to a discussion of *pros and cons* of global projective dynamics as compared to *local projective dynamics* (see *Nagurney and Zhang*, Chap. 8). The dynamic adjustment models focus on adjustments of cost perceptions reflecting day-to-day variations that are slow with respect to the within-day variations of route and departure time choices (see also *Bernstein and Friesz*, Chap. 7). A dynamic adjustment model is then embedded in a network design model. This problem is formulated in terms of continuous capacity enhancements and in terms of an objective reflecting maximisation of net benefits. A budget constraint for capacity enhancements is added. The resulting optimal control problem is intrinsically difficult due to non-smooth equations of state dynamics. By discretisation the optimal control problem may be transformed into a finite dimensional mathematical programme.

A comprehensive introduction into the field of projected dynamic systems for traffic network equilibrium problems is presented by *Nagurney and Zhang* (Chap. 8). Starting from static network equilibrium problems formulated as finite-dimensional variational inequalities, there is a naturally associated dynamic system (*a projected dynamical system*), the stationary points of which correspond precisely to the solutions to the variational inequality problem. *Nagurney and Zhang* synthesise recent advances in research on formulation, stability analysis and computation of dynamic traffic network models with both elastic and fixed travel demands. Existence and uniqueness of the solution to the projected dynamic system can be proved under certain conditions. The stability analysis deals with questions like: Do behavioural adjustments lead to an equilibrium? Will the dynamic system stay close to an equilibrium? Does the system return to equilibrium after a perturbation? Conditions are stated for stability theorems in general and examples are worked out in detail for elastic travel demand models with known travel disutility functions or known travel demand functions as well as for the case of fixed travel demand. The stability results are also crucial for establishing convergence of discrete time algorithms for computation of equilibria and adjustment trajectories.

Also *Bernstein and Friesz* (Chap. 7) develop dynamic network models focusing on the traffic assignment problem. Early application oriented dynamic traffic assignment models had the most difficulty in the formulation of the physical model of traffic flow dynamics. Time-based simulation and models based on systems of differential equations have been suggested for improving the treatment of traffic flow dynamics. *Bernstein and Friesz* suggest a third approach, motivated by event-based simulation, which is formulated as an infinite dimensional fixed-point problem for determining travel times in the case of first-in-first-out travel delay functions. This formulation is then applied to dynamic traffic assignment problems based on departure-times or departure-rates as well as to a simultaneous route and departure-time choice equilibrium model. Finally, disequilibrium adjustments of route choices and travel time perceptions are discussed on the basis of the dynamic traffic assignment models developed. The discussion is closely related to *Friesz et al.* (Chap. 6).

1.4.2 Spatial Equilibrium Models

In spite of the clear need for dynamic models of network infrastructure and the urban environment, there is still scope for new and elaborated approaches to spatial equilibria. Development of evaluation criteria in land-use analysis and estimation methods for combined traffic network equilibrium models are two examples of research of great interest for applications in urban planning. Spatial price equilibrium models and spatial computable general equilibrium models constitute elaborate tools for advanced analysis of regional systems. The research area is very active and the potential for development of operational models based on these approaches, integrating analyses of spatial markets and models of network infrastructure usage, looks promising.

Hårsman and Quigley (Chap. 9) analyse the spatial employment and residential location pattern by industry and occupation in Stockholm between 1970 and 1990. In a similar way, the spatial distribution of residential location by industry, occupation and education is studied during the same period. For the year 1990 a cross-sectional comparison of the joint residential and workplace distribution of workers could be carried out. The aim is to investigate the common, but often implicit, assumption that the diversity of jobs and human capital does not influence commuting patterns (see *Mills* (Chap. 4) and *Gordon et al.* (Chap. 5) for related studies). Appropriately decomposed entropy measures are used as indicators of spatial segregation. As for workplace composition, segregation has declined in terms of occupation but increased in terms of industry and education. The workplace segregation by sex is low. The corresponding results for residential composition show declining segregation by occupation and industry and increasing segregation by education. There is almost no residential segregation by sex. Similar results are obtained when the distribution of categories over locations are compared. Both workplaces and residences have decentralised (workplaces at higher speed) but still residences are more dispersed. The overwhelming impression is that the urban processes have led to less segregation in most respects except in terms of education. It is also the high education category that exhibits the most pronounced relative concentration of commuting patterns in 1990.

Combined network equilibrium models for destination, mode and route choices in the urban commuting context are discussed by *Boyce and Zhang* (Chap. 10). In particular their contribution addresses the estimation of parameters in situations when origin-destination travel times of car traffic are not observed or are considered unreliable. A second aim is to estimate and evaluate different formulations of the nested logit models of travel demand. The estimation problem has the character of a bilevel optimisation problem. A heuristic solution procedure is developed and applied to Chicago data. The estimations tend to favour the simultaneous model of destination and mode choices and the estimation procedure has appeared to be operational. A plea is made for introducing state-of-the-art research methods into transportation planning practice by providing information, training programmes, software and data sets for best practice techniques.

The classical spatial price equilibrium problem deals with shipment patterns between price sensitive supply and demand on a spatial market. *Erlander and Lundgren* (Chap. 11) consider a discrete version of the spatial price equilibrium problem, where unit shipments are assumed. The equivalent optimisation problem for the discrete version can be shown to have a behavioural interpretation in terms of the simultaneous cost minimising behaviour of each shipper. Furthermore, by using the efficiency principle of spatial interaction, it can be proved that the most probable shipment patterns are discrete spatial price equilibria. Aggregating unit shipments of the homogeneous commodity, the probability of trade patterns can be stated. Finding the most probable trade pattern is a very difficult optimisation problem. By relaxing the integer conditions on shipments, a continuous approximation of the discrete spatial price equilibrium can be obtained. The standard spatial price equilibrium problem emerges. In the continuous case, the most probable trade pattern can be computed as a continuous dispersed spatial price equilibrium.

Friesz et al. (Chap. 12) discuss a generalised spatial price equilibrium problem, where demand and supply functions are replaced by explicit modelling of production and consumption characteristics. A spatial computable general equilibrium model (CGE) is formulated as a nonlinear complementarity problem and is then integrated with a Wardropian user equilibrium of transportation flows. The computational difficulties of such integrated models are large indeed, and even more so for more general and realistic model formulations. One such complicating factor is the possibility of imperfect competition between carriers and in terms of spatial arbitrage. Some computational approaches are briefly outlined and, finally, the development of a computable general *disequilibrium* model is envisaged (cf. Chaps 6-8).

The contribution by *Miyagi* (Chap. 13) discusses a spatial computable general equilibrium model for studying multiregional impacts of network infrastructure. The formulation of the regional economic equilibrium model is reported in great detail. It is based on the assumption that transportation costs (from a separate transportation submodel) are known to actors. The behaviour of firms, traders and households is analysed and procedures for calibration of parameters and for solving the model are outlined. The transportation submodel is used to analyse the impacts of the demand decisions (shippers, households) from the regional economic equilibrium model on mode and route choices in the network. It can be viewed as a fast process of adaptation to the demand from the economic model, while the impact of transportation costs on the economic equilibrium is delayed (cf. the delayed structure of land-use/transportation interactions in e.g. the SUSTAIN model system, see *Roy et al.* (Chap. 19)). For the case of mutual and simultaneous demand-supply interactions, integrated approaches of the type discussed by *Friesz et al.* (Chap. 12) need to be developed.

1.4.3 Urban Land-Use/Transportation Modelling

The third part of this volume reports on the development and use of operational urban models and on regional development analysis. Seven contributions treat state-of-the-art approaches to urban land-use/transportation modelling. They cover the whole range from a comprehensive survey, via examples of applied urban modelling approaches to compact computer systems for analysis and interactive design.

Wegener (Chap. 14) provides an up-to-date overview of applied urban models with special emphasis on their potential for modelling interactions between urban development and the environment. The public interest in comprehensive policy packages for meeting environmental threats has led to a new demand for land-use/transportation models. *Wegener* considers interactions between nine urban subsystems and evaluates fourteen contemporary operational models in terms of the following criteria: comprehensiveness, overall structure, theoretical foundations, modelling techniques, dynamics, data requirements, calibration and validation, and operationality and applicability. The aim for sustainable development means new challenges for urban modellers to take interactions with the urban environment into account. A table of cross-impacts between urban land-use/transport and environmental resources, emissions and immissions is taken as the basis for a thorough discussion. New subsystems with different dynamics and spatial resolution need to be incorporated. A mini-survey of 24 models showed that most urban models far from deserve the name land-use/transport/environment (LTE) models. Whereas most models are capable of computing at least some environmental impacts, the majority still lack a feed-back from environmental conditions to the behaviour of urban actors. Land consumption, energy consumption and air pollution are treated in more than half of the models.

Network equilibrium modelling is a fundamental ingredient of any system for integrated analysis of activity location and transportation. Another basic tradition is spatial interaction theory based on micro-economic principles, exemplified by random utility theory, or on macro-oriented approaches, such as entropy maximisation. Theoretical elaborations have contributed to consistency and rigour in models that earlier have been formulated on a rather ad hoc basis. Random utility theory, for example, has prompted a reinterpretation of certain models in a micro-economic framework admitting a consistent formulation of benefit measures. *de la Barra* (Chap. 16) discusses some new developments of random utility theory in order to cope with problems related to different scales on the one hand and the possibility of negative composite costs on the other hand. The need for such developments is related to applications in hierarchical and aggregate land-use/transport/environment models and *de la Barra* has incorporated these improvements in the TRANUS model system.

Three contributions report on land-use/transportation models with applications to U.S. and European cities. *Putman* (Chap. 15) reviews recent results from implementing the land-use models DRAM and EMPAL together with transport modelling packages (TRANPLAN, EMME/2) in two metropolitan regions. Based on forecasts of regional totals and the base year spatial pattern, EMPAL projects employment location by type and zone and DRAM subsequently projects household location and land use by type and zone. Travel demand models and traffic assignment models produce link flows in the transportation networks, which underlie a new employment location by EMPAL, etc. *Putman* presents developments of the submodels and of the linking of land-use and transport submodels. The applications show smooth and quick convergence. Linking land-use and transport models produces results that are different from unlinked models. The possibility of calibration based on endogenously generated travel costs is mentioned as a future option (cf. *Boyce and Zhang*, Chap. 10).

Anderstig and Mattsson (Chap. 17) describe the structure of the IMREL model of residential and employment location, which is linked to the EMME/2 model of traffic assignment. The housing submodel is a normative surplus maximising model, taking local density externalities and travel demand patterns into account. The employment location model is a positive model replicating the location behaviour of employers. The model has been extensively used in Stockholm. Two strategic planning applications are reported. A recent development of a model system for the Öresund region is also outlined. Both the model systems of *Putman* and *Anderstig and Mattsson* have been applied in several urban regions. Their common feature is that they do not require excessive amounts of data and they rely on prevailing commercial transportation software packages. There are differences in model structure, in estimation techniques and in the way of linking submodels, however.

A simultaneous treatment of activity location and transportation flows is the main characteristic of the TRANSACT model described by *Lundqvist* (Chap. 18). The model can be seen as an extension of combined network equilibrium models (see *Boyce and Zhang*, Chap. 10) to also include land-uses of production activities and households. It can be related to the modelling of input-output interactions in regional systems. Building on many prominent theoretical traditions, the model determines activity volumes and densities by zone and commodity, export volumes by zone and commodity, and interzonal shipments by commodity and mode in such a way that the sum of the (user equilibrium) equivalent transport cost and opportunity costs of land and capital are minimised. The main constraints apply to export requirements, input-output balances by zone, and land availability. The model has been applied to Chicago (Kim, 1989) and preparations have been done for an application to Stockholm. The integrated nature of the model is similar to the ambition of *Friesz et al.* (Chap.12). TRANSACT can be seen as an urban version of Leontief type computable general equilibrium models integrated with a transport network equilibrium model.

The two contributions by *Roy et al.* (Chap. 19) and *Roy and Snickars* (Chap. 20) both present interactive computer systems for experimental use in education and planning contexts. *Roy et al.* (Chap. 19) develop a Lowry-type employment and housing location model combined with a doubly constrained gravity model of commuting trips, a logit model for modal split and a user equilibrium traffic assignment model. Housing location is modelled by a rent bidding utility maximising procedure. The location submodels are based on transportation costs from the previous period. The geometry is radially symmetric in order to simplify the analysis. This opens for some endogenous network design capabilities. The model is linked to the MAPINFO geographical information system.

MacNetwork described by *Roy and Snickars* (Chap. 20) is a desktop computer package intended for the modelling and analysis of urban and regional systems. It is considered to be a compromise between a GIS, a spread-sheet package and a comprehensive land-use/transportation model. MacNetwork includes editing facilities for networks and development zones. Import and export of data in spreadsheet form are supported. The transportation analysis options include interactive shortest paths and contour mapping and the land-use analysis includes interactive computation and mapping of accessibilities.

1.4.4 Regional Development Analysis

The last two chapters of the volume deal with infrastructure and regional development issues on a supra-regional or national scale. *Hanley and Kim* (Chap. 21) have similar ambitions as the two preceding chapters: developing a user friendly model system for desktop computers based on commercially available software. A cost minimising interregional input-output model is linked to a user equilibrium traffic assignment model. The computer system combines the spread sheet programme LOTUS, the mathematical programming package GAMS and the transportation package TRANPLAN. An application to Jawa, Indonesia, is reported.

The impact of network infrastructure on transport corridors has attracted increasing attention. Availability of data with high spatial detail has increased and made such studies feasible. *Rienstra et al.* (Chap. 22) discuss impacts of highway development on corridors in the Netherlands using data based on statistical regions. An overview of relations between transport infrastructure and spatial development is presented with special emphasis on transport infrastructure as a production factor and as a prerequisite for trade. Three approaches were used to study the impact of highway investment on regional employment: a study of accessibility and economic growth, a reference region approach and a labour market approach. The result is that there is no evidence that the construction of main road infrastructure has had a clear impact on overall employment in the regions of the corridor. Productivity and transport cost effects were not dealt with in this analysis and the impact of congestion on accessibility was not modelled either (cf. *Hanley and Kim*, Chap. 21).

1.5 Emerging Research Directions

The contributions to this volume provide an up-to-date summary of research issues, innovative modelling approaches and advances in development and application of urban and regional models. From this rich menu a number of interesting and interrelated research directions emerge:

- *Land-use/Transport/Environment integration.* The importance of a thorough modelling of relations between land-use, mobility and environmental systems in response to changes in the urban context is well documented (see Chaps 3 and 14).
- *Dynamic modelling.* Operational models are mainly of the equilibrium type, sometimes with possibilities for treating the development over time in a recursive manner. Many factors including new ITS options raise the need for introducing a truly dynamic analysis. Chaps 6-8 show clearly that this is a feasible task.
- *Equilibrium modelling.* Operational urban and regional models mainly handle interactions between activity location and transportation by linking separate submodels. Chaps 11-13 and 18 provide examples of how spatial equilibria might be computed based on a true integration of transportation market models with models of economic activities.
- *Network design and market behaviour.* Chaps 6 and 19 touch upon the design of network infrastructure in cases of equilibrium or disequilibrium transportation markets. Such problems can be seen as bilevel optimisation formulations, which are notoriously difficult to handle.
- *Spatial models and GIS.* The increasing availability of disaggregate spatial data and GIS in planning will profoundly affect the urban modelling context. The integration of land-use/transport/environment models in GIS is a viable research programme in its own right.
- *Complexity of transport patterns.* Many models presented in this book cover both person transport and freight. The complexity of modern logistics in freight traffic and household trip patterns (including trip chaining) and extended policy instruments require refined or renewed transport models. Activity based approaches and microsimulation techniques are examples of such developments, which also may provide alternatives to the rationality assumption usually made.
- *Model validation.* This important issue is discussed in many chapters (e.g. Chaps. 14 and 17). Validation of models of long-term spatial processes (sometimes with normative components) is a difficult undertaking that should be paid more attention. Good before-and-after data related to major network infrastructure decisions may assist such developments.
- *Comparative spatial modelling.* Applying many models to one spatial setting or applying one model to many spatial settings was the main idea of the ISGLUTI study (Webster et al., 1988). Such work should be continued with new generations of models in order to gain insights in the mechanisms of the models and their policy relevance.

- *Policy packages and policy evaluation.* In the current planning context, policy packages employing many simultaneous instruments are developed in order to meet multidimensional objectives of e.g. sustainable development. This will increase the required richness of spatial models, both in terms of the range of instruments and in terms of the range of criteria for policy evaluation.
- *Infrastructure and sustainable development.* A variety of sophisticated models are available for analysing how infrastructure affects economic growth and for studying the relationship between transportation, land-use and the environment. This knowledge should be used in studies on sustainable transportation systems and in designing and evaluating scenarios for sustainable development.

References

Dupuis, P. and Nagurney, A. (1993), 'Dynamical systems and variational inequalities', *Annals of Operations Research*, vol. 44, pp. 9–42.

Hayashi, Y. and Roy, J. (eds) (1996), *Transport, Land Use and the Environment*, Kluwer Academic Publishers, Dordrecht.

Kim, T. J. (1989), *Integrated Urban Systems Modeling: Theory and Applications*, Kluwer Academic Publishers, Boston, MA.

Lee, D.B. (1973), 'Requiem for large-scale models', *Journal of the American Institute of Planners*, vol. 39, pp. 163–178.

Rietveld, P. (1989), 'Infrastructure and regional development', *The Annals of Regional Science*, vol. 3, pp. 255–274.

Webster, F.V., Bly, P.H. and Paulley, N.J. (eds) (1988), *Urban Land-Use and Transport Interaction: Policies and Models,* Report of the International Study Group on Land-Use/Transport Interaction (ISGLUTI*)*, Avebury, Aldershot.

Wegener, M. (1994), 'Operational urban models: State of the art', *Journal of the American Planning Association*, vol. 60, pp. 17–29.

Wilson, A.G. (1998), 'Land-use/transport interaction models: Past and future', *Journal of Transport Economics and Policy*, vol. 32, pp. 3–26.

Part I

Issues

2 Infrastructure and Economic Milieu: Swedish Contributions 1960–1995

Börje Johansson
Jönköping International Business School
P.O. Box 1026, S-551 11 Jönköping, Sweden, and
Department of Infrastructure and Planning
Royal Institute of Technology
S-100 44 Stockholm, Sweden

2.1 The Self-Organised Swedish ILA-Program

Production and economic growth have at every point in time an uneven distribution across countries, regions, and zones within regions. Such differentials can be interpreted as a consequence of attributes of the economic milieu of each region. We shall call them location attributes. For each type of economic activities one can identify certain constellations of attributes that support these activities better than other location attributes. Some milieu properties are gifts of nature, others are created by means of investments in physical and human capital with a fixed or semi-fixed location. This chapter intends to present Swedish contributions to this field of research during the period 1960–1995. Much of this work has not been published in English and is remained unknown outside Sweden. Our aim is to outline the profile of a rather coherent series of research efforts.

2.1.1 Introduction

A regional economic milieu can be described by its location attributes. Standard economic theory has devoted little interest to regional differences as regards location attributes, productivity and growth. Consequently, research with this focus has been forced to create its own platform, and to establish its own conventions which specify relevant and challenging research problems. Using the terminology of Lakatos (1968; 1970), such a cohesive effort can be interpreted as a research programme.

This essay claims that one can identify a self-organised research programme in Sweden, based on the work of economists, geographers and other regional scientists who have emphasised the above phenomena during a period extending from the early 1950s to the early 1990s. A main focus has been how regional differences in economic performance and in the distribution of economic activities can be related to spatial differentiation of resources with a fixed location (trapped resources), and accessibility to such resources. In this presentation the implicit research programme is called ILA with reference to (i) infrastructure, (ii) location attributes, and (iii) accessibility.

Internationally, similar types of aims were initiated by Walter Isard and promoted by the International Regional Science Association. However, this international regional science movement covers a broader scope than the ILA-programme. Moreover, in Sweden the ILA-programme has, to a certain extent, remained a rivalling effort vis-à-vis mainstream economic thought. By emphasising spatial networks, it has also promoted models with non-linearities and non-convexities.

The following presentation describes Swedish theoretical and empirical studies within the indicated tradition. In addition, one may identify a series of associated planning and forecasting models which reflect basic assumptions of the same tradition (e.g. Lundqvist, 1978; 1981; Snickars and Granholm, 1981). However, the presentation does not examine such contributions in any detail. Instead it tries to assess pertinent empirical studies as well as associated theoretical models adhering to the ILA-programme. These studies focus on how location attributes are formed and how they influence the economic activity pattern and productivity in space. The group of Swedish scholars within this field include Andersson, Anderstig, Holmberg, Hårsman, Karlqvist, Lundqvist, Mattsson, Snickars, Strömqvist, Törnqvist, Weibull, Westin, Wigren, Åberg and many others. The reference list provides an overview of their contributions.

2.1.2 Economic Milieu and its Component Parts

Rietveld (1989) provides an overview of approaches to model the relation between infrastructure, economic milieu and regional development. Swedish overviews that present discussions of how to define and include infrastructure in economic models are found in Forslund and Karlsson (1991; 1992), Karlsson (1994), and Batten and Karlsson (1996). Transportation infrastructure is discussed in Lundqvist (1991).

The following presentation concentrates on the economic milieu of a functional region and examines how a region's infrastructure influences the development of the regional milieu. Formally, the economic milieu comprises those properties (location attributes) of a region which are durable, and (i) which the individual firm cannot control, (ii) for which there are no market prices and no direct charges, (iii) which influence the firm's input deliveries, production activity, distribution and sales activities, management and innovation activities. In order to be recognisable, the location attributes have to vary across regions.

A region with a clear specialisation has a quite different economic milieu as compared with a region that has a diversified economy. Strong specialisation may rely on the availability of a particular natural resource, on economies of scale or on localisation economies. In all these cases the economic milieu and its infrastructure will be less general and less multipurpose in nature than in the case of a diversified economy. Localisation economies can develop when a cluster of similar activities are located together. According to Marshall (1920), localisation economies is a phenomenon based on (i) a pooled market for workers with specialised skills, (ii) provision of non traded inputs of collective nature, and (iii) spillovers of market and technological knowledge which can spread more easily and accurately in a local environment.

The milieu of a diversified regional economy is characterised by agglomeration economies which provides (i) a creative milieu (Andersson, 1985), (ii) a diversified supply of various producers services, (iii) a differentiated supply of employment categories with a higher than average knowledge intensity, (iv) information flows about products, production techniques, suppliers, and customers.

Johansson and Snickars (1992) recognise physical infrastructure as properties of the built environment in a region. Built environment is durable capital that embodies and hosts location attributes. This environment comprises (i) networks for flow of people, materials (goods and residuals) and messages, (ii) plants and systems for provisions of public services, and (iii) housing infrastructure. Built environment may have infrastructure properties by generating services which influence the regional economic milieu. Table 2.1 specifies three fundamental properties which together imply that built environment becomes infrastructure.

Table 2.1. The three characteristics of physical or material infrastructure

TEMPORAL SCALE	Infrastructure is physical capital that adjusts at a slow pace both as regards capacity and geographical pattern. The speed of change is slow in comparison with the fast adjustments of variables describing social and economic activities. In the short-term it is an arena for fast processes.
SPATIAL COLLECTIVITY	The services from infrastructure are supplied to a collective of users, the spatial extension is limited, and the benefits reduce as accessibility declines.
FUNCTION	Infrastructure is characterised by satisfying at least one of the following criteria: (i) polyvalence, (ii) generality over time, and (iii) a systems or network function that generates accessibility.

Source: Adapted from Johansson and Snickars (1992).

The economic milieu and its location attributes are strongly dependent on infrastructure characteristics as defined in Table 2.1. Scholars like Andersson (1985) also identify non-material infrastructure consisting of collective, durable and spatially fixed location attributes, e.g. "permanently" located knowledge-intensive labour and other R&D-related resources. The slow time scale is essential. The durability and generality over time of location attributes imply that the allocation of productive resources has sufficient time to adjust to given spatial differentials. Although Table 2.1 was formulated in the 1990s, it summarises some basic assumptions that implicitly or explicitly can be found among scholars from the very beginning of the research tradition examined here. This tradition is signified by the acronym ILA-programme or ILA-group; it should be remembered that it represents an informal network of interacting individuals.

The main reference to Lakatos in this essay is his concept of a methodology of scientific research programmes (Lakatos, 1970). Such a programme informs its participants about what are, and what are not important research problems. It also contains a hard core of beliefs which should not be questioned or tested within the programme (de Marchi, 1991). In particular we shall emphasise the following two aspects of a Lakatos programme as important for the ILA-group (de Marchi, 1991; Lakatos, 1978):

> A programme is characterised by its evolution of heuristics, or perhaps a combined development of heuristics and a hard core; the hard core hardens only slowly by trial and error: "... it is the progressing problematic frontiers of knowledge, and not its relatively solid core [unproblematic background knowledge], which gives science its scientific character".
>
> There is a strong relation between heuristics as the techniques of problem-solving and program-related guidelines for modelling. Reformulating Lakatos' message we may emphasise that progress is rather related to the mathematical and econometric novelty of problem-solving techniques. These may lead to and form a part of a progressive programme.

Given the above characterisation we may observe that many Swedish ILA-scholars have their origin in disciplines like theoretical and economic geography, applied statistical analysis and demography, regional economics and economics related to applied mathematics, and operations research and optimisation theory. Can we find jointly shared theoretical approaches among these scholars? In their models we find that systems for interaction are structured in networks – both regarding interaction within urban regions and interregional interdependencies. In various forms they have also elaborated measures of time distance and friction for links in a network. Moreover, each accessibility measure is related to a specific resource or attraction phenomenon which is located in a node. In this way alternative and problem-oriented measures have been developed. In addition, infrastructure is frequently referred to as accessibility networks.

2.1.3 Elements of a Programme

The research efforts that are associated with the ILA-programme in this presentation are centred around the concept of interaction. To a large extent market behaviour is modelled with explicit recognition of the interaction between actors or groups of actors. In this spirit the economic milieu of a region is analysed with regard to the interaction conditions it offers. This focus also implies a concentration on variables such as accessibility, contact frequency, interaction patterns and contact opportunities. These in turn are related to infrastructure via transportation, communication and density. One may trace the central role played by interaction to the influence of Hägerstrand and his "time geography" based on network descriptions of space and interaction opportunities (Hägerstrand, 1974).

Another common feature to the implicit programme is that the models depict a discrete geographical space, using network formulation both for intersectoral and interregional flows and relations. Discrete space with nodes and zones may again be attributed to the Hägerstrand tradition. However, equally important is the regional comparative advantage analysis adhering to the Heckscher-Ohlin approach, which represents a second Swedish influence. In this context regional location attributes have a significant analytical role. In Sweden, this tradition is contrasted by a parallel approach, with continuous-space models by Palander (1935), Puu (1982; 1984) and Zhang (1988).

In addition, ILA-models stressing interaction often attempt to model market solutions as well as combined market and planning solutions as equilibria with constrained market clearing. This can for example be observed for a series of intraurban and interregional allocation models, both with a static and dynamic framework (e.g. Andersson and Plaut, 1977; Karlqvist and Snickars, 1977; Gustavsson et al., 1977; Hårsman and Marksjö, 1977). In this context one can also observe the use of optimisation models to describe market phenomena, in particular entropy maximisation models referring to a minimum information theory (Snickars and Weibull, 1977).

The theoretical background referred to above gives priority to accessibility. Following theoretical assessments by Karlqvist (1975) and Weibull (1976; 1980), accessibility has frequently been specified as in formula (1) within the ILA-framework. For each kind of interaction along a path in a network we can identify a generalised cost. Generalised travel, transport or interaction costs are usually the base for calculations of accessibility values, T_r^A, of node r and with respect to a resource of type A.

$$T_r^A = \ln \sum_s A_s \exp\{-\lambda t_{rs}\} \tag{1}$$

where t_{rs} is a distance measure of the link (r,s) and λ represents the distance sensitivity, where A_s represents the size or the capacity of resource A in node s. In this way T_r^A denotes the compound accessibility to resource A from node r. Obviously, for certain types of interaction one has to specify both displacement costs, time consumption (time distance) as well as other forms of accessibility costs.

As noted above, one cornerstone in the ILA-programme is that variations in location attributes of regions give them individual opportunities to specialise in accordance with comparative advantages. This theory refers back to Ricardo and the refined HO-model as developed by Heckscher and later Ohlin (1933). This model defines an equilibrium in which every region has a relative advantage, given the fixed or semi-fixed (trapped) resources that constitute the invariant (slowly changing) economic milieu of each particular region. This may be compared with the minimal set of assumptions used in the models due to Beckmann and Puu (1985). In their framework interaction (or accessibility) conditions are enough to generate the differentiation in the land use structure.

A long sequence of ILA-studies have investigated variations between regions as regards regional product per capita, i.e., differences in aggregate productivity.

Such results can be derived from SPE-models as a partial equilibrium solution (SPE = spatial price equilibrium). The SPE-framework adheres to the Heckscher-Ohlin tradition. It was outlined by Samuelson (1952) and further developed by Takayama and Judge (1964; 1971). In its pure form the model explains location as a consequence of transportation costs in combination with a fixed pattern of located demand (Beckmann and Thisse, 1986). In Swedish research the SPE-model has been associated with putty-clay assumptions in models where each production unit has a specific technology vintage, and where already located, as well as optional, units compete for a multiregional market (Johansson and Westin, 1987; Westin, 1990). Such formulations can also reflect location dynamics. Related earlier and empirically applied contributions are, e.g. Lundqvist (1975; 1981) and Johansson and Strömqvist (1981).

Another fundamental track in the implicit ILA-programme is the dynamic theory of spatial product cycles as introduced by Vernon (1966) and Hirsch (1967), with Thorngren (1967) as an early Swedish follower. In this theory a basic distinction is made between the development of new (young) products and the production of mature, standardised products with routinised production. Non-standardised products (goods and services) comprise customised deliveries as well as young products. According to the theoretical framework referred to here, firms with such products find it advantageous to locate in urban regions that are characterised by accessibility to a broad set of various customer categories, as well as to R&D resources and suppliers of knowledge services. Other desirable location features are a high purchasing power and rich contact opportunities. When a product gradually matures and a growing market develops it becomes possible to standardise its design and automate and routinise its production process. The comparative advantages are in this case found in regions that have lower land values and costs of premises, lower costs of routinised labour inputs (Andersson and Johansson, 1984; Snickars et al., 1989; Johansson 1993b). Additional empirical support to this model can be found in e.g. Anderstig and Hårsman (1986), Johansson and Strömqvist (1986), Karlsson and Larsson (1990).

Swedish scholars have frequently related this analytical approach to the static von Thünen class of location models, for which an equilibrium solution is such that contact-intensive and non-standardised activities locate in zones or subnodes with high accessibility values, whereas goods-handling and standardised activities are forced to select less accessible locations.

2.1.4 Outline of the Chapter

Section 2.1 has outlined a framework in which many of the studies can be included and interpreted on a common basis. This effort includes a short introduction to Lakatos's ideas. Section 2.2 describes and examines studies of the Swedish urban hierarchy and associated location advantages. In particular the studies presented in this section relate regional variations of interaction costs with location decisions. Section 2.3 presents models and empirical work of location dynamics and how such change processes are governed by differences in regional attributes. Section

2.4 presents a series of studies employing regional production functions. These are aggregate sectoral production functions, using regional attributes as arguments – together with ordinary production factors. The section provides an assessment of econometric studies which relate regional differentials as regards production cost, productivity and growth rate to pertinent location attributes. Section 2.5 gives a brief sketch of the hard core and the frontier of unsolved problems within the ILA-programme.

2.2 Urban Hierarchy, Location Decisions and Interaction Costs

The Swedish governmental organisation ERU (Expert Group for Regional Development) was established in 1965. It has initiated a long series of studies investigating conflicts between rural and urban regions, between Sweden's North and South, as well as studies of intraurban and interurban location dynamics as regards firms and households, the importance of transportation and other interaction costs, etc. A main focus has been the stability of the urban hierarchy (e.g. Bäcklund, 1991) and equity aspects of the regional distribution of employment, income and welfare. This section provides an assessment of these studies.

2.2.1 The Metropolitan Issue

In 1970 a metropolitan investigation was presented, composed of many volumes and by different authors. In one of these, Andersson (1970) stresses the importance of dense urban regions: "*High population density in a commuting region is a precondition for the economic feasibility of certain production processes, characterised by scale economies.*" This argument was combined with observations of scale economies in the consumption of different services, e.g. exhibitions, theatre performances, etc. Messages of this type represent one of the sides in the Swedish "metropolitan issue", in which the policy goal of "regional balance" was matched against the goal to promote economic growth and factor mobility. Andersson emphasises the scope advantages a metropolitan region may offer in comparisons with other locations.

Andersson claims that the metropolitan advantage is critical for firms demanding different types of customised production services, which require a rich supply of such services. The second basic advantage reveals itself in the labour market – a rich variation as regards jobs and supply of competence can only be realised in a metropolitan region. During the 1960s these phenomena became popular under the name of agglomeration factors (Törnqvist, 1963; 1970; Ramström, 1967; Kristensson, 1972). According to their theorising, firms benefit from a diversified environment and this requires an urban infrastructure that combines density and short time-distances. Table 2.2 shows how the metropolitan study from 1970 tried to demonstrate the above characteristics by reporting on income differentials and, in particular, differences regarding labour market participation.

Table 2.2. Income, unemployment and job vacancies in A-regions (labour market regions)

A-regions in size groups (1000)	Income per capita in 1967, thousand SEK	Income per person employed, thousand SEK	Number of persons unemployed divided by the number of job vacancies in 1965, %
Less than 60	8.2	19.3	96
60 – 95	8.2	19.4	87
95 – 130	8.6	19.7	49
130 – 180	9.1	20.6	52
Göteborg, Malmö-Landskrona	11.0	23.8	22
Stockholm region	12.7	27.7	15

Source: SOU 1970:15 Appendix 7.

In the 1980s the metropolitan issue became a focal point for political debates (Beckman, 1987; Snickars, 1984; 1989; Snickars et al., 1989). On the one hand it was argued that metropolitan regions were growing at the expense of more peripheral regions, on the other hand it was claimed that the Swedish economy was lagging behind internationally because the renewal of the economy in the metropolitan regions (and especially in Stockholm) was not allowed to gain full momentum – due to concern for other regions. A new large metropolitan project was organised and produced a series of reports (SOU 1990: 32–36). In some of these reports Malmö-Copenhagen is described as a potential future Scandinavian centre for international-cosmopolitan creative activities (see also Wichmann-Mathiessen and Andersson, 1993). For Sweden the Mälardal region (including the Stockholm region) is analysed as a gateway for import of knowledge, technical solutions and equipment. Moreover, empirical data show how new ideas, R&D-results and equipment are diffused from the Mälardal region in commercial networks throughout the country (see also Johansson and Strömqvist, 1986).

To a certain extent the metropolitan study from 1990 offers a conflict resolution to "the metropolitan issue" by assigning different roles to different regions in an overall interpretation of spatial product cycles in the Swedish and Scandinavian context (see also Oscarsson, 1989). Another influential book from 1984, "Regional Diversity and Pluralism in Order to Favour the Country" by four leading Swedish regional scientists gave inspiration to such interpretations. A leading idea is that urban regions and settlements of varying size form an interdependent system with opportunities for mutual benefits. The picture may be summarised by the following two components (Andersson et al., 1984):

(i) Certain urban regions constitute environments for creative activities and frequent initiation of expanding product cycles.

(ii) As product cycles leave the experimental stage and enter their phases of growth, a demand develops for locations which provide competitive advantages for routine production of standardised products. Such location decisions may concern the maturing product itself or various components from which the product is assembled.

This dynamic model emphasises that regions characterised by creativity and product development and places dominated by cost efficient processing of standardised products interact in an intricate way. Maturing products relocate to export regions which can offer more favourable cost conditions. Development-oriented regions host firms that continue to deliver new technical solutions to regions with standardised manufacturing and other such activities. In this schematised picture the productivity of the entire economy is determined by how well different parts of this interdependent pluralism can work together.

The analysis in the book by Andersson et al. (1984) concentrates on how the infrastructure of a region provides conditions for transportation, communication, face-to-face meetings and logistically balanced resource flows. The authors stress that the infrastructure changes by means of extremely slow adjustment processes. In this way it can form a collective platform or arena for the faster processes of the economy. Contributions with similar messages can be found in Snickars (1984), Andersson (1985), Karlsson (1988; 1994), Snickars et al. (1989), Karlqvist (1990).

2.2.2 Metropolitan Land Use and Economic Density

In Andersson (1970) the location of households and firms is analysed. His background references, Alonso and Muth, are today represented by Fujita (1989). Analyses in the latter contribution appear in a governmental report on the infrastructure of metropolitan regions (SOU 1989:12). In this type of model each household selects the size and location of its dwelling, given the accessibility to workplaces, public facilities and the supply of services. In the standard case this yields a unique equilibrium relation between travel costs and the willingness to pay for alternative locations. Land rents and transportation costs are inversely related.

Andersson makes direct use of von Thünen (1966/1826) when modelling the location behaviour of firms. With given output price, every firm is assumed to maximise its profit function π such that

$$\pi = I(r, m) - C(I, r, m) - P(r)m \tag{2}$$

where r denotes the distance from the location to the urban centre; m the size of the selected premises (production space); I the firm's revenues (output value) as a function of space and distance; C the firm's costs as a function of output, space and distance; P the price of a unit production space as a function of the distance r. This price reflects the land value.

Hypothetically a firm can maximise its profit function π for each distance r, in order to find out its maximal bid price in that location. Different kinds of firms (different types of activities) have different bid-price curves as illustrated by Fig. 2.1 This reflects that every location provides different productivity conditions for distinct activities. In equilibrium each firm has found a location in which its own willingness to pay is higher than that of all other firms. Relocation friction delays such adjustments. Hence, at any point in time one should expect deviations from the equilibrium identified by Andersson. In more recent analyses it has been emphasised that this type of friction is lower in urban regions with a general, flexible infrastructure.

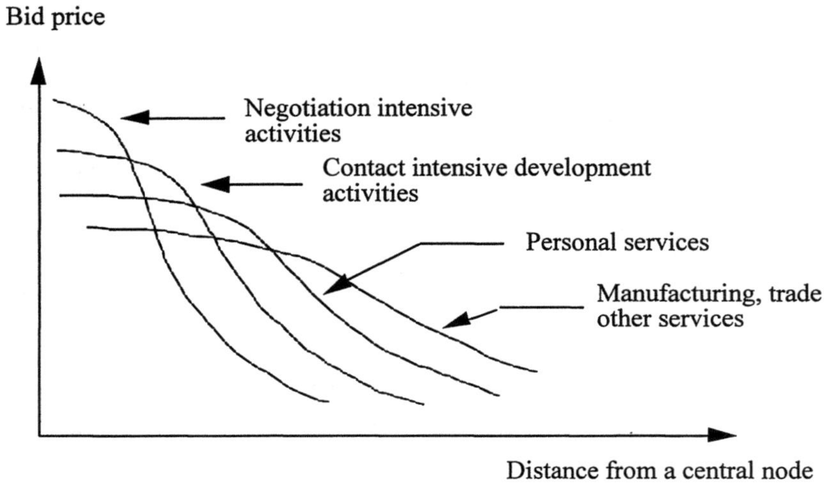

Fig. 2.1. Bid price and land rent for different categories of activities

Andersson's study has a follower in Anderstig et al. (1990). They examine with a fine spatial resolution the influence of transport systems on the intraurban accessibility patterns and its consequences for location decisions made by households and firms. Anderstig and Mattsson (1992) formulate a model in which a household selects a dwelling and workplace in response to commuting costs, the local population density and price level in the zone where the dwelling is sited. They estimate econometrically how a zone's attractiveness is determined by the following accessibility factors of the zone: (i) availability of public transport, (ii) conditions for walking and riding a bicycle, (iii) travel costs of automobile, (iv) travel time of public transport, (v) distances for walking and riding a bicycle.

Anderstig and Mattsson also consider how the location of workplaces is affected by the attractiveness of individual zones. According to their presentation the following accessibility factors play a significant role: (i) accessibility to labour supply within the region, (ii) local accessibility to subway terminals, (iii) local accessibility to commuting trains, (iv) location in sequential concentric zones (rings), (v) availability of land.

A supplementary empirical investigation is found in Johansson and Strömqvist (1979). They examine the profitability of manufacturing firms in zones of Swedish municipalities. The share of unprofitable establishments and the share of persons employed in such establishments are shown to depend on the age (vintage) of each zone's built environment, its supply of land, its distance to infrastructure for long distance transportation. Moreover, zones exhibit higher productivity and profits if they have been included in the municipality's land use planning for industries.

2.2.3 Interaction Costs and Location Decisions

The importance of infrastructure in functional regions can be examined from two angles. The first option is to ask: how do the production possibilities differ for two similar firms in two unequal regions? Does one of the regions provide superior preconditions to obtain high productivity, large profitability or fast expansion? The second option is to examine a region's attractiveness when firms and their establishments select locations and decide about where to expand and contract existing activities. Fig. 2.2 illustrates how these two issues may be interpreted as two strongly interdependent phenomena.

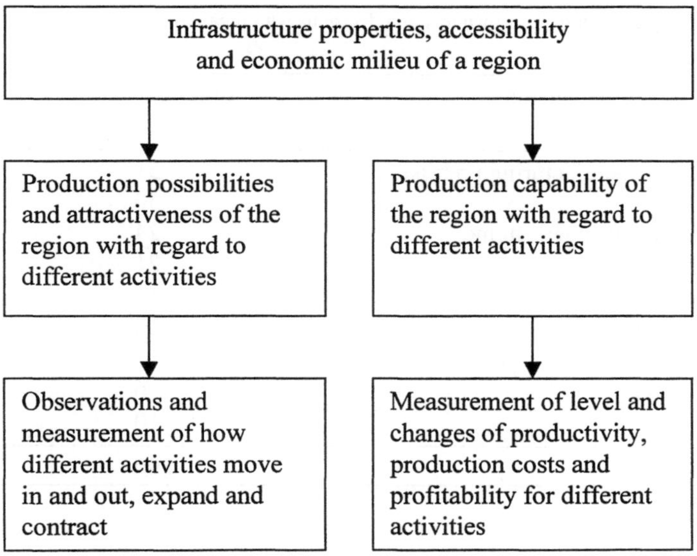

Fig. 2.2. Interdependence between location attributes and production capability of a region

The above theoretical perspective seems to have influenced Törnqvist (1963; 1974) in his analysis of transportation conditions of various industries. His study from 1963 focuses on freight costs and how they affect location decisions of manufacturing firms. He investigates direct transportation costs, and the analysis deals with both time distances and other forms of economic distances. In his study transportation costs are measured as a share of production costs for different industries, based on direct observations of individual firms. In this way he can present a best and a worst location example. He concludes that several types of industries appear to be completely unconstrained in their siting decision, if the size of transportation costs is the only criterion.

For other industries transportation costs must be regarded as decisive location factor. An alternative measure is transportation costs as share of the profit margin. The latter varies strongly between different types of goods as a consequence of the size of intermediary inputs in the production of each good.

During the 1980s one finds several studies that reveal for both manufacturing and other industries that the transportation of goods is often a limited share of a firm's total interaction costs (e.g. Andersson and Strömqvist, 1988). This conclusion can be drawn already in Törnqvist's investigation from 1963. Table 2.3 shows information from a questionnaire in his investigation. By examining the table we can extract four factors that all refer to accessibility and interaction conditions. These factors are (2) intraregional accessibility to labour supply, (4) communication conditions, (5) accessibility to customer markets, (6) accessibility to input deliverers. The four factors represent 49 % of the sum of index values.

Table 2.3. Managers' explanation of their own location decisions with regard to industrial establishments

Factor influencing location decision	Index reflecting importance
(1) Available land for manufacturing premises	51
(2) Supply of labour	50
(3) The location of the manager's home	24
(4) Communication conditions	21
(5) Accessibility to customer markets	15
(6) Accessibility to input deliverers	12
(7) Support from the pertinent municipality	11
(8) Supply of services	6
(9) Other factors	10
Sum of index values	200

Source: Törnqvist (1963)

Törnqvist (1970) may be seen as a complement to and a comment on his earlier study. The new focus is on the importance of contacts between individuals. The interaction frequency is shown to be largest for negotiating organisations, and these are followed by R&D activities, consulting and financial activities. These observations imply that transportation conditions for passenger flows and personal mobility determine the costs of nearby, face-to-face contacts in the economy. This will in turn influence the long-term composition of industries in a region and hence its potential aggregate productivity. Improved accessibility as regards nearby contacts makes it possible for a region to host a greater variety of firms and types of activities. We may call this a potential advantage.

Törnqvist (1970) is a part of the governmental report "Urbanisation in Sweden". In the same report Thorngren analyses location of establishments within the framework of spatial product cycle theory. In several of his studies he concludes that new product cycles frequently are initiated in metropolitan regions, relying on a varied composition of supporting activities (Thorngren, 1967; 1972). During the introduction phase a new product is dependent on extra-market flows of information between the producer, input suppliers, competitors, technical expertise, customers, etc. At this stage the price elasticity of demand is generally low. As a consequence, the information and communication conditions have basic importance for location decisions.

When the product has passed the introduction phase it tends to be more standardised. Moreover, production and interaction tend to be more routinised. At this stage production and goods transportation costs increase in importance. Contributions from the 1980s continue this analysis and identify how specific knowledge and development services support product development in regions (Andersson, 1985; Johansson and Strömqvist, 1986). These services may be classified as follows:

(i) *Expanding development services* comprise technology/equipment trade, R&D services, special consulting services, media and marketing services.

(ii) *Slowly growing development services* comprise technical consultants, architects, construction consultants, and organisation consultants.

These two groups refer to firms that deliver support services to the product renewal of other firms. In the 1980s one could observe an extreme skewness in the spatial distribution of these services, as illustrated in Table 2.4. The measure "regional concentration" expresses for an industry a region's share of the total employment in that sector divided by the region's share of the entire employment in the country.

Table 2.4. Regional concentration of development services 1987. Index (Stockholm region = 100)

	Expanding development services	Slowly growing development services
Stockholm region	100	100
Metropolitan regions	78	90
Other urban regions with above average density	32	54
Non-peripheral municipalities with a below average density	23	51
Non-dense peripheral municipalities	14	27

Source: Johansson and Karlsson (1991)

2.3 Location Dynamics and Regional Attributes

2.3.1 Gradual Adjustment of Households and Jobs

Going back to Fig. 2.2 we can identify how differences between regions regarding their location attributes have two observable expressions. In a region with favourable conditions many activities may operate more efficiently than elsewhere, which we can observe in the form of a higher value added per employee in given industries. Secondly, over time favourable conditions may attract new activities to the region and stimulate existing ones to expand. This second group of processes implies that regions with location advantages in the long run increase their share of the total employment in the country. Such processes also include forces that have an equilibrating tendency. An important element in this dynamics is the stimulus from increasing productivity and economic expansion to bring about growing wage levels and land values. As a consequence the decline of obsolete activities may accelerate. In a study of changing employment shares of municipalities during the period 1965–1985, Wigren (1988) shows that the growth rates of muncipality-shares vary considerably.

Swedish empirical models of the above process of change have been inspired by a study by Mills and Carlino (1989) in which they collect a vector of structural variables from all counties in the U.S. (ca 2500) for the years 1970 and 1980. In a quasi-dynamic econometric model they use these structural characteristics of each region to explain the population change across counties during the decade (cf. Tiebout, 1956). A study by Holmberg and Strömqvist (1988) follows the described approach. In the Swedish analysis the following equilibrium conditions are introduced:

$$\begin{aligned} L^* &= H(B, x) \\ B^* &= G(L, x) \end{aligned} \tag{3}$$

where L^* and B^* denote the equilibrium of jobs and population in a municipality, given the infrastructure (location attributes) of the municipality as specified by the vector x and the current values of B (population) and L (jobs). According to model assumptions, the system in (3) continues to change as long as $B^* \neq B$ and $L^* \neq L$. The adjustment towards equilibrium is assumed to be gradual and slow. In order to detect this kind of change process with limited data, the econometric analysis is based on observations of B and L from year $t = 1980$ and $t+\tau = 1988$. The model is specified as follows:

$$\begin{aligned} L(t+\tau) &= L(t) + \lambda(L^* - L(t)) \\ B(t+\tau) &= B(t) + \mu(B^* - B(t)) \end{aligned} \tag{4}$$

where $L^* = H(B, x)$, $B^* = G(L, x)$, λ and μ are dependent variables and hence estimated, and where λ and μ denote the speed at which the gaps $L^* - L(t)$ and $B^* - B(t)$ are closed, given that the process is allowed to come to rest without further disturbances or incitement. We may first note that the econometric results imply that both λ and μ have low values, corresponding to an adjustment period longer than two decades. The strongest influence on L comes from B and vice versa. Table 2.5 summarises the results.

Table 2.5. Overview of results from the study by Holmberg and Strömqvist (1988)

Location attributes that influence the adjustment processes in municipalities, and that appear as significant in one or several alternative model specifications are specified below:

CHANGE OF POPULATION	CHANGE OF JOBS
Number of jobs in the municipality	Number of inhabitants in the municipality
Number of jobs in neighbouring municipalities	Number of inhabitants in neighbouring municipalities
Supply of public transport	Supply of public transport
Local supply of single-family houses	Supply of air traffic
Accessibility by car to population concentrations	Frequency of intercity train departures
Flow capacity of main roads in the municipality	Flow capacity of main roads in the municipality

2.3.2 Infrastructure and Industrial Dynamics

The previous subsection presents an aggregate study, investigating the location of all jobs or workplaces. What can be said about the subset of workplaces in the manufacturing industry? The total number of these workplaces has diminished in almost all municipalities during the period 1980–1988. Holmberg and Johansson (1992) examined if there was any covariation between the change of manufacturing output and the change of jobs in the manufacturing industry. Following their analysis, let Q_r and L_r denote manufacturing output and the number of manufacturing jobs in municipality r, and let ΔQ_r and ΔL_r denote the corresponding change during the period. The result is that $\Delta Q_r / Q_r$, is negatively correlated with $\Delta L_r / L_r$. The same pattern also obtains if we match the productivity change, $\Delta(Q_r / L_r)$, and ΔL_r. Hence, in municipalities where the manufacturing production and productivity is growing, the number of jobs in the manufacturing industry shows a stronger tendency to diminish than elsewhere. The empirical results in Holmberg and Johansson (1992) indicate that in municipalities with location attributes that favour manufacturing production and productivity, the number of manufacturing jobs tend to reduce at a faster speed than in the average municipality.

The reported findings are consistent with various aggregate observations of economic development across regions in Europe during the period 1970–1990 (e.g. Cheshire and Hay, 1989; Rodwin and Sazanami, 1991). Essentially these observations indicate that vital regions with a high expected regional efficiency have a faster than average reduction of persons employed in the manufacturing industries. However, the study by Holmberg and Johansson (1992) indicates that an opposite pattern applies for other industries. They examine the following two aggregates of service activities in detail: (i) the WT-sector comprising wholesales and transport, and (ii) the CB-sector comprising various forms of consulting and all kinds of bank services. For each of these sectors, j, a variable, V_{rj}, is introduced to express the sector's location intensity in municipality r. The sectoral intensity is specified as follows:

$$V_{rj} = W_{rj} / H_r \tag{5}$$

where W_{rj} denotes the sum of wages in sector j and municipality r, and where H_r signifies the potential labour supply in municipality r. By using the wage sum instead of the number of persons employed, the measure also reflects the competence (and expected productivity) of the labour force in each region. The study by Holmberg and Johansson (1992) indicates that the examined service sectors are concentrated on municipalities in which the infrastructure facilitates interpersonal contacts and mobility. Moreover, a large share of these municipalities have experienced a higher than average growth rate as regards the total number of jobs. These conclusions are based on the following econometric equation:

$$V_{rj} = aA(x_r) \tag{6}$$

where a denotes the intercept and where $A(x_r)$ is a function of the following x_r-variables describing the production milieu of r: (i) flow capacity of the road system, (ii) value of premises per job, (iii) accessibility to labour supply with college education, (iv) frequency of interregional trains, (v) size of the public sector relative to the number of jobs, and (vi) travel time to Arlanda airport outside Stockholm. The last of these variables is the only one that is not significant at the 95 % level.

Given the results presented above, Holmberg and Johansson continue to study the change in the total number of jobs, $\Delta L_r = L_r(t+\tau) - L_r(t)$, in each municipality between $t+\tau$=1990 and t=1980 and show that the following econometric model can describe the spatial variation in the change of jobs:

$$\Delta L_r = \lambda(L_r^* - L_r(t))$$
$$L_r^* = A(x_r)^\alpha H_r(t)^\beta \qquad (7)$$

where L_r^* can be interpreted as the equilibrium number of jobs, $H_r(t)$ denotes the potential total labour supply in the initial year, and where λ shows at which speed the gap $L_r^* - L_r(t)$ is being closed. $A(x_r)$ refers to the function in (6) applied to the WT and the CB-sector, respectively. In both cases the estimation of (7) is characterised by high F-values and statistically significant values of α, β and λ. Hence, location attributes which, according to the estimation of (6), favour the concentration of these two service sectors also seem to stimulate location of jobs in general over time.

One can shed further light on the associated process of location changes by using results from a study by Wigren (1995). He subdivides the economy into 19 sectors and collects observations for the period 1965–1990. In this case a region is a cluster of several municipalities, where each cluster represents a labour market, an "A-region". In this investigation the location variable is the relative sector share, l_{rj}, defined for each sector j and region r as follows:

$$l_{rj} = L_{rj} / L_j \qquad (8)$$

where L_{rj} denotes the employment in sector j and region r, and L_j denotes the employment in sector j in the whole country. As dependent variable in his regression equations Wigren introduces the change, $\delta_{rj} = \Delta l_{rj}$. The regression equation may be summarised as follows:

$$\delta_{rj} = H(x_r, s_{rj}, \varepsilon_{rj}) \qquad (9)$$

where the vector x_r denotes characteristics of the regional economic milieu, the vector s_{rj} denotes sector j's initial competitive power in region r and contains variables describing the relative mix of various sectors, and where ε_{rj} is a random error term. The x_r-vector comprises variables reflecting (i) accessibility, (ii) type of region, and (iii) the size of regional public support sectors. These variables are described below:

Accessibility: (i) travel time to Stockholm, (ii) changed travel time to Stockholm between 1965 and 1985, (iii) regions adjacent to Stockholm, (iv) travel time to the closest metropolitan region, (v) regions adjacent to a metropolitan region, (vi) peripheral regions, (vii) accessibility to regional income concentrations.

Type of region: (i) the region's share of Sweden's total population, (ii) the region's share of urban population, (iii) the share of the region's population living in its largest urban node, (iv) the region's share of persons below 15 and above 65 years of age, (v) metropolitan region, (vi) university region, (vii) primary centre, (viii) regional centre.

Regional support: (i) relative concentration of education and research, (ii) relative concentration of energy, culture and public administration, (iii) relative concentration of regional commercial employment, (iv) relative public production, i.e., administration, health care and administration.

Of the 19 sectors initially studied, Wigren excludes agriculture, forestry and mining. According to the regression analysis, the size of a region and its density had, surely, a significant influence on the change pattern of 11 sectors. No influence can be verified for public administration, education and research, and health care. As many as 6 of the 16 sectors studied seem to develop invariantly with respect to the region's accessibility, whereas the change process of all national sectors such as manufacturing, wholesale, hotel and restaurant, bank and insurance, and consulting activities are statistically affected by the region's accessibility features.

2.4 Regional Economic Milieu and Productivity

Wigren (1976) identifies the following four separate but interlinked categories of activities of a firm: (i) organisation of input flows and personnel management, (ii) production, (iii) distribution and sales activities, (iv) administration and management.

The efficiency of a firm is determined by the efficiency with which it performs the activities enumerated above. How could we explain that one firm is more efficient, and has a higher total productivity, than another competitor? The first firm may have a more clever management, a better composition of its labour competence, younger vintages of the production technique embodied in the production capital and organisation, a larger scale of the activities, and a better capacity utilisation. In the individual case such a favourable position may be due to random phenomena. However, in a statistical sense we could expect all the above conditions to be positively influenced by the economic milieu (called "production milieu" by Wigren). This fourth section of the paper presents econometric studies that illustrate and support this conclusion.

2.4.1 Productivity Differentials

In the following presentation regional efficiency is a central concept. We may think of a firm's efficiency (or total productivity) as its capability to make use of its internally controlled resources. Given this formulation, the regional efficiency is that part of the firm's efficiency which depends on the characteristics of the economic milieu. This approach could be compared with Marshall's external economies (or localisation economies) as introduced in Section 2.1.2.

The Marshall arguments from Section 2.1.2 imply that a specialised industrial centre can self-reinforce its location advantages for a specific industry. They also mean that a vital and multifaceted urban agglomeration can do even better. It can provide a creative milieu (Andersson, 1985), a diversified supply of various producer services, an intraregional network for world-wide flows of information about new production techniques, products, suppliers and customers (Johansson, 1991), and a differentiated supply of labour categories. This Marshall-inspired theory implies that firms mutually constitute each other's economic milieu. Still, Swedish scholars continue to stress the additional importance of interregional interaction links and pertinent accessibility properties (e.g. Lundmark and Malmberg, 1988; Lundqvist and Olander, 1992).

In Hägerstrand (1970), the general milieu advantages in urban regions are described by emphasising that it facilitates considerably the contacts between decision makers, between experts, and between sellers and buyers. Many of the results presented in this section verifiy that the costs associated with contacts between persons are a main efficiency factor for the manufacturing industries. We start by presenting econometric studies by Wigren that focus on a large set of sectors within the manufacturing industry, and show how they vary in their dependence on the characteristics of the production milieu. Next, we describe the results from a study that utilises an aggregate production function referring to the total production in a region (Andersson et al., 1990). As a final step we present a quasi-dynamic model that has been applied to estimate how the economic milieu in municipalities influences the production in different manufacturing industries (Johansson, 1993a).

2.4.2 Sectoral Production Functions and Economic Milieu

In Wigren (1976; 1984b; 1985) the manufacturing industry is differentiated into about 15 sectors. A similar approach is also found in Wigren (1984a) which focuses specifically on one single sector, i.e., fabricated metals. The output, Q_r, of a sector in region r is assumed to depend on (i) the inputs of production factors $v_r = (v_{1r},...,v_{mr})$, (ii) firm specific characteristics $u_r = (u_{1r},...,u_{nr})$, and (iii) regional characteristics $x_r = (x_{1r},...,x_{hr})$. The components of the last vector refer to characteristics of the production milieu. They are assumed to be common to all firms in the same region. With these assumptions a regional production function is specified as follows:

$$Q_r = F(v_r, u_r, x_r) = f(v_r, u_r)A_r$$
$$A_r = A(x_r) \tag{10}$$

In (10) it is assumed that the F-function can be decomposed such that a regional efficiency factor A_r may be identified. Observe also that the functions f and A are explicitly assumed to be the same in all regions. Consider now $A^* = A(x^*)$, where x^* represents the most favourable composition of location attributes observed in any region. In this way A^* signifies an "optimal" value and we can define the relative regional efficiency as

$$a_r = A(x_r) / A^* \tag{11}$$

One may note that (11) is based on a condition of Hick's neutrality. Suppose that we want to express regional efficiency without assuming Hick's neutrality. Then formula (12) should be substituted for (11) as follows

$$\hat{a}_r = F(v_r, u_r, x_r) / F(v_r, u_r, x_r^*) \tag{12}$$

The available information also makes it possible to distinguish between complete and incomplete firms. An incomplete firm is a "producer-establishment" in the sense that it lacks a sales function, whereas a complete firm has its own marketing and sales organisation. An incomplete firm may either be a subsidiary or just have all its output contracted to one dominating buyer.

Firm-specific characteristics and the inputs of factors will be described in the next subsection. With regard to the regional characteristics we note that they comprise as many as 62 variables. In order to transform this rich information set to useful arguments in the production function introduced in (10), factor analysis was carried out (principal components). The first seven factors collected above 73 % of the variation with regard to the complete firms and just around 70 % for the incomplete firms. As regards the complete firms, the first factor collected 40 %, the second another 12 %, and the third an addition of 11 % of the variance. As a consequence, in the next subsection we focus on the content of these three factors.

Table 2.6 describes the three most important regional factors with regard to complete firms. These may be compared with the x_r-vector in formula (9). The same type of factors are relevant for the incomplete firms, although in a less complex constellation. Each factor is characterised by those x-variables (location attributes) that determine the properties of the factors by having high factor loadings. In practice, the table enumerates all x-variables with factor loadings larger than 0.8 in absolute values. One should observe that some of the location attributes are measured for functional regions. Each municipality belongs to a functional region (A-region), often formed by 3-4 municipalities, although the metropolitan regions are formed by a larger number of municipalities.

The export landscape is described by market potentials with regard to (i) the Nordic countries, (ii) UK, USA and countries where Germanic languages are spoken, and (iii) other countries. The potential expresses the size of the demand in each group discounted by the travel-time distance. The market landscape is described by distances to harbours, and domestic market potentials. In particular the following variables characterise profoundly the contact landscape for both complete and incomplete firms: (i) travel time to Stockholm, (ii) travel time to the closest of the three metropolitan regions Stockholm, Göteborg and Malmö, (iii) travel time to the town hosting the provincial government, (iv) travel time to the location of the nearest university, road distance in kilometres from the centre of a municipality to the centre of its functional region.

To a large extent the incomplete firms are dependent on the same type of factors as the complete ones. However, the variables that describe the accessibility conditions are fewer for the incomplete firms. A noteworthy difference is that the link distance between an incomplete firm and its superordinate complete firm (often head office) matters.

Table 2.6. The three most important factors referring to the regional production milieu for complete firms

ACCESSIBILITY PROPERTIES	(I) 10 variables which describe the location in the Swedish market landscape, (II) 3 variables which describe the location of the municipality in the export landscape, (III) relative supply of office (white collar) personnel in functional regions, (IV) relative supply of business services measured for functional regions, (V) travel time to the nearest of Stockholm, Göteborg and Malmö, (VI) the total number of persons employed in the municipality.
THE SIZE OF THE MUNICIPALITY BUDGET	(I) the municipality's expenditures on streets and buildings, (II) its expenditures on harbours, communications and industry
THE DEGREE OF INDUSTRIALI-SATION	(I) the relative supply of female workers in the functional region

Source: Wigren (1984a; 1985)

2.4.3 Milieu Sensitivity of Sectors

In his early study Wigren (1976) used a rather coarse spatial resolution with 24 counties. In his later analyses (Wigren, 1984a; 1985) the counties are subdivided into 275 municipalities. The observation period is the three years 1973–1975. Wigren's material is unique in the sense that the observation units are individual establishments, with information about type of output, consumption of production resources, sales value, costs of intermediary inputs, value added, wages etc. The regional efficiency is that part of the total factor productivity that can be explained by the characteristics of the production milieu (location attributes). Other factors which affect the productivity are the scale of output, the technology vintage of the capital equipment, and factor allocation in general. For all firms the analysis is initiated by estimating sector-specific production functions, based on observations of individual firms, j, in particular regions, r. Hence, the random error term is observed with the specification ε_{jr}.

The firm-specific variables in the production function include several labour input variables, energy input, and capital input (capital income). The estimation is made separately for complete and incomplete firms. In the following step the observed residuals $\tilde{\varepsilon}_{jr}$ are confronted with the factor values, k_{sr}, based in the variables in Table 2.6 and obtained from the principal components analysis mentioned above. This confrontation is accomplished by means of the following estimation, specified for each firm *j* in each region *r*:

$$\tilde{\varepsilon}_{jr} = \beta_o + \sum_s \beta_s k_{sr} + \gamma_{jr} \tag{13}$$

The regression in (13) is carried out for each individual sector. When the parameters of (13) have been estimated the following index, $\hat{a}_r$, can be calculated to express the relative value of the regional efficiency in *r*:

$$\hat{a}_r = 100 \sum_s \beta_s k_{sr} \tag{14}$$

Note that the type of index described in (14) is calculated with respect to each industrial sector. Hence, one obtains a sector-specific information about the regional efficiency. On the basis of statistically significant results one may compare different manufacturing sectors. Compiling the results one can divide the sectors into three groups with roughly one third of the industries in each group: one reveals a strong dependence on the production milieu, a second a medium sensitivity, and a third shows no sensitivity at all vis-a-vis the regional milieu. It is likely that this type of sectoral sensitivity will gradually change as time goes by and the technology develops. The important observation is that the sensitivity varies. Moreover, industries characterised by standardised products and routine production methods tend to be less dependent on the production milieu than others.

2.4.4 Production Functions Estimated on A-Region Data

Wigren's investigations in the 1970s were initiated in a climate of equity discussions like – how can the economic milieu be improved in disadvantaged regions by means of public policy? In the 1980s the Swedish policy focus changed and researchers were asked to answer questions of the type – in which regions should the production milieu be improved in order to stimulate national economic growth and increase overall productivity. Another difference is that in the 1980s there is a stronger focus on R&D and knowledge-intensive sectors, inside and outside the manufacturing industry. In this intellectual atmosphere Andersson et al. (1987; 1990) estimated production functions using characteristics of the regional milieu as explanatory variables. In Andersson et al. (1992) one finds an econometric investigation based on similar objectives. Their study is spatially more aggregated and examines three Nordic countries – Denmark, Finland and Sweden. In Andersson et al. (1990) the production function is specified as follows:

$$Q_r = A(x_r) f(v_r) \tag{15}$$

where Q_r represents the value of production in a functional region $r = 1, ..., 70$, v_r is a vector representing the size of the labour input and its knowledge intensity, x_r is a vector referring to the regional milieu. In the estimations, the value of production and other variables (except the accessibility measure) are normalised by means of division by the size of the entire area of the functional region. The *A*-function has Cobb-Douglas form, while the *f*-function is assumed to adhere to Frisch's regular ultra passum law which implies that when inputs increase proportionally the output increases along a sigmoid path.

The initial work by Andersson et al. (1987) contains studies of both the entire economy and individual sectors across regions. The presentation here is restrained to the contribution from 1990 which focuses on the production capability of the economy as a whole in regions, represented by the observable "total regional income". The authors argue that the economic milieu in a functional region should be thought of as regional infrastructure. As a basic aspect in the definition of infrastructure they require publicness in space and time. All capital that influences conditions of production beyond the spatial and dynamic limits of the decision-making firm is regarded as infrastructure. In addition they add that land without infrastructure is useless.

One particular feature of the *f*-function is that it distinguishes between standard and knowledge-intensive labour. Table 2.7 presents the econometric results in qualitative form in 1970 and 1980. Variable 2, which is a true accessibility variable, is strongly significant both years. The length of roads and railroads divided by land area indicate the capacity for intraregional interaction and, to a certain extent, also long-distance interaction. The airport and R&D variables reflect in their own ways each region's capacity to interact as regards long-distance person contacts and knowledge contacts.

The authors present six other estimations in which the variables are not normalised by the land area. Instead the developed (built) land area is included as a new variable. In all essence these alternative specifications support the result in Table 2.7. Moreover, these equations all include an additional accessibility variable expressing interregional accessibility to population concentrations across the country. In a follow-up paper Anderstig and Mattsson (1989) examine the investment options and the associated returns from alternative infrastructure investment programmes.

Table 2.7. The economic milieu described by infrastructure variables included in the *A*-function in formula (15)

INFRASTRUCTURE OF EACH REGION	1970	1980
1. Value of the building capital	+	+
2. Time distance to the nearest metropolitan region	-	-
3. Airport capacity	(0)	(+)
4. R&D capacity	-	(-)
5. Airport capacity multiplied by R&D capacity	+	+
6. Length of main roads	(-)	(+)
7. Length of railroads	+	(+)
LABOUR INPUTS		
8. Standard employment categories	+	+
9. Knowledge-intensive employment categories	+	+

Remark: Parameters that are not significant at the 95 % level are given within parentheses. For 1970 variable 6 has the wrong sign. The negative sign for variable 4 is compensated by the positive sign for variable 5.

2.4.5 Infrastructure, Accessibility and Production Potential

Sections 2.4.2–2.4.4 present two different econometric approaches which emphasise similar characteristics of the economic milieu. Further support can be found in a study by Johansson et al. (1991) based on information about aggregate sectors in municipalities. In the sequel we follow the presentation in Johansson (1993a) and Forslund and Johansson (1995). The study employs a quasi-dynamic econometric model in which the production potential in each municipality is a function of the municipality's accessibility to different resources. These accessibility measures are assumed to reflect the economic milieu of existing and potential firms in that municipality.

A set of alternative functions are estimated for several different sectors of the manufacturing industry (Johansson, 1991; 1993a). In this presentation we will focus on the analysis of the manufacturing industry as a whole. The general form of the function referred to is:

$$\begin{aligned} \widetilde{Q}_r &= A(x_r)F(K_r, L_r) \\ F(K_r, L_r) &= f(K_r, L_r)L_r \end{aligned} \qquad (16)$$

where $\widetilde{Q}_r$ represents the potential output in municipality r, x_r is a vector of variables characterising the infrastructure in r, K_r and L_r refer to the recorded capital and labour inputs to the industry of municipality r respectively. As described in Forslund and Johansson (1995) this type of aggregate function refers to an "average" or "typical" firm. Alternatively, one may assume that formula (16) represents a "typical composition of firms". The function A is increasing in its arguments. Thus, when the x-variables increase by Δx_r, the value $A(x_r)$ increases to $A(x_r + \Delta x_r)$ and in this way $\tilde{Q}_r / L_r$ moves upwards as described in Fig. 2.3.

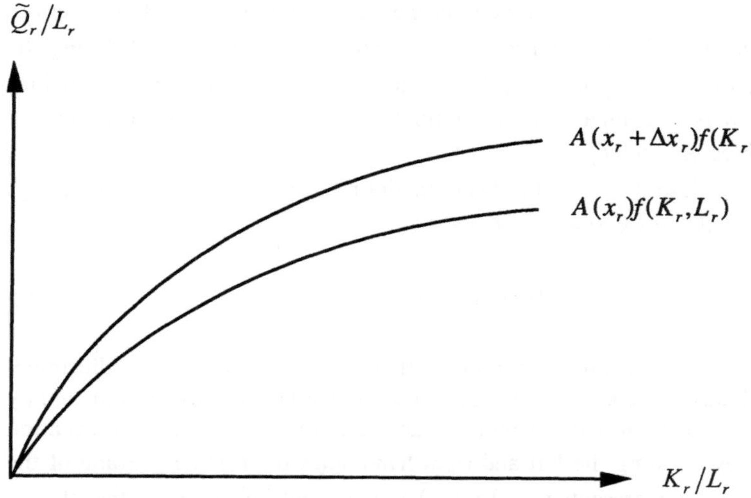

Fig. 2.3. Effects of Δx_r on potential output per labour input

The characteristics of the economic milieu of a municipality is described by infrastructure variables referring to (i) the quality of the local built environment, (ii) intraregional accessibility, and (iii) interregional accessibility. The general form for depicting the accessibility to a resource j with capacity or value, A_{sj}, in municipality s is given by the formula introduced in (1). Among recorded variables of A_{sj}-type one should observe (i) capacity of international harbours, (ii) potential labour supply for various education levels, and (iii) population concentrations.

The regression of (16) is based on the overall assumption that the production potential $\tilde{Q}_r$ will increase when Δx_r is augmented. As time goes by the realised and observed production Q_r is assumed to gradually approach the value $\tilde{Q}_r$.

In order to capture this type of dynamic adjustment, the functions f and A are estimated by means of a quasi-dynamic model called DYN (Johansson, 1991; 1993a). This formulation has been inspired by Mills and Carlino (1989) and Holmberg and Strömqvist (1988). Formally, the regression analysis attempts to explain the difference $\Delta Q_r = Q_r(t+\tau) - Q_r(t)$ in each municipality, where $Q_r(t+\tau)$ and $Q_r(t)$ denote the observed value added in year $t+\tau = 1988$ and $t = 1980$, respectively. The strategy is to estimate how ΔQ_r depends on the potential $\tilde{Q}_r = A(x_r) F(K_r(t+\tau), L_r(t+\tau))$, where

$$F(K_r(t+\tau), L_r(t+\tau)) = K_r(t+\tau)^{\alpha_1} L_r(t+\tau)^{\alpha_2}$$
$$A(x_r) = x_{r1}^{\beta_1} \dots x_{rm}^{\beta_m} \tag{17}$$

The variables $K_r(t+\tau)$ and $L_r(t+\tau)$ are observed at time $t+\tau$. The value of $A(x_r)$ should represent the infrastructure properties during the entire 8-year period. Hence, the vector x_r is assumed to remain approximately unchanged during the estimation period. This approach reflects the assumption that the x_r-variables adjust at a slow pace compared to the normal adjustment speed of the somewhat "faster" variables K_r and L_r.

Formula (18) describes how the DYN-model relates the observed change ΔQ_r to the potential $A(x_r)F(K_r(t+\tau), L_r(t+\tau))$:

$$\Delta Q_r = \lambda[F(K_r(t+\tau), L_r(t+\tau)A(x_r) - Q_r(t)] \tag{18}$$

where $0 < \lambda \leq 1$ is an estimated parameter that indicates how far the adjustment has developed during the 8 years. We may observe that (18) has to be estimated by means of non-linear regression techniques. In the estimation process the equation is normalised by dividing the left and right hand sides by $L_r(t+\tau)$. Some of the estimation results are presented in Table 2.8. It should be observed that the α-coefficients of the F-function are almost completely invariant across the various alternative specifications of the A-function.

Table 2.8. Alternative specifications of the A-function associated with robust estimation results

	EQUATION NUMBER					
LOCATION ATTRIBUTE	1	2	3	4	5	6
I. Flow capacity of the road system in the municipality		X		X	X	X
II. Capacity of regional public transport	X		X			X
III. Value and density of built environment				X		
IV. Population density	X		X			
V. Accessibility by car to population concentrations	X		X			X
VI. Accessibility by truck to international harbours		X		X	X	
VII. Accessibility to airport capacity	(X)	X				

Remark: A parenthesis signifies that a parameter is not significant at the 95 % level. The notion "robust results" refer to the fact that α-coefficients remain invariant when the A-function is varied.

We may emphasise that the implicit dynamics in (18) are driven by the size of the gap between $\widetilde{Q}_r$ and $Q_r(t)$. The change process is a movement which gradually closes the gap. If we divide λ by τ the value λ / τ indicates the average annual speed of this process. In a series of alternative specifications of the production potential function, the average value of λ comes out as $\lambda \approx 0.5$, which corresponds to 16 years of adjustment (e.g. Johansson, 1993a).

2.5 Conclusions

This paper provides a survey of Swedish research of the regional economic milieu with two related strands. One of them is focused on theoretical and empirical relations between infrastructure, regional economic milieu and regional disparities as regards productivity and production capacity. The other strand relates location dynamics of households to the attractiveness of functional regions and their subzones, based on each region's infrastructure and the associated location attributes. A general conclusion is that all studies tend to associate location attributes with local, intraregional and interregional accessibility – where accessibility is related to infrastructure networks for interaction (e.g. Karlqvist, 1990; Persson and Wiberg, 1991; Batten et al., 1995).

2.5.1 The ILA Network of Scholars

The presentation of contributions in this paper is not based on a systematic survey. It rather relies on the author's intuition, reading habits, personal contact pattern (see also Lorendahl, 1974). As a consequence, it is biased. Given this confession, which clusters of scholars and departments can be associated with the ILA-programme of Lakatos type? Remembering that ILA refers to infrastructure, location attributes and accessibility, we may start our sketch with a regional picture.

The Department of Geography at the University of Lund has in the period 1970–1995 hosted ILA-scholars, who in this survey are represented by Torsten Hägerstrand and Gunnar Törnqvist. Another gravitation point has been Åke E. Andersson. One may trace his interaction pattern with two focal points: (i) the regional research at the Department of Economics in the University of Gothenburg in the 1970s, and (ii) a Stockholm milieu consisting of the research at the Regional Planning Office (RPO) together with research at the Royal Institute of Technology, which was started in the 1970s by Anders Karlqvist as leader of the Research Group for Urban and Regional Planning and later placed in the Division of Regional Planning. In the 1970s and 1980s this group included Lars Lundqvist, Lars-Göran Mattsson, Folke Snickars and Jörgen Weibull, among others. Moreover, during the 1980s Åke Andersson and Tönu Puu inspired a series of contributions from economists and geographers of the University of Umeå. They were also associated with the Centre for Regional Science at the same university together with scholars such as David Batten, Börje Johansson, Folke Snickars, Gunnar Törnqvist and Sture Öberg.

A lot of the research reported on here has been initiated and financed by a small set of "centres". During the years examined, ERU has been one of these centres. As the national government's expert group for regional development, ERU has initiated many studies on its own but also influenced the government to establish research commissions and other investigation projects. Other focal points for ILA-studies have been RPO in Stockholm and NordREFO, an internordic research organisation for regional issues (e.g. Lundqvist and Persson, 1994).

2.5.2 Is There an ILA-Core?

The reference in this paper to an ILA-group comprises scholars from many, partly disparate, disciplines. Hence, one should not expect any superficial cohesion among the ILA-scholars. However, a closer inspection reveals that results from econometric and other statistical investigations are incorporated in associated planning, forecasting and optimisation models. Geographers and economists have to a large extent focused on the same variables and mutually examined both labour supply and product cycle dynamics.

During the thirty years, various ILA-scholars have continuously returned to investigations of how the economic performance and production potential in regions are affected by the economic milieu of each region. An invariant assumption has been that the characteristics of the regional milieu play a significant role in the development of the long term economic performance of each region. Moreover, the milieu is described as strongly associated with a region's infrastructure and the pertinent accessibility to different input and output markets – both in cross-section and time-series perspectives. Local interaction conditions are often assumed to be vital for ongoing economic renewal and evolution. Moreover, the importance of face-to-face contacts has remained unquestioned.

Another cohesive message is that the milieu-influence on the economy comes from the combination of infrastructure resources rather than from any particular element. The regional economic arena is generally assumed to constitute synergy-effects. A typical assumption is that it is the accessibility in a region that matters, rather than just transportation costs. And accessibility features are viewed as multidimensional.

For example, it is emphasised that accessibility to labour supply with university education is affected by (i) the number of persons with such education in the nearby environment, and (ii) local transportation networks and travel conditions which connect residential areas with workplace areas.

Thus, systems properties and the accessibility to distinct resources and to concentrations of purchasing power are stressed as fundamental variables. The concept of an economic milieu comprises such systems properties. However, there are certain soft parts of the ILA-core. I cannot claim that I have found among all ILA-scholars a clear consensus about how the economic milieu should be theoretically defined. There seems to be a common ground, but its nature is fairly eclectic. Of course, this also influences the attitudes to how fundamental or generic components of the milieu should be selected.

Thus, there is no full agreement about how to observe and measure the constituents of the economic milieu. At the same time one can note that ILA-scholars seem to accept that the impacts of the milieu over time should be investigated from two interacting perspectives. One is the cross-section approach revealing productivity differentials and differences in regional efficiency. The other represents a dynamic approach with a focus on location propensities and location processes. The implicit ILA-programme also includes a further suggestion – the two approaches do combine into models of how regions differ as regards economic growth and specialisation development.

Evidently, the ILA-programme has not completed the development of an uncontroversial hard core. Still, one may stress some further programme elements.

The empirically observed covariation between (i) economic growth and renewal, and (ii) infrastructure and economic milieu is frequently interpreted as an equilibrium-related phenomenon, for which the dynamics are characterised by mutual adjustments, although the time scale is slow for infrastructure and faster for variables referring to economic performance and activity location. With this perspective one is confronted with complex adjustments towards stable or unstable equilibria and other attractors instead of simple causality issues. In Andersson and Mantsinen (1980) a new research path is outlined. They investigate a multiregional system in which located physical and knowledge capital influence both the current spatial production pattern at each point in time and the growth trajectories across regions – with possibilities of converging, cyclic and diverging development.

An evolutionary theory is outlined in Andersson (1985) and Andersson and Strömqvist (1988). These two contributions stress that economic development options are offered as a consequence of logistical revolutions based on infrastructural discontinuities that alter accessibility patterns and interaction conditions. Aspects of their interpretation of the economic history of Sweden 1850–1970 is examined and tested in Westlund (1992). This confrontation shows that new research problems arise and new heuristics are likely to develop within the ILA-field – in particular new approaches to empirically analyse interdependent adjustment processes that operate on different time scales.

Acknowledgments

The author is grateful for valuable comments from two anonymous referees and the editors.

References

Andersson, Å.E. (1970), 'Storstadsproblematiken' (The metropolis problem), Reprinted from *SOU 1970:15*, Allmänna Förlaget, Stockholm.

Andersson, Å.E. (1985), *Kreativtitet: Storstadens framtid* (Creativity: The Future of the Metropolis), Prisma, Stockholm.

Andersson, Å.E., Anderstig, C. and Hårsman, B. (1987), 'Knowledge and communication infrastructure and regional economic change', CWP-1987:25, CERUM, University of Umeå.

Andersson, Å.E., Anderstig, C. and Hårsman, B. (1990), 'Knowledge and communication infrastructure and regional economic change', *Regional Science and Urban Economics*, vol. 20, pp. 359–376.

Andersson, A., Holmberg, I. and Ohlsson, O. (1992), 'Vägsystemet, produktivitet och inkomster' (The road system, productivity and income), in P-O Hesselborn (ed.), *Infrastruktur och samhällsekonomi*, TFB-rapport 1992:21, Transportforskningsberedningen, Stockholm.

Andersson, Å.E. and Johansson, B. (1984), 'Knowledge intensity and product cycles in metropolitan regions', WP-84-13, IIASA, Laxenburg.

Andersson, Å.E. and Mantsinen, J. (1980), 'Mobility of resources: Accessibility of knowledge, and economic growth', *Behavioral Science*, vol. 25, pp. 353–366.

Andersson, Å.E. and Plaut, T. (1977), 'Economic analysis of the supply and demand for labor in regions', in Andersson, Å.E. and Holmberg, I. (eds), *Demographic, Economic, and Social Interaction*, Ballinger Publishing Company, Cambridge, MA.

Andersson, Å.E. and Strömqvist, U. (1988), *K-samhällets framtid* (The Future of the C-Society), Prisma, Stockholm.

Andersson, Å.E., Törnqvist, G., Snickars, F. and Öberg, S. (1984), *Regional mångfald till rikets gagn* (Regional Diversity and Pluralism in Order to Favour the Country), Liber Förlag, Stockholm.

Anderstig, C., Fridén, L. and Mattsson, L.-G. (1990), 'Dennispaketets regionala effekter' (Regional consequences of the Dennis package), in Konsekvenser av "Dennispaketet" för trafik- och regionalekonomi, Regionplane- och trafikkontoret, Promemoria 3, Stockholm.

Anderstig, C. and Hårsman, B. (1986), 'On occupation structure and location pattern in the Stockholm region', *Regional Science and Urban Economics,* vol. 16, pp. 97–122.

Anderstig, C. and Mattsson L.-G. (1989), 'Interregional allocation models of infrastructure investments', *The Annals of Regional Science*, vol. 23, pp. 287–298.

Anderstig, C. and Mattsson L.-G. (1992), 'Appraising large-scale investments in a metropolitan transportation system', *Transportation*, vol. 19, pp. 267–283.

Batten, D., Casti, J. and Thord, R. (eds) (1995), *Networks in Action*, Springer-Verlag, Berlin.

Beckman, B., Holm, E., Lenntorp, B. and Tapper, H. (1987), *Geografin i politiken* (The Geography of National Public Policy), Ds I 1987:6, Ministry of Industry, Allmänna Förlaget, Stockholm.

Beckmann, M. and Puu, T. (1985), *Spatial Economics: Density, Potential and Flow*, North-Holland, Amsterdam.

Beckmann, M. and Thisse, J.-F. (1986), 'The location of production activities', in Nijkamp, P. (ed.), *Handbook of Regional and Urban Economics*, vol. I, North-Holland, Amsterdam, pp. 21–95.

Bäcklund, D. (1991), 'Ortssystemet i Sverige efter 1820' (The Swedish urban system since 1820), in Persson, L.-O. and Wiberg, U. (eds), *Regioner och nätverk*, ERU, Ministry of Industry, Ds 1991:37, Allmänna Förlaget, Stockholm.

Cheshire, P.C. and Hay, D.G. (1989), *Urban Problems in Western Europe*, Unwin Hyman, London.

Forslund, U.M. and Johansson, B. (1995), 'Assessing road investments: Accessibility changes, cost benefit and production effects', *The Annals of Regional Science*, vol. 29, pp. 155–174.

Forslund, U.M. and Karlsson, C. (1991), *Infrastrukturens regionala effekter* (Regional Effects of Infrastructure), ERU, Ministry of Industry, Ds 1991:55, Allmänna Förlaget, Stockholm.

Forslund, U.M. and Karlsson, C. (1992), 'Infrastructure and its contribution to regional development', in Andersson, L. and Blom, T. (eds), *The Medium-Sized City*, Research Report 92:3, University of Karlstad.

Fujita, M. (1989), *Urban Economic Theory: Land Use and City Size*, Cambridge University Press, Cambridge.

Gustavsson, J.R., Hårsman, B. and Snickars, F. (1977), 'Interregional migration and the housing market', in Andersson, Å.E. and Holmberg, I. (eds), *Demographic, Economic, and Social Interaction*, Ballinger Publishing Company, Cambridge, MA.

Hirsch, S. (1967), *Location of Industry and International Competitiveness*, Oxford University Press, Oxford.

Holmberg, I. and Johansson, B. (1992), 'Growth of production, migration of jobs and spatial infrastructure', Working Paper 1992-6, Regional Planning, Royal Institute of Technology, Stockholm.

Holmberg, I. and Strömqvist, U. (1988), *LOSE – Länstrafik och samhällsekonomi* (Economic Assessment of Public Transport in Counties), Appendix 3, Transportforskningsberedningen, Stockholm.

Hårsman, B. and Marksjö, B. (1977), 'Modelling household changes by "efficient information adding"', in Andersson, Å.E. and Holmberg, I. (eds), *Demographic, Economic, and Social Interaction*, Ballinger Publishing Company, Cambridge, MA.

Hägerstrand, T. (1970), 'Tidsanvändning och omgivningsstruktur' (The allocation of time and local environment), Appendix 4 in *SOU 1970:14*, Allmänna Förlaget, Stockholm.

Hägerstrand, T. (1974), 'Tidsgeografisk beskrivning' (Time geography modelling), *Svensk geografisk årsbok*, Lund.

Johansson, B. (1991), 'Economic networks and self-organisation', in Bergman, E.-M., Maier, G. and Tödtling, F. (eds), *Regions Reconsidered – Economic Networks, Innovation and Local Development in Industrialized Countries*, Mansell, London.

Johansson, B. (1993a), 'Infrastructure, accessibility and economic growth', *International Journal of Transport Economics*, vol. XX-2, pp.131–156.

Johansson, B. (1993b), 'Economic evolution and urban infrastructure dynamics', in Andersson, Å.E., Batten, D.F., Kobayashi, K. and Yoshikawa, K. (eds), *The Cosmo-Creative Society*, Springer-Verlag, Berlin.

Johansson, B., Anderstig, A., Holmberg, I. and Strömqvist, U. (1991), 'Infrastruktur och produktivitet' (Infrastructure and productivity), *Expertutredning Nr 9 till Produktivitetsdelegationen*, Allmänna Förlaget, Stockholm.

Johansson, B. and Karlsson, C. (1991), *Från brukssamhällets exportnät till kunskapssamhällets innovationsnät* (From Export Networks of the Industrial Region towards Innovation Networks of the Knowledge Region), Länsstyrelsen i Värmlands län, Karlstad.

Johansson, B. and Snickars, F. (1992), *Infrastruktur* (Infrastructure), T33: 1992, Byggforskningsrådet, Stockholm.

Johansson, B. and Strömqvist, U. (1979), *Arbetsområden med industri i svenska tätorter* (Industrial Zones in Swedish Urban Regions), Rapport R24: 1979, Byggforskningsrådet, Stockholm.

Johansson, B. and Strömqvist, U. (1981), 'Regional rigidities in the process of economic structural development', *Regional Science and Urban Economics*, vol. 11, pp. 363–375.

Johansson, B. and Strömqvist, U. (1986), 'Teknikspridning och importsubstitution – Stockholmsregionens roll för svensk teknikförnyelse' (Diffusion of technology and import substitution – The role of the Stockholm region), Länsstyrelsen i Stockholms län, Report 1986 No. 7:2, Stockholm.

Johansson, B. and Westin, L. (1987), 'Technical change, location and trade', *Papers of the Regional Science Association*, vol. 62, pp. 13–25.

Karlqvist, A. (1975), 'Some theoretical aspects of accessibility-based location models', in Karlqvist, A., Lundqvist, L. and Snickars, F. (eds), *Dynamic Allocation of Urban Space*, Saxon House, Westmead.

Karlqvist, A. (ed.) (1990), *Nätverk* (Networks), Gidlunds bokförlag, Värnamo.

Karlqvist, A. and Snickars, F. (1977), 'Regional migration and labor markets', in Andersson, Å.E. and Holmberg, I. (eds), *Demographic, Economic, and Social Interaction*, Ballinger Publishing Company, Cambridge, MA.

Karlsson, C. (1988), *Innovation Adoption and the Product Life Cycle*, Umeå Economic Studies No. 185, University of Umeå.

Karlsson, C. (1994), 'Infrastruktur och samhällsekonomiska konsekvenser – en kunskapsöversikt' (Infrastructure and economic consequences – a survey), in Maskell, P., Jean-Hansen, V. and Johansson, B. (eds), *Infrastruktur, lokaliseringsegenskaper och produktivitet*, TemaNord 1994:624, Nordiska ministerrådet, Copenhagen.

Karlsson, C. and Larsson, J. (1990), 'Product and price competition in a regional context', *Papers of the Regional Science Association*, vol. 69, pp. 83–99.

Kristensson, F. (1972), *Människor, företag och regioner* (People, Firms, and Regions), Surte.

Lakatos, I. (1968), 'Changes in the problem of inductive logic', in Lakatos, I. (ed.), *The Problem of Inductive Logic*, North-Holland, Amsterdam.

Lakatos, I. (1970), 'Falsification and the methodology of scientific research programmes', in Lakatos, I. and Musgrave, A. (eds), *Criticism and the Growth of Knowledge*, Cambridge University Press, Cambridge, pp. 99–196.

Lakatos, I. (1978), *Philosophical Papers, Vol I-II*, J Worrall and Currie, G. (eds), Cambridge University Press, Cambridge.

Lorendahl, B. (1974), *Nordisk regionalekonomi: Lokalisering i teori, praktik och politik* (Nordic Regional Economics: Location in Theory, Practice and Policy), Föreningarna Nordens förbund, Stockholm.

Lundmark, M. and Malmberg, A. (1988), 'Industrilokalisering i Sverige – en regional och strukturell förändring' (Location of the manufacturing industry in Sweden – regional and structural change), Geografiska regionstudier Nr 19, Department of Geography, University of Uppsala.

Lundqvist, K.-J. and Olander, L.-O. (1992), 'Svenska regioner i internationell konkurrens' (Swedish regions in international competition), Rapporter och notiser 107, Department of Geography, University of Lund.

Lundqvist, L. (1975), 'Integrated location-transportation analysis – a decomposition approach', *Regional and Urban Economics*, vol. 3, pp. 233–262.

Lundqvist, L. (1978), 'Planning for freedom of action', in Karlqvist, A., Lundqvist, L., Snickars, F. and Weibull, J. (eds), *Spatial Interaction Theory and Planning Models*, North-Holland, Amsterdam.

Lundqvist, L. (1981), 'A dynamic multiregional input-output model for analyzing regional development, employment and energy use', TRITA-MAT-1980-20, Royal Institute of Technology, Stockholm.

Lundqvist, L. (1991), 'Interregionala effekter av transportinvesteringar' (Interregional effects of investments in transportation systems), in Persson, L.-O. and Wiberg, U. (eds), *Regioner och nätverk*, ERU, Ministry of Industry, Ds 1991:37, Allmänna Förlaget, Stockholm.

Lundqvist, L. and Persson L.-O. (eds) (1994), *Northern Perspectives on European Integration*, NordREFO 1994:1, Stockholm.

Marchi de, N. (1991), 'Introduction: Rethinking Lakatos', in de Marchi, N. and Blaug, M. (eds), *Appraising Economic Theories – Studies in the Methodology of Research Programmes*, Edward Elgar, Aldershot.

Marshall, A. (1920), *Principles of Economics*, Macmillan, London.

Mills, E. and Carlino, G. (1989), 'Dynamics of county growth', in Andersson, Å.E., Batten, D., Johansson, B. and Nijkamp, P. (eds), *Advances in Spatial Theory and Dynamics*, North-Holland, Amsterdam, pp. 195–205.

Ohlin, B. (1933), *Interregional and International Trade*, Harvard University Press, Cambridge, MA.

Oscarsson, G. (ed.) (1989), *De nordiska huvudstäderna – Drivkrafter eller skapare av regional obalans* (The Nordic Capitals – Causes of Regional Inbalances), NordREFO, Helsingfors.

Palander, T.F. (1935), *Beiträge zur Standortstheorie*, Almqvist and Wicksell, Uppsala.

Persson, L.-O. and Wiberg, U. (eds) (1991), *Regioner och nätverk* (Regions in Networks), ERU, Ministry of Industry, Ds 1991:37, Allmänna Förlaget, Stockholm.

Puu, T. (1982), 'Continuous flow modelling in regional science', in Albegov, M., Andersson, Å.E. and Snickars, F. (eds), *Regional Development Modelling: Theory and Practice*, North-Holland, Amsterdam.

Puu, T. (1984), 'The long-run equilibirum of a continuous spatial exchange economy', in Andersson, Å.E., Isard, W. and Puu, T. (eds), *Regional and Industrial Development Theories, Models and Empirical Evidence*, North-Holland, Amsterdam.

Ramström, D. (1967), *The Efficiency of Control Strategies. Communication and Decision-Making in Organizations,* Uppsala.

Rietveld, P. (1989), 'Infrastructure and regional development', *The Annals of Regional Science,* vol. 23, pp. 255–274.

Rodwin, J. and Sazanami, H. (1991), *Industrial Change and Regional Economic Transformation,* Harper Collins Academic, London.

Samuelson, P. (1952), 'Spatial price equilibrium and linear programming', *American Economic Review,* vol. 42, pp. 283–303.

Snickars, F. (ed.) (1984), *Beslut för regional förnyelse* (Decisions for Regional Renewal), Allmänna Förlaget, Stockholm.

Snickars, F. (1989), 'On cores and peripheries in the network economy', in Oscarsson, G. (ed.), *A European Resource-based Periphery in a Knowledge Economy,* NordREFO 1989:3, Helsingfors.

Snickars, F. (1990), 'Samhällsekonomins transport- och kommunikationsförsörjning' (Transportation and communication systems supporting the economy), Appendix 16 in *Långtidsutredningen 90.*

Snickars, F. and Granholm A. (1981), 'A multiregional planning and forecasting model with special regard to the public sector', *Regional Science and Urban Economics,* vol. 11, pp. 377–404.

Snickars, F., Hjern B., Johansson, B., Lindmark, L. and Åberg, R. (1989), *Chans för Norrbotten* (Opportunities for Norrbotten), Bothnica 10, Norrbottens Museum, Luleå.

Snickars, F. and Weibull, J.W. (1977), 'A minimum information principle: Theory and practice', *Regional Science and Urban Economics,* vol. 7, pp. 137–168.

SOU (1970), *Balanserad regional utveckling* (Balanced Regional Development), SOU 1970:3, *Urbaniseringen i Sverige* (Urbanisation in Sweden), SOU 1970:14, *Regionalekonomisk utveckling* (Regional Development), SOU 1970:15, Allmänna Förlaget, Stockholm.

SOU (1974), *Produktionskostnad och regionala produktionssystem* (Production Costs and Regional Production Systems), SOU 1974:3, Allmänna Förlaget, Stockholm.

SOU (1989), *Storstädernas infrastruktur* (Infrastructure of Metropolitan Regions), SOU 1989:12, Allmänna Förlaget, Stockholm.

SOU (1990), *Staden* (The City), SOU 1990:32, *Urban Challenges,* SOU 1990:33, *Stadsregioner i Europa* (Urban Regions in Europe), SOU 1990:34, *Storstädernas ekonomi* (The Economy of Metropolitan Regions), SOU 1990:35, *Storstadsliv* (Metropolitan Life), SOU 1990:36, Allmänna Förlaget, Stockholm.

Takayama, T. and Judge, G.G. (1964), 'Equilibrium among spatially separated markets: A reformulation', *Econometrica,* vol. 32, pp. 510–524.

Takayama, T. and Judge, G.G. (1971), *Spatial and Temporal Price and Allocation Models,* North-Holland, Amsterdam.

Thorngren, B. (1967), *Regional External Economies,* EFI, Stockholm.

Thorngren, B. (1972), *Studier i lokalisering: Regional strukturanalys* (Studies in Location: An Analysis of Regional Structures), EFI, Stockholm.

Thünen, J. H. von (1966/1826), *The Isolated State,* Pergamon Press, New York, NY.

Tiebout, C. M. (1956), 'A pure theory of local expenditures', *Journal of Political Economy,* vol. 64, pp. 416–424.

Törnqvist, G. (1963), *Studier i industrilokalisering* (Studies in Industrial Location), SOU 1963:49, Allmänna Förlaget, Stockholm.

Törnqvist, G. (1970), 'Personkontakter och lokalisering' (Face-to-face contacts and location), Appendix 5 in *SOU 1970:14,* Allmänna Förlaget, Stockholm.

Törnqvist, G. (1974), 'Kommunikationer och verksamhetslokalisering' (Communication and the location of activities), Appendix 9 in *SOU 1974:3,* Allmänna Förlaget, Stockholm.

Vernon, R. (1966), 'International investment and international trade in the product cycle', *Quarterly Journal of Economics,* vol. 80, pp. 190–207.

Weibull, J. W. (1976), 'An axiomatic approach to the measurement of accessibility', *Regional Science and Urban Economics,* vol. 6, pp. 357–379.

Weibull, J. W. (1980), 'On the numerical measurement of accessibility', *Environment and Planning A*, vol. 12, pp. 53–67.

Westin, L. (1990), *Vintage Models of Spatial Structural Change*, Umeå Economic Studies No. 227, University of Umeå.

Westlund, H. (1992), *Kommunikationer, tillgänglighet, omvandling: En studie av samspelet mellan kommunikationsnät och näringsstruktur i Sveriges mellanstora städer 1850–1970* (Communications, Accessibility, Change: A Study of the Interaction between Communication Networks in Sweden's Medium-Sized Towns 1850–1970), Umeå Studies in Economic History 16, University of Umeå.

Wichmann-Mathiessen, C. and Andersson, Å.E. (1993), *Öresundsregionen – Kreativitet, integration og vaegst,* (The Öresund Region – Creativity, Integration and Growth), Munksgaard, Copenhagen.

Wigren, R. (1976), *Analys av regionala effektivitetsskillnader inom industribranscher: En teori med tillämpning på svenska förhållanden* (Analysis of Regional Efficiency Differentials within Industrial Sectors), Memorandum 58, Department of Economics, University of Gothenburg.

Wigren, R. (1984a), 'Measuring regional efficiency – a method tested on fabricated metal products in Sweden 1973–75', *Regional Science and Urban Economics*, vol. 14, pp. 363–379.

Wigren, R. (1984b), 'Regional efficiency in the Swedish manufacturing industry', mimeo, National Swedish Institute for Building Research, Gävle.

Wigren, R. (1985), 'Productivity and infrastructure: An empirical study of Swedish manufacturing industries and their dependence on the regional production milieu', in Snickars, F., Johansson, B. and Lakshmanan, T.R. (eds), *Economic Faces of the Building Sector,* Document D20:1985, Swedish Council for Building Research, Stockholm.

Wigren, R. (1988), 'Tillväxtpoler och krisorter' (Growth pools and decline regions), SIB, Gävle.

Wigren, R. (1995), 'Trender och trendbrott – Sysselsättningen 1965–2015 i Gävle/Sandvikens A-region' (Trends and switching trends – Employment patterns in the labour market region Gävle/Sandviken), mimeo, Department of Housing Research, University of Uppsala.

Zhang, W.-B. (1988), 'The pattern formation of an urban system', *Geographical Analysis*, vol. 20, pp. 75–84.

Åberg, Y. (1973), 'Regional productivity differences in Swedish manufacturing', *Regional and Urban Economics*, vol. 3, pp. 131–156.

Åberg, Y. (1974), 'Regionala produktivitetsskillnader' (Regional productivity differentials), Appendix 9 in *SOU 1974:3*, Allmänna Förlaget, Stockholm.

3 The Changing Context of Transportation Modeling: Implications of the New Economy, Intermodalism and the Drive for Environmental Quality

T.R. Lakshmanan
Bureau of Transportation Statistics
U.S. Department of Transportation
Washington, DC 20590, USA

3.1 Introduction

Over the last decade and a half, there has been a dramatic transformation of the environment in which transport infrastructure planning takes place. This transformation derives from three developments which have been gathering steam over the decade – first a structural change in the industrial production and market order and two broad changes in the policy and institutional context of transportation planning in the U.S.

The contemporary literature on technical change and industrial evolution underway in the industrialized economies of North America, Europe and East Asia highlights several dimensions of this structural transition: a high level of continuous technical innovations; an increasing capacity to take advantage of the economies of scope instead of the earlier concentration on exploiting economies of scale; a shift towards high value added, knowledge intensive industrial structure; an evolution towards higher levels of embodied knowledge and flexibility in the labor force; an increasingly global competition among industries; and the development of networks and networking among firms, in order to cope with the pervasive technical, and market, uncertainties (Piore and Sabel, 1984; Perez, 1985; DeBresson and Amesse, 1991; Lakshmanan, 1992; Lakshmanan et al., 1993; Lakshmanan and Okumura, 1995).

The economic networks developed in the process of such competition in time and space include transportation and communication networks. In the case of transportation networks, new logistical systems such as 'Just-in-time' appear to be promoting more efficient seamless multimodal transportation systems with expanded capabilities in terms of speed, reliability, and safety.

Second, there has been an extensive national public policy debate in the last two decades on the context and scope of transportation planning. Major outcomes of this debate have been (i) an increasing deregulation of and a significant expansion of market incentives in the transport sectors and (ii) a broadening of the objectives of transportation in the U.S.

Additional objectives of transportation policy include now the promotion of economic competitiveness, improved energy efficiency and environmental quality.

Consequently, the assessment of transportation investments has become a multicriteria problem where the claims of mobility, economic efficiency, natural resource use efficiency, and environmental quality have to be balanced.

In this context, two recent Acts of U.S. Congress, *have altered significantly the policy and institutional context of transportation planning.* The first is the Intermodal Surface Transportation Efficiency Act (ISTEA) of 1991, which mandated a major shift in transportation priorities and altered greatly *the context and scope* of transportation policy making at the state and metropolitan levels. ISTEA's declared policy is to create a coordinated, flexible network of diverse but complementary forms of transportation to move efficiently and safely persons and goods in an environmentally sound manner. Further, the transportation policy tool kit is also being expanded.

The second major change in the institutional context of transportation policy apparatus is the Clean Air Act (CAA) as amended in 1990. This act seeks to reduce vehicular emissions not only through cleaner vehicles and fuel, but also through transportation control measures. As many metropolitan areas fail to meet the National Ambient Air Quality Standards (NAAQS), the requirements to determine conformity of transportation plans, programs and projects with state air quality plans, dictate consideration of air quality concerns in transportation investment planning.

The purpose of this paper is threefold: first, it explores in Section 3.2 in some detail, the above mentioned factors which are changing the economic and policy environment of transportation. As structural changes in the economy and deregulation of the transport sector converge to promote intermodalism and new logistical systems, the public policy response has been in the form of Congressional legislation (ISTEA and CAA) which has expanded the scope of transportation planning beyond infrastructure capacity expansion. The new transportation policy kit includes additionally a variety of measures for transport control and travel demand management measures (many of which are market based), new management systems (e.g. for congestion reduction, promotion of intermodalism etc.), and close links between transportation and air quality planning.

Second, Section 3.3 of the paper proceeds to a discussion of the knowledge base – analytical models and information – necessary to support strategic decisionmaking in the new policy environment at *the national and metropolitan scales.* This knowledge base will comprise of a broad range of new analytical capabilities and information appropriate to metropolitan assessments of a) more integrated transportation and air quality planning, b) of responsiveness of travel behavior to the new transportation demand and supply side measures c) of management systems for congestion and intermodalism d) of new technologies (e.g. Meglev, Intelligent Transport Systems (ITS)). At the national or interregional level, the knowledge base will include analytical models and information relevant to the existing and future extent, conditions, services, flows, performance and economic/environmental/safety impacts of the multimodal national transportation networks and facilities.

Third, the paper surveys the potential that existing and emerging analytical or modeling frameworks have in responding to the challenges of creating such a knowledge base. It outlines in particular emerging approaches such as activity-based travel modeling and microsimulation of air emissions.

3.2 The Transformation in the Context and Scope of Transportation Planning

Three developments have contributed, as noted above, in the last decade or more to the rapid transformation of the context of strategic transportation decision-making.

The first of these derives from the remarkable structural changes in the economy initiated by the emergence of Information Technologies (IT). IT has ushered in a period of rapid technological progress, which, in turn, has led to a very high level of technological uncertainty, and market volatility. To cope with these dynamic uncertainties in an era of flexible specialization (Piore and Sabel, 1984) and increasing globalization of production (Porter, 1987), many production establishments have developed an organizational response in the form of networks and networking (DeBresson and Amesse, 1991; Lakshmanan and Okumura, 1995). The emerging national economy with firms actively competing in time and space, has made possible and has been made possible by a major logistical revolution embracing transportation networks. In fast environments, buffers and inventories in the network reduce speed and allow time for adjustment. However, they impose costs resulting from inventories and missed opportunities. Consequently, we have now logistical systems where the producer-customer relationship based on speedy delivery of quality products and services is developed towards 'Just-in-time' deliveries of customized commodities at the moment of consumption. This concept implies fast and exact transportation, information handling, and decision-making in networks comprising of flexible firms exploiting economies of scope. The resulting change forces and the restructuring of the transport sector have promoted efficient connections among different modes of transportation, choices among alternate modes, communication and cooperation and competition within the transportation community. Both in the U.S. and elsewhere in OECD countries, there has been a remarkable set of intermodal investments (e.g. double stack cars) which appear to move vast volumes of freight speedily long distances at low prices seamlessly over different modes. Now information technologies which permit fast and flexible trading, routing, and management of freight over many transport modes greatly facilitate these seamless intermodal movements.

Such an evolution of the transportation system from one capable of serving a material-intensive production system based on economies of scale towards one serving the emerging flexible, knowledge-intensive production system exploiting economics of scope is a process only dimly understood. Yet such a knowledge base in terms of changes in the types and quantity of commodities moved, their characteristics in terms of cost, speed of movement, reliability, safety, and mode is necessary to guide strategic decision-making in the transportation sector.

More than three decades of investment on the U.S. Interstate Highway System (authorized by the 1956 Highway Act) has created not only a high level of highway-based mobility, but also a serious disaffection with the neglect of other modes of transport, with various inefficiencies, congestion, and pollution and indeed with the singleminded focus on the mobility (or automobility) goal itself. Public policy response to these disaffections over the last two decades has taken three forms: first, increasing deregulation of and an expansion of market incentives in the transport sector to unleash competitive forces and improve efficiencies; and second, promote transit, rail, and other modes of transport; and third, broadening of the public policy objectives of strategic transportation decision-making.

Over the last two decades various transportation sectors have been deregulated in the U.S. thereby spurring market competition and economic efficiency.[1]

The second response is to provide incentives for promoting transit, rail or other possible transportation modes. Several motivations underline this response: transit and rail sources offer transportation choices to groups such as the carless, the aged, the poor, and the commuter; further, these services, by removing some traffic off highways, reduce congestion, energy use, and improve air quality.

The third component of public policy response is the growing realization that the objectives of transportation policies are not only to promote mobility but also to support other valued national goals such as economic growth, natural resource use efficiency, environmental quality, and safety.

These broader and multiple objectives of strategic decision-making are enshrined in two recent Acts of the U.S. Congress. The first is the Intermodal Surface Transportation Efficiency Act (ISTEA) of 1991 which declares a new intermodal transportation policy orientation, provides for a greater role for market incentives in the transportation sector, and promotes mechanisms which would obtain increased use of existing transportation capital stock and systems for managing congestion, intermodalism, etc. Further, ISTEA requires Metropolitan Planning Organizations and States to develop intermodal transportation plans.

ISTEA has reintroduced a variety of market incentives into the transport sector. The 19th century, when the U.S. was industrializing, was the heyday of private transportation with turnpikes, paved roads, and toll roads built and managed by about 2,000 companies – an era which came to an end in the Progressive Era and the Federal Aid Highway Act of 1916, which barred the use of tolls on any highway receiving federal money. ISTEA reversed that 76-year policy against private roads and allows federal agencies to support toll roads and to participate in their financing. California has permitted the testing of the feasibility of building four privately funded transportation facilities (Fielding, 1993). Since ISTEA also allows for experimentation in congestion pricing, the prospects for road pricing policies in the U.S. may be improving (Small et al., 1989).

---

[1] In addition, deregulation primarily in the transportation sector has benefited U.S. by 36-46 billion (1990 dollars) annually – the bulk of the benefits going to consumers, but not at the expense of labor or producers; the latter as well, have benefited, on net, from reform (Winston, 1993).

The Clean Air Act (CAA) of 1990 places emphasis on transportation control measures to reduce emissions of ozone, CO, particulate matter of 10 microns in size (PM-10) in metro areas which fail to meet National Ambient Air Quality Standards (NAAQS). While there have been significant reductions in mobile emissions from cleaner vehicles and cleaner fuels, the anticipated continuing growth in VMT (Vehicle Miles Traveled) poses problems. Since U.S. Environmental Protection Agency (EPA) is required to apply sanctions against metro areas which are not in conformity with state plans for achieving air quality, CAA implies increased coordination requirements for transportation and air quality decisionmaking, entailing greater complexity and requiring more complex analysis in transport planning.

ISTEA complements CAA by providing metro and state transportation decisionmakers with flexibility to develop environmentally sound, balanced intermodal transportation systems. ISTEA and CAA encourage the use of transportation control measures (TCMs) whose objectives are to promote switching to environmentally beneficial modes, improve traffic flow, bring about workplace/workweek changes or modify vehicle performance. The tool kit to address these objectives include market mechanisms such as road pricing, parking tax, smog/VMT tax, regulations (on parking, road use, vehicle use, etc.), facility improvements (e.g. HOV lanes, bicycle paths, etc.) or information devices. ISTEA also encourages the use of transport demand management market based measures which could be directed to reduce congestion and promote greater use of network capacity.

3.3 Requisite Knowledgebase in a Changing Decision Environment

To the degree the foregoing review of the emerging world of transportation and its public and private institutional structure is valid, the analytical challenges posed by this world are formidable. There will be an expansion of the scope of transportation modeling and policy assessment in two ways: *expansion of its functional scope and spatial scope.*

Transportation planning makers will be called up to address trends and issues relating not only to mobility but also to economic and ecological sustainability in a period when the production system is undergoing structural change and significant changes in demography and lifestyles are afoot. Second, such broadranging analytical efforts should be directed not only at the traditional metropolitan spatial scale but also at the national scale.

3.3.1 Metropolitan Scale Knowledge Base

The particular institutional and policy requirements of transportation public investment planning authorized by the Federal Highway Aid Act of 1956 (which created the Interstate System) over the next thirty-five years have been met by the creation and progressive improvement of a vast knowledge base. The latter comprised of a variety of land use and transportation models and databases on land use patterns and land use forecasts, trip generation, trip distribution, modal split and traffic assignment, based on the familiar 5 step Urban Transportation Planning System (UTPS). UTPS uses aggregate data for small subareas (traffic zones) in the metropolis to estimate personal travel on current and future networks in order to guide strategic decisions on future additions to highway and transit system capacity.

The land use and transportation models available for use in the UTPS in the early nineties have been built upon the pioneering efforts of the creative modeling era in the late fifties and early sixties (Voorhees, 1959; Harris, 1965), expanding their theoretical foundations and operational capabilities. Such improvements reflect a quarter century of experimentation and operational modeling not only in many U.S. metropolitan areas but also in other institutional contexts (e.g. UK, Rest of Europe, Australia, Japan, Latin America, etc.). Theoretical developments in discrete choice theory, Transportation and the New Urban Economics, entropy-maximizing modeling, and Network equilibrium theory have been incorporated into these operational models to strengthen their theoretical foundations, to provide greater transparency to their structures, and to improve calibration and validation procedures. Wegener (1994) provides a comprehensive and welcome survey of these models. (See also the relevant discussion in the next part of this paper).

These are welcome developments towards the knowledge base necessary for assessing decisions pertaining to transportation capital investments. However, the implementation of ISTEA and CAA require *two types of capabilities* that are not offered by the extant land use and transportation models. These two types of capabilities are:

a) *Modeling Impacts of New Policy Tools*: This would permit the assessment of a broader range of policy options than those offered by the currently available land use and transportation models. Specifically, such models would generate five different types of information listed later in this paper.

b) *Decision Support for Cross Modal Choices*: The utilization of analytical information provided by the models (including those referred to in (a)) in a decision framework which permits the comparison of investments in different modes of transportation or in intermodal facilities. ISTEA requires the use of criteria for crossmodal choices of investments. We discuss this issue in detail in the next section.

3.3.2 New Policy Modeling

Improvements in modeling are necessary to provide *five new classes of information* required in the ISTEA policy environment.

First, the need to assure conformity of transportation investment plans with Environmental State Implementation Plans (SIPs) require information in great spatial and temporal detail on vehicular flow and associated criteria pollutant emissions. Such information can be generated by new linkages between the traditionally separate modeling traditions of transportation modeling and air quality modeling. The obvious linkages occur when the (traffic flow) output of transportation models are converted through the use of EPA's models (e.g. MOBILE 5 emission factors) into emissions of ozone (produced by combination of hydro carbons and monoxide, and particulate matter of under 10 microns in size PM-10). The more subtle and complex linkages arise from the requirements of air quality conformity assessment and air quality episode management. The latter require real time outputs of vehicle flow by time of day, speed of travel and the like a far cry from the traditional output of transportation models in the form of average daily or peaktime flows.

Second, the need to reduce vehicular emissions through the use of Transport Control Measures (TCMs) suggest that the transportation modeling system should be capable of measuring the different types of likely responses of the metropolitan transportation system to the introduction of such transport control measures (TCMs). TCMs seek several kinds of changes: switching of modes by car commuters towards transit vehicles or high occupancy vehicles (HOV) in order to reduce vehicle trips; freeway traffic flow improvement to reduce emissions; and reduction of vehicle trips by changes in workweek or workplaces. They attempt to make these changes by *altering the market incentives* (congestion or roadpricing, parking taxes, smog/VMT tax, etc.) or *regulations* (on parking, driving times and zones, or on land use) or *by facility improvements* (HOV lanes, transit improvements, bicycle paths etc.) or by *information strategies.*

Third, the models need to describe the effects of transportation demand management (TDM) measures – (which are a complement to the supply side measures) on travel behavior and the occurrence of congestion. The emergence of such transportation demand management (TDM) measures (e.g. changes in the departure time, in destination choice, in mode of travel, etc.) requires that transportation models should be able to assess such combinations of demand and supply side strategies.

Assessment of such TDM and TCMs requires extensive modeling of travel behavior patterns and the underlying variables which could respond to the relevant policy instruments. Table 3.1 provides an illustration of the types of new policy issues and the requisite modeling and information needs for addressing such issues in the future. Marketbased TDM and TCMs, which can target specific users and facilities and specific times call for more precision and more detailed impact measurements. Precision is warranted in the estimation of mobile emissions, since overestimation or underestimation will lead either to the application of drastic mitigation approaches or to air quality deterioration. The emphasis on such detail in impact assessment stems from our need to understand the process of dynamic adaptations of the traveler to the changes likely to be generated by the various TDMs and TCMs. Not only do we need elasticities of response to transportation price changes, both in the short and longrun, also we need such elasticities *by user type, by facility, by time of day and by market segment*. Such information presupposes that we are able to estimate traffic vehicles by time of day. In addition, freight flow modeling becomes important along specific corridors in the metropolis both for multimodal congestion management and for planning pollution reductions. Further, management of air quality episodes requires knowledge base on the emission levels under conditions when air quality standard violations are likely to occur. The new models need to provide time-of-day sensitivity and variability in travel and driving patterns. In general, time path of emissions during a day or week is a requisite piece of knowledge.

Table 3.1. The new transportation modeling agenda (metropolitan level)

Illustrative Issues	Requisite Knowledge Base
ISTEA	
Transportation Demand Market Instruments (pricing/user fees)	Short and longrun response elasticities by user type, by facility, by time of day or week and market segment.
Congestion and Capacity-Related Travel Behavior	Traffic volumes by time of day inducted demand. Freight flow models, intermodal analysis along corridors
CLEAN AIR ACT	
Emission Targets	Type of emission by source and time of day
Episode	Time of day or week emission estimates, spatial-temporal management.
Transportation Control Measures (TCM) (e.g., market based regulation, facility improvement exhortation)	A comprehensive analysis of travel behavior
New Technologies	
Telecommuting IVHS/ITS Clean Vehicles	Factors influencing decision making by technology producers and consumers, and technology diffusion processes.
Other Sustainable Development	Models of economic growth, land use density and travel behavior

Fourth, the models should generate information on the likely consequences of the broad social and structural changes in the emerging network economy. Specifically, what are implications on urban development of (a) demographic changes and (b) industrial restructuring and the IT revolution? The models should capture the effects of the evolution in the household structure and lifestyles on the activity patterns of household members and their changing activity programs and the related travel patterns of the individuals. As firms coordinate their various functional activities (e.g. R&D, input logistics, output logistics, production, service etc.) on a global basis (Porter, 1987), locations and relocations of employment centers change in response to changes in rents, quality and quantity of labor supply, and in costs and access. How well do the models capture such complex factors governing travel generation and urban development patterns in the emerging world economy?

Fifth, the models must be able to describe the consequences of new information technologies such as ITS on travel behavior, traffic patterns, congestion, and safety.

3.3.3 Cross Modal Decision Support

Since ISTEA planning framework encourages states and localities to make tradeoffs between choices of investments in different modes or intermodal facilities in a particular corridor, the land use and transportation models must generate the types of *outcome measures*, which facilitate intermodal tradeoffs. Such measures would inform cross-modal criteria, which can compare mass transit projects, expressways, and other unconventional TCMs such as HOV lanes or bicycle routes systems. These *outcome measures* would differ in two significant ways from the traditional mode-based criteria. (Highway criteria pertain to number of vehicles moved, while mass transit criteria focus on people movement). First, cross-modal criteria provide a basis for comparing quantitatively an investment project's ability to meet transportation objectives, such as mobility, economic performance, safety, cost effectiveness, and environmental quality.

Second, these outcome measures would also help MPOs and states to select the most effective mix of projects – drawn from different modes if necessary – to address the region's mobility and air quality objectives.

Examples of such measures of performance are: people moved, savings of travel time/cost savings per traveler, etc.; productivity trends, income effects etc. for measuring economic performance; rates for accidents, fatalities, or injuries per million passenger/ton miles as measures of safety; additional benefits (in the form of mobility, environmental quality etc./per unit cost); air quality improvements per unit investment, etc. Consequently, the designers of improved models of travel must include model outputs which could serve as the basis for generating such cross-modal *transportation performance indicators*.

3.3.4 National Level Assessment

U.S. Department of Transportation (DOT) is engaged in a strategic analysis related to a National Transportation System (NTS)[2]. The strategic initiative is an assessment of the condition, performance, and economic, safety, energy and environmental consequences of the multimodal NTS. DOT believes that the integrated system perspective it employs in this process will help identify the most

[2] More precisely, NTS comprises not only of the physical infrastructure of airports, highways, rail lines, transit systems, pipelines, ports and waterways but also the nonmaterial infrastructure. The latter pertains to the regulatory environment, the institutional structures and the market based incentives governing private and public choices in transportation investments, strategic management and operational activities.

strategic and effective uses of scarce resources available for advancing goals of mobility, safety, economic growth, and environmental quality. Viewed this way, NTS is not just a map but an electronically specified multimodal network described and analyzed in the following manner:

- By information on the location, extent and connectivity of major transportation facilities and services;
- By information on the safety and patterns of use of major transportation facilities and services;
- Information on the condition, performance and economic, energy, and environmental) consequences of major transportation facilities and services and;
- Tools to analyze and interpret changes in the use, conditions, performance and consequences of major transportation facilities and services – in other words, the ability to design and play a variety of 'what-if' policy exercises.

This strategic initiative is jointly pursued by the Office of the Secretary (policy) and the Bureau of Transportation Statistics (BTS).

BTS has completed a major national commodity flow survey (CFS) and is engaged in an American Travel Survey (ACTS) to track intercity personal travel in order to provide the information on the current patterns of the use of NTS. CFS, conducted jointly with the U.S. Bureau of the Census, has surveyed approximately 200,000 establishments to capture information on about 12 million shipments – described in terms of tons, tonmiles, value, mode of transportation (including intermodal combinations), shipment distance, shipment size, and combinations of origins and destinations – all in considerable commodity and spatial detail. The results will be available at the level of the nation, 50 states and for 89 National Transportation Analysis Regions (NTARS)- combinations of BEA Economic Areas).

The American Travel Survey (ACTS) tracks personal travel (of over 75 miles in length) of 83,000 households (based on diary records) by purpose, distance, mode or combinations thereof), origins and destinations with intermediate stops and some expenditures.

The Geographic Information System (GIS) Center at the Bureau of Transportation Statistics (BTS) prepares and maintains the existing multimodal National Transportation System – specified in terms of networks and facilities, the conditions, the services, and commodity and passenger flows. Efforts are underway to develop new analytical models of passenger and freight flow and other attributes of the multimodal network. Parallel research efforts at BTS (a) to develop a transportation satellite account (in cooperation with Bureau of Economic Analysis) in the National Economic Accounts, (b) to cooperate with Bureau of Labor Statistics in developing improved output measures in transport productivity analysis and (c) Transportation System Performance Indicators – all of these will in turn support the NTS Strategic Initiative.

3.4 Emerging Metropolitan Modeling Frameworks

The ongoing logistical revolution, ISTEA, and CAA are clearly transforming the transportation policy landscape in the U.S. in the nineties. Policy horizons and tool kits are being expanded, public-private interactions and a new institutional context of strategic transportation decisionmaking are being defined. The critical research policy questions is: How robust are the extant land use and transportation models (which have been developed in response to the policy needs of an earlier era) in their ability to respond to the sharply expanded scope of transportation policy development? Are there any modeling frameworks on the horizon, which hold a better promise?

3.4.1 New Informational Requirements

Two types of developments appear to be necessary if the extant analytical models and the information they provide are to be relevant to the assessment of the new policies promoted by ISTEA and CAA. First, the new policy models must acquire the *five types of new analytical capabilities* outlined in the last section of this paper.

These new types of analytical capabilities will be reflected by the models' provision of information on:

- Vehicle flows and pollutant emissions in considerable temporal and spatial detail,
- The travel response elasticities of different travel control measures and transport demand management measures – regulations, facility improvements, pricing, taxing and other market incentives, and information strategies (ITS)-specified by user type, by facility, by time of day and by market segment,
- The consequences of broad technical, structural and demographic changes in the emerging networks and travel patterns,
- Markets for and impacts of new technologies (eg. Maglev, Intelligent Transport Systems (ITS)).

Second, in the process of doing so, they should draw upon and *link together* the two research traditions which have developed separately in the past: namely land use and transportation modeling and vehicle emission modeling.

The initiatives to link transport and emission modeling is exemplified by Environmental Protection Agency's (EPA) Mobile Source Emission Model (MOBILE).[3] MOBILE 5, the latest version, is developed to estimate motor vehicle emissions of ozone precursors (HC and NOX), and Carbon Monoxide, and to determine conformity of emissions with the environmental implementation plans.

[3] Another relevant model is the Regional Oxidant model which includes chemical reactions among pollutants and is organized on a broader geographical scale such as multistage corridors. We do not consider it here.

It is an integrated set of equations which develop emission rates for the fleet (organized into eight different types)[4] of motor vehicles in a given community. The model accounts for a variety of community parameters such as vehicle age and mileage, vehicle speed, ambient temperature, driving conditions.

In this approach, there are several sources of error. The errors may be in the basic emission rates for individual vehicles and/or in the correction factors used to adjust these measured rates.[5] Since basic emission rates are developed from data intended to represent average urban conditions and may not necessarily match the mix of speeds or other operating conditions in a community, correction factors are used to account for such differences may be erroneous. EPA has expressed concerns with the Federal Test Procedure's (FTP) treatment of cold starts, driving behavior, speed, acceleration behavior, and hot soaks[6] (DOT and EPA, 1993). While EPA is engaged in some of the issues relating to test cycles, there is room for creativity on the part of the transport modeling community.

3.4.2 Potential of Existing Models

Michael Wegener (1994) has identified about a score of land use and transportation models institutionally located at Universities, private firms or public agencies. About a dozen of these are operational models applied to the assessment of transportation investment policies. He has described briefly the theoretical foundations, model structure, policy scope, calibration procedures, and operationality of these models. Wegener (1994) identifies the degree to which a few of these models have introduced some environmental versions in the form of selected pollutant emissions. While these are welcome developments, the gap between the transportation and environmental information needs of the ISTEA/CAA process and the information provided by existing models is too wide for comfort.

We anticipate a quickening of analytical development and creative modeling in the next decade in response to the challenges of ISTEA and CAA, comparable to the creative modeling output generated in the first decade after the 1956 Federal Highway Act (Voorhees, 1959; Harris, 1965). While it is foolhardy to predict the specific forms the new creative phase of land use and transportation modeling may

---

[4] The 8 vehicle categories are: light duty gasoline vehicles, two types of light duty gasoline trucks (under 6,000lbs and 6,000-8,500lbs truck weight), heavy duty gasoline vehicles (over 8,500lbs), light duty diesel vehicles, light duty diesel trucks, heavy duty diesel trucks (over 8,500lbs) and motor cycles, different from those of the test-cycle, control programs etc.) and produces grams of pollutant emitted per mile driver (Sierra Research, 1994).

[5] Emission rates are developed from test measurements of in-use vehicles at various odometer readings. They have two components: an intercept (zero-mile level, and a slope (deterioration rate by vehicle age).

[6] Fuel exposed to the hot engine when it is turned off may evaporate and escape to the atmosphere. These are the so-called "hot soak" emissions modeled by MOBILE as grams/event or grams/mile.

take, they are highly likely to grow out of the existing broad research traditions and some emerging concepts.

Our survey of the theoretical and substantive developments and existing operational modeling in land use and transportation analysis in the last three decades suggest *several modeling traditions.* Each of these traditions represents a particular theoretical-operational-policy nexus (in the sense that the models in that tradition are built around a specific theoretically coherent view of the metropolis), a set of functional specifications, estimation procedures, and policy assessment format. Each tradition also includes a group of scholars and professionals who have a shared understanding of decisionmaking in land use and transportation investments and the way that is to be represented in mathematical models. The researchers in each group have in all likelihood heavy intellectual and other investments in that tradition including a large publication record and/or a commercially available model. However, some of the models in each group have selectively borrowed and used concepts and techniques from another group, resulting in considerable overlap and degree of convergence among the models – still retaining, however, the dominant concepts/techniques of their class. The rationale for recognizing these classes of models lies not in their orthogonality but because we believe that many modelers would (in responding to the ISTEA modeling challenges) likely continue in their tradition or use it as a point of departure.

While analyzing land use and location, some of these models are *Lowry derivatives,* e.g. California Urban Futures Model – Landis (1992; 1993); others use *a discrete choice framework,* e.g. LILT Model of Leeds by Mackett (1983; 1990) and IRPUD Model by Wegener (1985; 1986) etc; yet others develop transportation- locations models viewing *the Urban System as a set of markets,* e.g. Kim's (1989) model, Echenique's et al. (1990) MEPLAN model, HUDS Model (Kain and Apgar, 1985). Such land use models are combined with some State-of-the-art multimodal transportation network models, often with network equilibrium as the dominant trip assignment method – this development stimulated by the pioneering work of Beckmann et al. (1956). A few transport models employ a multiple-path assignment for allowing route-choice dispersions.

How these types of models will evolve in response to the analytical challenges posed by ISTEA and CAA is difficult to assess. Predictably, some of these models based on market framework will be more easily amenable to further development in directions congenial to developing information on the responsiveness of different classes of travelers to different travel demand management and traffic control measures. It is possible that some of the linkages being built between transport and air emission modeling can be expanded to a more satisfactory integration. The next decade will tell.

Our focus, however, in the last part of this paper is to describe briefly new approaches to travel and emission modeling. In the first approach, the activity-based approach to travel modeling analysis pursued over the last decade and a half – the focus of analytical attention is not the *trip*, but decisions governing activities which affect the demand for travel. Analysis of the interrelated set of decisions about each activity – *whether to, where, for how long, when and with whom to engage in* – yields the derived demand for travel or its substitute. The second approach is an ambitious new effort called TRANSIM, whose first phase takes advantage of certain developments in statistical methodology and computational speed to provide microdescriptions of vehicle locations and air emissions (Los Alamos National Laboratory, 1994).

3.4.3 Activity-Based Travel Modeling

Activity-based travel modeling is an intellectual descendent of the time-space geography ideas of Hägerstrand (1970), who viewed individual's movements as a series of passages through time and space. He described a time-space 'prism' where, locations fit within the available spatial-temporal opportunities for individuals as they engage in work and domestic activities. In the choice of such activities, they face three constraints: *capability* (e.g. personal resources, accessibility's etc.), *coupling* (activity requirements), and *authority* (rules of access to locations).

While Hägerstrand's prescient time-space framework has resisted operationalisation in transport modeling, its rich ideas underlie much of the activity-based modeling. The basic ideas of activity modeling can be summarized as follows:

> The activities necessary to satisfy household needs drive the decisions which result in travel. Travel is an outcome of the decision to engage in activities outside the home. Modeling the travel process requires that activities, not trips, be the object of analysis. Households, endowed with certain resources and structure, adapt incrementally (within budget and time constraints) to large changes in the urban economy or land use activities and transport infrastructure systems also evolve over time in response to changing household activity patterns.

A variety of transport analysts have formally explored these ideas. Such analyses include: grouping of households by activity patterns; analysis of impacts of life cycle stage on travel patterns; exploration of trip chaining patterns; hypotheses of adaptation processes of households to change in the household and outside; alternative approaches to activity simulation etc. (Axhausen and Gärling, 1992; Goodwin et al., 1990; Kitamura, 1988; Recker et al., 1986; Jones et al., 1983; 1990).

To assess this literature we outline the framework that is implicit in Activity Modeling literature (Fig. 3.1). The focus here is on activity choices leading to short and long term scheduling of household activities that meet their goals. Two sets of factors govern these choices:

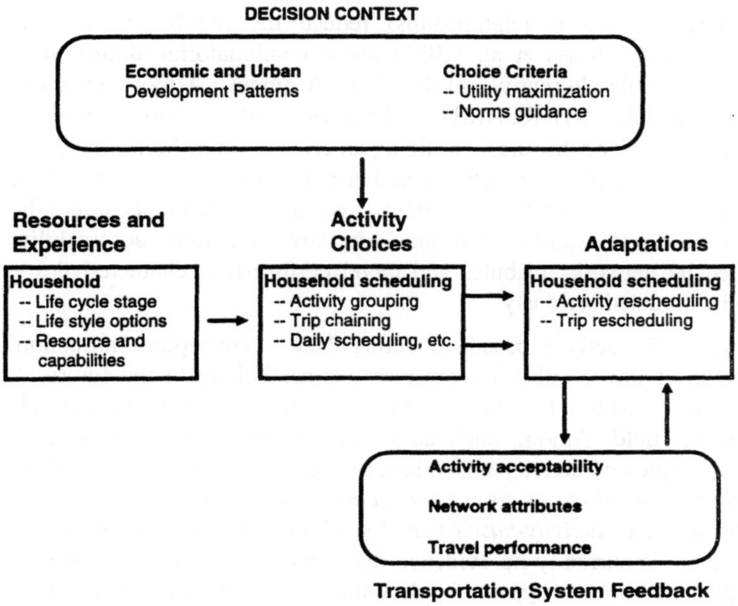

Fig. 3.1. Framework for activity-based travel analysis

Resources and Experience – defining the supply and demand sides of household activities. By *experience* we refer to the different attributes of the household that govern its activity demand e.g. its life cycle stage, its life style options, income, transport and other capital stock, location, and other social characteristics. R*esources* pertain to what is available to implement household activities (e.g. time, monetary resources, human capital, physical capital, etc.).

Decision Context: expressed in two forms – the broader context provided by economic/urban patterns in the area where the households reside; and what can be called *decision criteria,* which underlie the framework in which activity choices are made under constraints. One such framework is provided by the *theory of households' time allocation* (Becker, 1965; Winston, 1982), whose optimality assumptions bother many analysts in this area (Axhausen and Gärling, 1992). Another approach due to Fried, Havens and Thall (1978) holds that activity and travel and adaptations thereof are governed by social class and status, ethnicity, life cycle stage, and location. Such factors act on households through perception of *norms, and roles,* constrained by household resources – a problematic formulation. As contrasted with these two frameworks of *utility maximization and norms guidance,* there are a few eclectic frameworks which yield activity simulation models, e.g. taking advantage of the Hägerstrand-inspired insight, that a combination of spatial-temporal constraints (people's coupling in space and time through work and other activities, and limits on the use of space due to rules and power relationships) reduce the number of feasible alternatives greatly, Jones et al. (1983) use a combinatorial algorithm to simulate the feasible alternatives in the CARLA model. A more advanced simulation model, STARCHILD, identifies and chooses between representative activity schedules – utilizing an eclectic set of frameworks.

Adaptations take place in activity and trip rescheduling as a result of interaction with the transport environment. Such adaptations that capture the *feedback* from the transportation system – in terms of activity acceptability as mediated by network attributes and travel performance characteristics – are yet to be modeled seriously.

In summary, while activity-based modeling has a conceptual attraction, operational models of direct utility to transport planner belong to the future. A number of theoretical, methodological, and data issues need to be resolved in this young but vigorous field. Among such issues are: more research attention is needed on the principles of activity allocation; *the major weakness lies in these models not linking their findings to the transport network and thus failing to capture the interaction of activity-based travel modeling with the transportation system*; further, there is currently no evidence that activity-based travel modeling currently can generate the new types of policy-relevant information required by ISTEA and CAA; and finally, there are the vast problems of getting data on household activities. To the degree that this research stream addresses such issues in the next decade, its viability as a source of transport policy-relevant information in the ISTEA era will increase.

3.4.4 Microsimulation of Air Emissions

A major effort is underway to model in great spatial and temporal detail metropolitan vehicle flow and air emissions as part of the TRANSIM model being developed by Los Alamos Lab. This capability is attempted in the Microsimulation component of the TRANSIM architecture (Fig. 3.2).

The microsimulation module attempts to implement every trip planned by households through systemwide traffic microsimulation. Consequently, component submodels include driver and driver behavior representations, vehicle type representation, modal representation of infrastructure, mode transfer representations, traffic management, information systems, as well as relevant data bases and structures. This appears to be aided by some new techniques of data pooling, data mining, and analysis. The analysts' view that the microsimulation will provide outputs at three levels of analysis:

- Low fidelity simulation for planning and conceptual engineering
- High fidelity simulation for preliminary design and engineering
- Support for detailed traffic operations analyses.

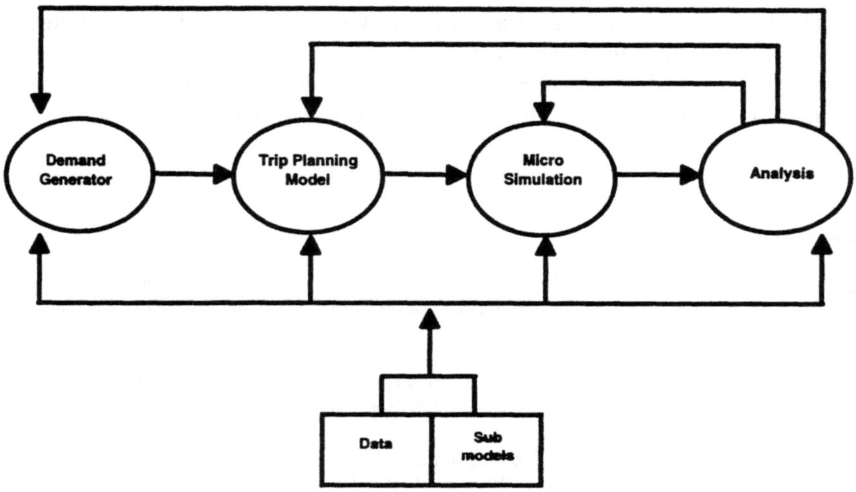

Fig 3.2 TRANSIM's architecture

As Fig. 3.2 shows, the full scope of TRANSIM extends well beyond its current capability in terms of simulating air emissions in extremely fine spatial temporal detail. Such future modules include 'demand generator' and 'a trip planning module' to be developed yet. At this point, there is little information about the theoretical basis, estimation methods or data sources on these modules. There is no basis yet to judge the relevance of these modules of TRANSIM to ISTEA-based planning.

3.5 Concluding Comments

As noted earlier, a major structural change in the American economy in the last two decades has made possible (and in turn been made possible by) a major revolution in the largely deregulated transport systems. The latter improvements (in terms of increasing efficiency, intermodalism etc.) and the public debate on transportation have in turn led to two Acts of U.S. Congress – ISTEA and CAA – which have altered the context and scope of transport planning.

Strategic planning in this new era at both the metropolitan and intercity or national levels requires new types of analytical methods and information. The paper has identified the scope of such an additional knowledge base. At the national level, much of the requisite analytical methods and data need to be created. The Bureau of Transportation Statistics (BTS) and the Office of the Assistant Secretary of Policy of the U.S. Department of Transportation have initiated initiatives in this area.

At the metropolitan level, policy modeling relevant to ISTEA and CAA may build on past transport modeling traditions; new research efforts in Activity-based travel modeling and microsimulation offer some potential. Only time can tell which analytical strands will be winnowed out. Given the importance of strategic decisions in transport in the public sector, the next decade promises to be a period of vigorous analytical research in land use, transportation, and environmental modeling – comparable to the creativity of the decade after The 1956 Highway Act.

References

Apogee Research, Inc. (1994), 'Costs and effectiveness of transportation control measures (TCMs)', Report prepared for the National Association of Regional Councils, Washington, DC.

Axhausen, K. and Gärling, T. (1992), 'Activity-based approaches to travel analysis: Conceptual frameworks, models, and research problems', *Transport Review,* vol. 12, pp. 323–341.

Batty, M. (1994), 'A chronicle of scientific planning', *Journal of the American Planning Association,* vol. 60, pp. 7–16.

Becker, G.S. (1965), 'A theory of the allocation of time', *Economic Journal*, vol. 75, pp. 493–517.

Beckmann, M. J., McGuire, G.B. and Winsten, C.B. (1956), *Studies in the Economics of Transportation,* Yale University Press, New Haven, CT.

Ben-Akiva, M.E. and Lerman, S. (1985), *Discrete Choice Analysis: Theory and Applications to Travel Demand,* The MIT Press, Cambridge, MA.

Boyce, D., Le Blanc, L.-J. and Chon, K.S. (1988), 'Network equilibrium models of urban location and travel choices: A retrospective survey', *Journal of Regional Science* vol. 28, pp. 159–183.

Damm, D. and Lerman, S.R. (1981), 'A theory of activity scheduling behavior', *Environment and Planning A*, vol. 13, pp. 703–718.

DeBresson, C. and Amesse, F. (1991), 'Networks of innovations: A review and introduction to the issue', *Research Policy*, vol. 20, pp. 363–379.

Echenique, M.H., Flowerdew, A.D.J., Hunt, J.D., Mayo, T.R., Skidmore, I.J. and Simmonds, D.C. (1990), 'The MEPLAN models of Bilbao, Leeds and Dortmund', *Transport Reviews*, vol. 10, pp. 309–322.

Fielding, G.J. (1993), 'Investigating toll roads in California', *Access*, vol. 2, pp. 22–24.

Fried, M.A., Havens, J.J. and Thall, M.T. (1978), 'Travel behavior synthesised theory: New applications to understanding travel behavior', Project 80–14, Phase I, NCHRP Program.

Goodwin, P.B., Kitamura, R. and Meurs, H. (1990), 'Some principles of dynamic analysis of travel behavior', in Jones, P.M. (ed.), *Developments in Dynamic and Activity-Based Approaches to Travel Analysis*, Gower, Aldershot, pp. 56–72.

Hägerstrand, T. (1970), 'What about people in regional science?', *Papers of the Regional Science Association*, vol. 24, pp. 7–21.

Harris, B. (1965), 'New tools for planning', *Journal of the American Institute of Planners*, vol. 31, pp. 90–95.

Jones, P.M., Dix, M.C., Clarke, M.I. and Heggie, I.G. (1983), *Understanding Travel Behavior*, Gower, Aldershot.

Jones, P.M., Koppelman, F. and Orfeuil, J.P. (1990), 'Activity analysis: State-of-the-art and future directions', in Jones, P.M. (ed.), *Developments in Dynamic and Activity-Based Approaches to Travel Analysis*, Gower, Aldershot, pp. 34–55.

Kain, J.F. and Apgar, W.C. Jr. (1985), *Housing and Neighborhood Dynamics: A Simulation Study*, Harvard University Press, Cambridge, MA.

Kim, T.J. (1989), *Integrated Urban Systems Modeling: Theory and Applications*, Kluwer Academic Publishers, Dordrecht.

Kitamura, R. (1988), 'An evaluation of activity-based travel analysis', *Transportation*, vol. 15, pp. 9–34.

Lakshmanan, T.R. (1992), 'Social change induced by technology', Invited paper presented at the Pacific Regional Science Summer Workshop, Taipei.

Lakshmanan, T.R., Han, X. and Liang, Y. (1993), 'The evolution of knowledge in the labor force during industrial restructuring in Japan', *The Annals of Regional Science*, vol. 27, pp. 41–60.

Lakshmanan, T.R. and Okumura, M. (1995), 'The nature and evolution of knowledge networks in Japanese manufacturing', *Papers in Regional Science*, vol. 74, pp. 63–86.

Landis, J.D. (1992), 'BASS II: A new generation of metropolitan simulation models', Working Paper 573, University of California at Berkeley, Institute of Urban and Regional Development, Berkeley, CA.

Landis, J.D. (1993), 'CUF model simulation results: Alternative futures for the Greater Bay Area Region', Working Paper 592, University of California at Berkeley, Institute of Urban and Regional Development, Berkeley, CA.

Los Alamos National Laboratory (1994), TRANSIM Program Description, April.

Mackett, R.L. (1983), 'The Leeds integrated land-use transport model (LILT)', Supplementary Report SR 805, Transport and Road Research Laboratory, Crowthorne.

Mackett, R.L. (1990), 'The systematic application of the LILT model to Dortmund, Leeds and Tokyo', *Transport Reviews*, vol. 10, pp. 323–338.

Pas, E.I. (1985), 'State-of-the-art and research opportunities in travel demand: Another perspective', *Transportation Research A*, vol. 19, pp. 460–464.

Pas, E.I., and Koppelman, F. (1986), 'An examination of the determinants of day-to-day variability in individuals' urban travel behavior', *Transportation*, vol. 13, pp. 183–200.

Perez, C. (1985), 'Microelectronics, long waves and world structural change: New perspectives for developing countries', *World Development*, vol. 13, pp. 441–463.

Piore, M.J. and Sabel, C.F. (1984), *The Second Industrial Divide: Possibilities for Prosperity*, Basic Books, New York, NY.

Porter, M.E. (1987), 'Changing patterns of international competition', in Teece, D. (ed.), *Competitive Challenge*, Ballinger Publishing, Cambridge, MA, pp. 27–58.

Recker, W.W., McNally, M.G. and Root, G.S. (1986), 'A model of complex travel behavior: An operational model', *Transportation Research A*, vol. 20, pp. 307–318.

Sierra Research, Inc. (1994), 'Evaluation of mobile vehicle emission model', Report prepared for Volpe National Transportation Systems Center, Sacramento, CA.

Small, K.A., Winston, C. and Evans, C.A. (1989), *Road Work: A New Highway Pricing and Investment Policy*, Brookings Institution, Washington, DC.

Transportation Research Board (1992), *Data for Decisions: Requirements for National Transportation Policy Making,* Special Report 234, National Research Council, Washington, DC.

U.S. Department of Transportation and U.S. Environmental Protection Agency (1993), *Clean Air Through Transportation*, Washington, DC.

U.S. General Accounting Office (1990), *Air Pollution: Reliability of EPA's Mobil Source Emission Model Could Be Improved,* Washington, DC.

U.S. House of Representatives (1991), *Intermodal Surface Transportation Efficiency Act of 1991*, Conference Report to Accompany HR2950, pp. 102–104.

Voorhees, A.M. (1959), 'The nature and uses of models in city planning', *Journal of the American Institute of Planners*, vol. 25, pp. 57–60.

Wegener, M. (1985), 'The Dortmund housing market model: A Monte Carlo simulation of a regional housing market', in Stahl, K. (ed.), *Microeconomic Models of Housing Markets*, Lecture Notes in Economic and Mathematical Systems 239, Springer-Verlag, Berlin, pp. 144–191.

Wegener, M. (1986), 'Transport network equilibrium and regional deconcentration', *Environment and Planning A*, vol. 18, pp. 437–456.

Wegener, M. (1994), 'Operational urban models: State of the art', *Journal of the American Planning Association*, vol. 60, pp. 17–29.

Winston, C. (1993), 'Economic deregulation: Days of reckoning for microeconomists', *Journal of Economic Literature*, vol. 31, pp. 1263–1289.

4 Excess Commuting in U.S. Metropolitan Areas

Edwin S. Mills
Real Estate and Finance
Kellogg School of Management
Northwestern University
2001 Sheridan Road
Evanston, IL 60208, USA

4.1 Introduction

Most social scientists in U.S. universities (not to mention federal, state and local government agencies) believe that workers in U.S. metropolitan areas make excessive use of cars for commuting.

Excessive use consists of excessively long automobile commuting trips and of deficient use of transit systems. Resource misallocation can also result because of deficient carpooling and because of excessive concentration of travel during peak periods. As an issue of government policy, excess commuting by car first rose to a significant number of decibels in the high oil price decade 1973 – 1982. Discussion by those who write and speak on government policy issues faded during the 1980s, but government has charged in like the proverbial bull in the china shop in the 1990s. Reductions in automobile commuting are being mandated by the national government on environmental grounds.

Serious interest among academic urban transportation specialists started in 1982 with Hamilton's (1982) "Wasteful Commuting." During the following decade, several additional papers have clarified the issues both conceptually and empirically.

The academic discussion exudes an aroma of applied welfare economics. However, none of the sources I have read confronts the government policy issue explicitly. If there is excess commuting, presumably it means from a social welfare point of view. If so, what is the set of remedies available to government and which one(s) should be chosen? In the 1990s, the federal government has taken two prominent steps to curtail automobile commuting: mandated reductions in parking spaces at places of employment, and increasingly stringent conditions for financial assistance to state and local governments for highway construction. One rarely knows what other policy options have been considered. However, the federal government appears not to have tried to improve the dreadful quality of public transit service in metropolitan areas or to have given high priority to measures that would make driving more expensive. President Clinton proposed a substantial increase in fuel taxes in 1993, but was persuaded that the proposal was not politically viable. Nor has the government taken obvious steps to ensure that cars on the roads maintain legally required pollution control standards. My judgment is that mandated reductions in employer-provided parking spaces is no higher than fifth on a priority ranking of socially desirable government programs.

The problem, if it exists, is primarily restricted to metropolitan areas. There is no evidence of excess commuting, and not much of dirty air, outside metropolitan areas. Commuting times and distances are shorter in small than in large metropolitan areas. On a per-worker basis, congestion may be worse in some small than in some large metropolitan areas. Air quality is generally worse, but not necessarily very bad, in large metropolitan areas, and government attention has focused on large metropolitan areas. The analysis in this paper is relevant to all metropolitan areas, although the paper is motivated by conditions in large metropolitan areas, partly because I live in one.

4.2 Literature Summary

Hamilton's characterization of the problem is insightful and led to the first measurable concept of excessive commuting. He employed estimates of the pattern of decline of the densities of workers resident and jobs with distances from the centers of 14 metropolitan areas (1970s data) for which relevant data were available. Then he calculated the cumulative number of jobs and the cumulative number of workers resident from the center of the metropolitan area to each distance from the center. Typical cumulative functions are graphed in Fig. 4.1. Cumulative jobs and workers resident are zero at the center and equal to each other and to the total working population of the metropolitan area at the edge. At each distance, the excess of the number of jobs between there and the center over the numbers of workers resident between there and the center is the number of workers who must commute toward the center through that distance. Adding over the entire metropolitan area gives the minimum total worker/miles of commuting. This is the minimum amount of commuting needed in the metropolitan area given the locations of jobs and worker residences. Other things equal, workers are motivated to choose residential locations so as to achieve the minimum commuting distance for the metropolitan area. If the jobs curve is everywhere higher than the residences curve in Fig. 4.1, as it is in fact, all that is required in order to achieve the minimum commuting pattern is for each worker to live farther from the center than he/she works. No worker would rationally commute away from the center on the way to work because he/she would pay greater housing plus commuting costs than if he/she moved at least as far from the center as his/her job, effectively exchanging dwellings with a worker who lived beyond the first worker's job and worked closer to the center than the first worker's residence. Both workers would be made better off by the exchange if land and housing are priced competitively.

Hamilton found that the mean optimum commute was only 13 percent of mean actual commuting distance in his 14 metropolitan areas. Thus, his conclusion that 87 percent of commuting is "wasteful."

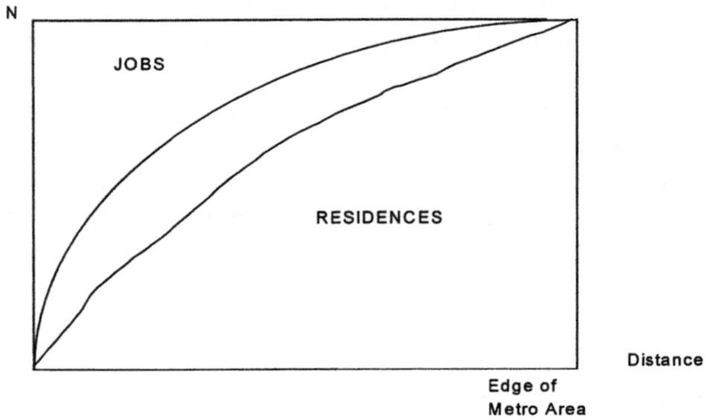

Fig. 4.1. Cumulative jobs and residences

Note that the model of the metropolitan area, for which the present author bears some responsibility, is two-dimensional. Densities and distances matter, but not directions. In fact, any commuting not directly toward the metropolitan center would be wasteful in the model. All residences at a given distance from the center are equally costly and a non-radial commuter could shorten the trip without increasing his/her housing cost by moving to a residence the same distance from the center as the original residence, but on the same radial line from the center as his/her job.

Real metropolitan areas do not resemble the model very closely. Most important for present purposes, clusters of jobs are located not only at the metropolitan center (the central business district) but also at various sub-centers that may be 5–35 miles from the center. Large sub-centers are sometimes referred to as "edge cities," although they are typically not near the edges of metropolitan areas. They are, however, important and may contain more jobs than many CBDs.

Residences cluster near sub-centers just as they cluster near CBDs, and for the same reason. It is easy to imagine that taking explicit account of sub-centers could have an important effect on calculations of excess commuting.

Implications for commuting distances of alternative spatial arrangements in a metropolitan area have also been discussed in the land use/transportation modeling literature since the mid-1980s. However, that literature does not embed the excess commuting concept in any more sophisticated welfare economics model than does the recent U.S. literature, to which attention is next addressed.

The foregoing induced White (1988) and others, most recently Small and Song (1992), to redo estimates of excess commuting. White has written several fine papers on sub-centers. White employs data from the *Journey-to-Work* volume of the 1980 census. It provides a complete origin-destination matrix showing community or county of residence and of work, along with data on minutes spent commuting for a sample of U.S. workers living in large metropolitan areas. Thus, employment sub-centers are accounted for realistically, at least at the level of geographical detail that the data permit. Using an assignment program, White's computer reallocates workers and their jobs among communities so as to minimize total commuting time subject to fixed numbers of jobs and residences in each community. She compares her calculated minimum with actual minutes of commuting time between each origin and destination obtained from the census volume. She concludes that only 11 percent of actual commuting time is wasteful.

Many workers live and work in the same community or county. The census volume does not provide commuting times for such workers and White assumed they were zero. Small and Song (1992) redid the commuting analysis for Los Angeles County (a very large county, covering 1289 square miles). They had data with much more spatial detail than White's, data provided by the State of California for 706 zones within the county, and travel time and distance data for 3.04 million workers who both lived and worked in the county in 1980. Using an assignment program similar to that used by White, Small and Song conclude that 65–69 percent of commuting is excess, depending on whether the commuting measure is time or distance. (Small has reported in correspondence that 1990 data support their conclusions from 1980 data.)

These are highly disparate calculations. An important part of the explanation is that intra-community commutes are a large part of the total. When Small and Song take them into account, they get results much closer to Hamilton's than to White's, whose data did not permit their inclusion. Evidence presented in the three papers suggests that, although actual commuting time and distance are correlated with metropolitan area population, the percent that is excess is not correlated with metropolitan area size. Impressionistic evidence suggests that excess commuting in Los Angeles might be a greater percentage of actual commuting than that in other metropolitan areas. Los Angeles has highly dispersed workplaces and residences, and it would appear that minimum commuting times and distances would be quite small. Yet major commuting highways are demonstrably congested during long periods of work days. All Angelinos believe that their commutes are interminable, yet Small and Song's data indicate an average commuting time of only 22 minutes and an average distance of only 10 miles, not far from national averages. These figures nevertheless lead to a large calculated percentage of excess commuting. Local evidence indicates that, commuting times, if not distances, lengthened somewhat between 1980 and 1990 in Los Angeles County.

In fact, employment has suburbanized rapidly in virtually all U.S. metropolitan areas during the post-World War II period. That should have reduced commuting times and distances. Commuting distances have actually increased by moderate amounts but quite steadily during most of the postwar period, and including the 1980s. Commuting times appear to have fallen during the same periods. There is independent evidence of increases in commuting travel speeds. Although contrary to popular impressions of increasing congestion, these conclusions are supported by all reliable data. See White (1994) for a full discussion. I believe there is a prima facie case that U.S. metropolitan workers engage in excessive commuting, but there is no substantial evidence that the situation worsened during the 1980s.

4.3 Characterization of the Problem

The papers summarized in the last section have provided valuable calculations concerning an important issue. Certainly, whether U.S. workers commute excessively is a justifiable social and government policy issue. The level of public concern on the subject is high. Many scholars have written on congestion (excess crowding) and Downs (1992) has put the subject in a broad context of government policy.

Nevertheless, the issue of excessive commuting has not yet been carefully posed as a subject of applied welfare economics or of government policy. The purpose of this section is to do so. In defense of the authors whose work was discussed above, their primary goal was to test a model of urban spatial structure, not to do applied welfare or government policy analysis. However, the three papers are the best starting point for an analysis of the applied welfare issue.

Nobody would think that jobs and dwellings are homogenous. Employers are far from indifferent among workers and workers are far from indifferent among jobs, employers or dwellings. Workers and their families care about dwelling characteristics, which vary among dwellings and among communities. In addition, in the United States more than in Europe, crime rates, school quality and local tax rates vary greatly among communities within a metropolitan area. A large literature of hedonic housing demand studies attests to and measures these preferences. There is no reason to presume that such tastes are minor or frivolous or that they are less legitimate concerns of welfare economics than are commuting time and distance. To do the assignment calculations using a detailed list of worker preferences among dwellings and communities, to be matched with a detailed list of worker capabilities and employer specifications of needed worker capabilities would be inconceivable. Any attempt to do so with conceivably available data would be criticized as having omitted or misspecified important variables. The excess commuting literature cannot possibly lead to any welfare economics conclusion. If that is true, it is unclear what goal other than model testing it might have or why the findings should be referred to as "excess" commuting. The answer probably is that it exposes a stone under which scholars should look for welfare implications.

Why might there be excess commuting in the welfare economics sense? Economists' parrot-like response is that there is excess commuting if the price of commuting is below its welfare optimum. I am very fond of that parrot and intend to teach it to analyze excess commuting in the remainder of this paper.

Most analysis of road use pricing in the economics of urban transportation literature concerns congestion pricing. The point of the congestion pricing model is that social efficiency requires that road users pay the full social cost of their use of the road. Road users automatically pay their user cost: their vehicle user cost and their time cost. Vehicle user cost includes fuel and related charges, maintenance and repair costs, insurance, (plus the actuarial value of uninsured accident costs), depreciation, interest on debt on the vehicle, and foregone interest on the owner's equity in the vehicle. In the very short run, if the driver owns the car and considers whether to take a particular trip, some of these costs are fixed. For my purposes, long run cost is appropriate and all the above costs should be included. The user's time cost arises because travel time could be used in other valuable ways if velocity were greater or the trip were shorter. Many estimates have been published of the dollar value that commuters actually or implicitly impute to the time cost of road travel.

In addition to the user costs borne by road users, vehicles on a congested road impose an external diseconomy on each other. Defining a congested road as one on which there are enough vehicles so that vehicle speed decreases as the number of vehicles per hour increases, it follows that each user of a congested road imposes a cost on other users equal to the additional time cost that each user causes others users because of their decreased speed. Although an additional user imposes only slight additional time cost on each other user, there are many other users. Estimates indicate that the external cost may be several times as large as the user costs borne by each user. Assuming that the number of users of the road is a decreasing function of the cost borne by each user (i.e. the demand curve for road use slopes downward), a well known theorem states that optimum road usage results when a congestion fee is imposed on all users equal to the value of the external diseconomy. The congestion fee is evaluated at the road usage at which the sum of the user cost plus the congestion fee equals the demand price. (See Mohring, 1976.)

Less widely appreciated is that, although the congestion pricing model is long run relative to vehicle costs, it is short run relative to right-of-way costs; it assumes that the capital cost and capacity of the road are fixed. In the long run, the capacity of roads can be and is varied. If the optimum congestion fee exceeds the capital cost of additional capacity, it implies underinvestment in roads. Optimum road capacity is that at which the marginal cost of additional capacity equals the demand price for road use.

At the socially optimum road capacity, there is no congestion. The fee per vehicle mile for road use equals the demand price equals the long run average capacity cost per vehicle mile equals the long run marginal cost per vehicle mile, valuing all inputs (especially land costs) at their opportunity cost. There is, however, crowding at the socially optimum capacity. Land values are much higher in large urban areas than elsewhere. In such places, land costs are a large part of capacity costs. Where land is valuable, it is optimum to use it intensively. The private sector does that by building tall buildings on valuable land. The analog in road transportation is to put large numbers of cars per lane mile on expensive land. This result merely states, in transportation terms, the usual welfare economics result that the optimum long run production quantity of a commodity is that which equates long run marginal (equals average) cost to the price that clears the market. The transportation application was first proved by Mohring (1976). For a recent survey see Small (1992).

Extending the analysis to a multi-modal model merely requires that the above condition apply for each mode. If one mode, specifically fixed rail transit, is subject to increasing returns beyond optimum capacity, things are more complex.

However, the best studies (see Small, 1992) conclude that additional fixed rail systems in U.S. metropolitan areas are unwarranted. All large metropolitan areas (those with populations in excess of three million people), with the exception of Detroit (where the issue is political), already have fixed rail systems. The low densities of smaller metropolitan areas and of most suburbs of even large metropolitan areas preclude the achievement of riderships that can justify investments in fixed rail systems. For example, the interest cost per ride for the capital cost of the D.C. fixed rail transit system is about $10.00, even though the D.C. metropolitan area population exceeds three million. Things are worse in Atlanta, Baltimore, Dallas and Los Angeles, where fixed rail systems are now being built, mostly with federal money.

Bus systems are different in that they use the same rights-of-way as cars. Scale economies are modest with bus systems (see Small, 1992). In principle, efficient express bus systems could be socially cheaper than cars for many drivers in many metropolitan areas, but no U.S. metropolitan area has a bus system that makes its use anywhere near socially superior to automobile commuting for any but low income commuters. A slow, unreliable or infrequent bus system makes the time costs of bus commuting large relative to those of a car-based system. Reserved bus lanes are a current enthusiasm among some transportation specialists. They cause scale economies of bus systems to be comparable with those of fixed rail systems. Large numbers of buses per lane mile are required to use a reserved bus lane efficiently, requiring more passengers per corridor than most U.S. metropolitan areas can generate. If the reserved bus lane is used well below capacity, the result is sheer social waste, since congestion is worsened in the remaining lanes available to cars.

Thus, an important question is the magnitude of the long run marginal cost of a car and highway commuting system.

Should users pay the fixed cost of a road system? Ever since Pigou, economists have argued that charging users the fixed cost of an uncongested facility is wasteful in that some potential users would thereby be deterred even though benefits of use to them would exceed social marginal costs. The argument cannot be faulted on its own terms. However, no one should feel guilty about peak load users paying the capacity cost that their use requires. Off-peak users are subjects of legitimate concern. With sophisticated metering of road use, they can be charged lower fees. However, even at the same fees paid by peak users, their cost per mile is less since the road is optimally crowded at peak times, but less crowded, and therefore faster at off-peak times. The same concern could be voiced about almost any private sector in the economy. Nearly all retail, and many business, facilities are subject to peak usage, yet off-peak users typically pay the same charges as peak users, and economists are not greatly exercised about the fact.

An important issue is how road users should be charged for road use. Many proposals have been made and some have recently been tried: tolls collected at toll booths, perhaps electronically; electronic metering of road use with bills sent to users at the end of the month; and fuel taxes. Each has its advantages and disadvantages; accuracy, capital costs and collection costs (both those of the collecting agency and those imposed on users) are the important ones. In the U.S., at least, an important issue would be illegal behavior to avoid fees. If users refused to pay their road use bills at the end of the month, the police would be forced to be collection agents; some people would demolish electronic gear in roads. In addition, users would become demoralized if there were many errors in charges in billing. Any sophisticated system would be several times more complex than the most complex activity of U.S. local governments.

As a practical matter, there is much to be said for fuel taxes. They are by far the cheapest taxes to collect; they are almost impossible to evade; they can easily be set at finely graded levels by metropolitan and non-metropolitan area; large road users, such as trucks, automatically pay more; and there is experience with their collection in every country. Their limitation is that, although they can be finely varied by time and place of sale, they cannot be finely varied by time and place of use. Off-peak users cannot be charged lower fuel taxes than peak users. A uniform national tax could be partially refunded on the national income tax form to those who neither lived nor worked in a metropolitan area, or in a large metropolitan area. As a good approximation, it would not be desirable to refund part of the fuel tax to those who work at home or take transit vehicles to work, since their fuel purchases would reflect that fact.

4.4 Estimates of Long Run Costs and Optimum Commuting

In this section I present rough estimates of socially optimum long run commuting and commuting cost in U.S. metropolitan areas. If commuting costs are below the social optimum, then there is excess commuting. Excess commuting may mean too little carpooling or public transit use. My guess, based on the excess commuting literature reviewed in Section 4.2, is that, in U.S. metropolitan areas, it mostly means that workers simply live too far from their jobs. Low fuel taxes permit workers and their families to base residential decisions on a large variety of relatively minor criteria, such as locations of friends' and relations' residences, religious meeting places, social institutions, favorite schools and familiar shopping facilities. These are not irrational concerns, but they entail resource misallocation if they are based on excessively low commuting costs. Incidentally, it would border on criminality if government agencies succeeded in their attempts to coerce or induce commuters to use public transit to a greater extent than now unless government also ensures that the quality of its service is greatly improved.

The calculations in this section are implicitly based on a one-worker household. Calculations for two-worker households would be more complex, but qualitative results hold unless the residence is located on the straight line connecting the two work places. U.S. data are very good about time and distance of commutes, but normally do not distinguish between one-and two-worker households. U.S. data show that women commute shorter distances than men, but that observation is of no help.

A final preliminary issue is that building more highways is probably not the best way to expand road capacity in most U.S. metropolitan areas. Most careful studies conclude that electronic control of traffic lights, controls on access, reverse direction lanes and streets and other such investments could increase metropolitan road use capacity by at least 25 percent. In Chicago, such things could be accomplished for a few tens of millions of dollars, whereas building or widening roads starts in the hundreds of millions or in the billions. Especially if existing structures must be demolished, building or widening metropolitan area highways is extremely expensive. However, I have no data on the cost of improved traffic controls that is of comparable quality to that on road building and widening, so I focus attention on the latter here.

I proceed as follows. I first ask what is the cost of commuting borne by a typical metropolitan commuter. I then ask what is the long run social cost per commuter mile and how high a fuel tax would be needed to bring the cost borne by the commuter up to the social opportunity cost. Finally, I ask what would be the likely effect of an optimum fuel tax on commuting distances. The data presented here are crude and approximate, intended to suggest likely magnitudes. Small (1992) has reviewed a large number of relevant studies, and much of the data I use is taken from his work and from the studies he reviews.

4.4.1 Present Costs

I take as representative an auto commuter on an arterial highway (not an expressway) in a U.S. metropolitan area.

At present, commuting cost per mile is about \$.42. Car cost is about \$.28 per mile and time cost is about \$.14 per mile.[1] Average fuel cost is about \$.04 per mile. (The other \$.01 in note 1 is oil cost.) Of the \$1.10 per gallon of fuel cost, total federal and state fuel taxes are about \$.20, or \$.01 per vehicle mile.

The average one way commute in a good size metropolitan area is about 10 miles. Assuming an average speed of 25 miles per hour, the trip takes .4 hour or 24 minutes. (That figure is a bit greater than the national average for 1980, but probably about right for a large metropolitan area in 1990.)

[1] An average car is worth about \$10,000, assuming it is a few years old. It gets about 21 mpg and is driven about 10,000 miles per year with fuel that averages \$1.10 per gallon, giving fuel and oil cost about \$.05 per mile. Depreciation is about \$1,000 per year, or \$.10 per mile. Adding insurance, licensing, maintenance and interest cost (8% of the value), we get \$.28. An average worker earns \$10 per hour and values commuting time at about one-third of the wage rate. An average commuting speed is about 25 miles per hour. These figures imply a time cost per mile of about \$.14. These data are from Small (1992) and Small and Gómez-Ibánez (1996).

The data in the Statistical Abstract (1993), Table 1037, p. 625 are from the Automobile Manufacturers Association. They are too high because depreciation is based on a new car. That excess cost is partly offset by the failure to include foregone interest on the car owner's equity. (The AMA data include interest paid on car loans.)

4.4.2 Optimum Fuel Charge

In this subsection, I compute the optimum fuel tax based on an optimum long run user charge for a metropolitan area car commuter. Again, I use data that appear to be typical for a large metropolitan area. Again, I employ fine data from Small (1992) for an arterial road. Although I calculate costs per lane mile, knowing the number of lanes is necessary, since all recent studies conclude that there are increasing returns with respect to the number of lanes for highway construction. Construction costs for expressways are considerably greater than those for the arterial highway considered here, but expressways carry more vehicles per lane, so capital costs per vehicle mile are similar to those calculated here.

Construction costs depend very much on where the highway is built in the metropolitan area. In a distant suburb, land is cheap and only few structures need to be bought and demolished. In a central city, land values may be 5 or more times as high as in a suburb, and expensive buildings may need to be demolished. The data employed here are for land that contains buildings of only small value.

Small (1992) estimated that the construction plus land cost per lane mile of a six lane arterial highway was $1.3 million (1989 prices) or about $1.4 million in 1992 prices. The rush hour traffic per lane might be 1,000 cars per hour, for four hours per five day work week (the assumption is that commuting is the peak burden that justifies the highway, and that rush hour traffic is about the same in each direction). Assuming 260 work days per year, (plus nominal usage on non-work days), the lane carries 1.4 million cars per year. Thus, capital cost per vehicle mile is about $1.00.

Assume that interest and depreciation imply capital user cost of 10 percent per year or $.10 per commuter mile. Add $.02 for maintenance, snow removal and police and the total use cost is $.12 per vehicle mile. Using the 21 mpg figure quoted above implies an optimum fuel tax of $2.50 per gallon, to be compared with present fuel taxes of $.20 per gallon, or $.01 per mile. In fact, a $2.50 per gallon fuel tax plus the present non-tax fuel cost of $.90 per gallon gives total fuel cost of $3.40 per gallon. That is close to prices in most European countries.

Raising the fuel tax from $.01 per mile to $.12 per mile would add $.11 to the cost per car mile, raising total commuting costs per mile from $.42 to $.53, or a little more than 25 percent. If the land to be acquired were centrally located and had valuable buildings on it, the calculated 25 percent increase could easily be 50 percent or more.

4.4.3 Effects on Commuting Distance

How much effect would a 25 percent increase in commuting cost per mile have on commuting? Needless to say, discussion of this question must be conjectural. As has been mentioned, some workers would be induced to carpool and some to substitute public transit, bicycles, walking or working at home for auto commuting. However, the great origin-destination diversity of workers who live and/or work in U.S. metropolitan suburbs implies that most such options are of only marginal practicality. My conjecture is that the largest effect of a 25 percent increase in commuting cost would be to shorten commuting trips, mostly by inducing workers to move their residences closer to their jobs. Most movement would probably be city-to-suburb and suburb-to-suburb. Suburb-to-city moves are seriously deterred by city problems of crime, schools and taxes.

How much reduced commuting might be induced by a large increase in fuel taxes appears not to have been studied previously. I present two approaches to the subject, based on existing data and analyses.

First, think of a worker who lives in one suburb and works in another that is 10 miles away. Both contain similar housing and similar amenities. Suppose that the main deterrent to moving is the transaction cost of finding a house, paying for the move, arranging the financing, and selling one house and buying another.

Some scholars have estimated that the transaction cost of such a move might be 15 percent of the house's value. To keep things simple, suppose that the house sold and the one bought each cost $100,000. That is close to the median house price in U.S. metropolitan areas and is a bit more than four times the annual median wage of $10 per hour. Fifteen percent of $100,000 is $15,000.

Suppose the fuel tax is raised to my calculation of the optimum, implying a 25 percent increase in commuting cost per mile, from $.42 to $.53. If the worker moves half way toward his/her job, $5.30 per day is saved by the 10 mile reduction in the daily round trip commute. Assuming a 240 day work year[2], the annual saving is $1270. That implies that the savings in commuting cost from the move will pay the transaction cost of the move in less than twelve years. (The average tenure of owner-occupants in the U.S. is a little less than 12 years.) If the move is viewed as being long term, the annual interest cost of the moving transaction cost, at an 8 percent interest rate, is $1,200, about the same as the annual saving in commuting cost. Of course, the saving from a move to a dwelling even closer to work would be greater. Indeed, even at the present $.42 cost per commuter mile, a move to a dwelling next to the work place would save $2016 per year, which would give a payback period of a little more than 7 years for the moving cost. We do not know the value to workers of dwellings that are long distances from work relative to the value of dwellings that are close to work. The calculations presented here suggest that whatever moves might result from my proposed increase in fuel tax would be moves to dwellings much closer to work. If 20 percent of commuters were induced to make moves that were 6 miles closer to work, the result would be a 60 percent reduction in commuting distance of movers and a 12 percent ($.6 \times .2$) reduction in total commuting distance.

Second, there are a few direct estimates of commuting demand elasticities in the literature (see Small, 1992). An ingenious estimate by Chan and Ou (1978) separates cost and service quality (as measured by time required). For Boston, the estimated elasticities are -.8 and -.5; for Louisville, they are -.4 and -1. If the elasticity for my above composite cost and time calculation were -.5, then the 25

[2] The 240 work days are for a given worker. The 260 day figure used previously is for number of days on which substantial commuting takes place.

percent increase in commuting cost implied by my calculation of the optimum pricing would reduce commuting distance by about 12 percent. That would be a 1.2 mile reduction in a 10 mile average trip.

Thus, my guess is that increasing the fuel tax to the social optimum might reduce car commuting by 10–15 percent by virtue of moves induced. If the fuel tax increase were accompanied by vigorous attempts to improve the quality of public transit, the effect of the two combined changes could easily reduce car commuting by 15–20 percent. Most of the difference between 10–15 percent and 15–20 percent would probably be in city-city and inner suburb-city commuting.

Finally, increasing fuel taxes by a factor of 10 would certainly reduce non-work travel, which is more miles of car travel than are work trips in U.S. metropolitan areas. That would be of value in reducing fuel use and auto emissions, but would not be of much value in reducing needed road capacity.

References

Chan, Y. and Ou, F. J. (1978), 'Tabulating demand elasticities for urban travel forecasting', *Transportation Research Record*, vol. 673, pp. 40–46.

Downs, A. (1992), *Stuck in Traffic*, Brookings Institution, Washington, DC.

Hamilton, B. (1982), 'Wasteful commuting', *Journal of Political Economy*, vol. 90, no. 5, pp. 1035–1053.

Mohring, H. (1976), *Transportation Economics*, Ballinger Publishing Co., Cambridge, MA.

Small, K. (1992), *Urban Transportation Economics*, Harwood Academic Publishers, Philadelphia, PA.

Small, K. and Song, S. (1992), ''Wasteful' commuting: A resolution', *Journal of Political Economy*, vol. 100, no. 4, pp. 888–898.

Small, K. and Gómez-Ibánez, J. (1996), 'Urban transportation', Mimeo.

U. S. Bureau of the Census (1993), *Statistical Abstract of the United States 1993*, U.S. Census Bureau, Washington, DC.

White, M. (1988), 'Urban commuting journeys are not 'wasteful'', *Journal of Political Economy*, vol. 96, no. 5, pp. 1097–1110.

White, M. (1992), 'Location choices and commuting behavior in cities with decentralized employment', *Journal of Urban Economics*, vol. 24, no. 2, pp. 129–152.

White, M. (1994), 'Housing and the journey to work in U.S. cities', in Noguchi, N. and Poterba, J. (eds), *Housing Markets in the United States and Japan*, University of Chicago Press, Chicago, IL.

5 Household Commuting: Implications of the Behavior of Two-Worker Households for Land-Use/Transportation Models

Peter Gordon
Yu-chun Liao
Harry Richardson
School of Urban Planning and Development
University of Southern California
Los Angeles, CA 90089, USA

5.1 Introduction

Single-worker households (and one worker's commuting trade-offs) populate the world of urban economic theory. Alonso (1964) elaborated the locational choices facing "an individual"; it made no difference if that individual was a member of a multi-worker household because all individual workers would commute to the same city center. More complex models (e.g. White, 1977) assume two work sites (the CBD and a "suburban ring") with all females commuting to the nearby suburban ring. Women's worktrip times would tend to be longer if by transit (their value of time is less than that of men) but shorter if by private automobile (because they generally earn less, and here they tend to economize on the monetary costs of commuting). Alonso (1980) wrote about the rise of the two-worker household in the U.S. and speculated that, " ... when there are several commuters rather than only one per household, and when non-working time is more valuable, ... locational factors may play a stronger role, ... this factor may result in more concentrated development to reduce travel time. The most convenient point to two distant suburban work places may in many cases be a location more near the center." (Alonso, 1980, p. 550). This view is challenged by Giuliano (1989) and Giuliano and Small (1993). Empirically, although the proportion of multi-worker households continues to expand, more concentrated development has not occurred. Census data on suburban (inside metropolitan areas but outside central cities) population (corrected for area redefinitions by the Population Division of the Bureau of the Census, Forstall (1994)) show that the U.S. population continues to suburbanize. Suburban residents were 23.3 percent of the population in 1950, 30.6 percent in 1960, 37.2 percent in 1970, 44.8 percent in 1980 and 46.2 percent in 1990. Even this trend may understate actual population dispersion since it does not account for settlement beyond metropolitan area boundaries and into the "exurbs". This peripheral expansion probably explains the slowing down of suburban ("inner ring") growth in the 1980s.

We restrict our analysis in this paper to private auto commuters because transit use in the U.S. remains negligible. We find that many male workers in two-worker households have the shorter commute.

The one-worker household and single CBD workplace assumptions drastically simplify urban land use models by eliminating the possibility of both multidirectional and cross-commuting. Of course, they increase the tractability of the model, and allow transportation (i.e. commuter) demand to be estimated along any route. But this analytical simplification is bought at a heavy price. If the majority of households has more than one worker, and only a small proportion of workers in multi-worker households are employed at the same location, convincing transportation/land-use models must recognize the importance of multidirectional commuting. The need for multiple workplace/multiple residential neighborhood approaches implies both 1) the final abandonment of continuous for discrete models (Gordon and Moore, 1989); and 2) the limitations of theoretical models that do not take account of real-world commuting behavior. If we are going to develop a better understanding of the many trade-offs involved in the locational choices made by multi-worker households, a first step involves a better understanding of actual household commuting.

In the U.S. single-worker households are now a minority (Table 5.1; the data are for families but approximate similar trends for households). In 1992, there were 44.1 million "families with earners" of which 22.1 million included two or more earners. Between 1980 and 1992, the number of one-earner families dropped by 17.6 percent, while the number of married couples working increased by 24.9 percent.

Table 5.1. Number of earners and median weekly earnings of families by type of family, 1980 to 1992

Characteristic	Number of Families (1,000)		Median Weekly Earnings (dollars)	
	1980	1992	1980	1992
Total Families with Earners[1]	41,162	44,137	400	688
Married-couple Families	33,825	34,280	433	779
One Earner	14,797	12,200	303	469
Husband	12,127	8,667	336	546
Wife	2,059	2,801	159	294
Other Family Member	611	732	163	282
Two of More Earners	19,028	22,080	535	942
Husband and Wife Only	12,990	16,225	507	913
Husband and Other Family Member(s)	2,369	1,617	557	848
Wife and Other Family Member(s)	426	597	350	592
Other Family Members Only	139	158	356	589

[1] Excludes families in which there is no wage or salary earner or in which the husband, wife, or other person maintaining the family is either self-employed or in the Armed Forces.

Source: U.S. Bureau of Labor Statistics, Bulletin 2307, and *Employment and Earnings*, monthly, January issues.

The increased participation of women in the labor force means that the majority of women, regardless of ethnic group, now work and that gender differences in labor participation are converging (Table 5.2). Extrapolations suggest that 63.5 percent of white, 61.7 percent of black and 58 percent of hispanic women will be in the labor force by the year 2005. This trend has numerous and profound consequences. It has been well studied in the transportation field, giving rise to several widely corroborated findings. Among these is that women, on average, have shorter commutes. Explanations for this difference abound, with many authors emphasizing that they reflect labor market hurdles facing females (Gordon et al., 1989).

Table 5.2. Civilian participation rates by ethnicity and sex 1970 to 1992 and projections 2000 to 2005

Race and sex	Participation Rate (percent)						
	1970	1980	1985	1990	1992	2000	2005
Total[1]	60.4	63.8	64.8	66.4	66.3	68.7	60.0
White	60.2	64.1	65.0	66.8	66.7	69.3	69.7
Male	80.0	78.2	77.0	76.9	76.4	76.7	76.2
Female	42.6	51.2	54.1	57.5	57.8	62.3	63.5
Black	61.8	61.0	62.9	63.3	63.3	65.7	65.6
Male	76.5	70.3	70.8	70.1	69.7	71.0	70.2
Female	49.5	53.1	56.5	57.8	58.0	61.2	61.7
Hispanic	(NA)	64.0	64.6	67.0	66.5	69.3	69.9
Male	(NA)	81.4	80.3	81.2	80.5	81.8	81.6
Female	(NA)	47.4	49.3	53.0	52.6	56.6	58.0

[1] Beginning 1980, includes other races not shown separately.

Source: U.S. Bureau of Labor Statistics, Bulletin 2307, *Employment and Earnings* monthly, January issues, *Monthly Labor Review,* November 1991, and unpublished data.

We examined data from the 1983 and the 1990 Nationwide Personal Transportation Surveys (NPTS) to learn more about households with two commuters. There are some differences between the 1990 Survey and preceding surveys, including the 1983 Survey which is the main basis for comparison here. The two most notable differences are that the 1990 Survey was a telephone survey rather than an in-person home interview and that the sample size in 1983 was much smaller, about 6,500 households as compared with 22,000 in 1990. With respect to the difference in the nature of the surveys, the questions were more or less identical. However, the use of computers in the 1990 Survey may have made that more accurate because on-line data consistency checking was possible. On the other hand, reliance on the telephone may have resulted in an undersampling of very low-income households.

The smaller sample of 1983 may have resulted in larger sampling errors, resulting in an underestimation of vehicle miles traveled, an underestimation of transit trips (especially in the larger metropolitan areas), and an undercount of low-income and minority households (one of the factors explaining the underestimation of transit use). The transit problem is not very serious in this research because of our focus on private vehicles.

Another difference is that the 1990 Survey permitted proxy interviews from other household members that are presumably less accurate than if the individual had been present. This could have contributed to the greater number of trips reported than in the earlier surveys, although the increased tripmaking is supported by other travel surveys.

Although these limits on comparability between 1983 and 1990 should be kept in mind, they do not seriously undermine the analysis. As far as the 1990 Survey results are concerned, the travel times reported are remarkably similar to those reported in the 1990 Population Census for comparable trip types. However, both Census and NPTS data are limited to counties. Hence, they cannot fully explore at the microlevel suburban development, reverse-commuting, and suburb-to-suburb commuting. Furthermore, the NPTS does not list specific destinations and the origin (the home) is not locationally specified (for more detail on all these points, see Hu and Young (1993)).

In order to investigate residential location (vis-à-vis the workplace) and its consequences, we focussed on *direct* (nonstop) worktrips only. Also, given that Liao (1993) has shown that the proportion of linked ("in-chain") worktrips increased from 14.7 percent of all worktrips in 1983 to 19.2 percent in 1990, intertemporal comparisons based on all worktrips would be misleading. The sample was also restricted to residents of metropolitan areas and to one and two-worker households where each worker conducted one round-trip to work each day. All the data reported in the tables are for one-way trips. Most of the analysis is also restricted to private vehicle drive-alone trips to remove the distorting effects of mode choice on distances (and especially times).

5.2 Findings

Recent studies of commuting behavior (using the 1990 Census and the NPTS) and of non-work travel (using the NPTS) have established that transit and carpool use have continued to decline while the use of single-occupied vehicles has continued to increase. The same appears to be true no matter how many workers are in the household (Table 5.3). The proportions of workers driving to work alone in private vehicles in two-worker and one-worker households in 1990 were almost the same. This reflects the rise of multi-car households; between 1960 and 1990, the vehicle-household ratio in metropolitan areas increased from 1.0 to 1.6 (Table 5.4). Both household types showed a decline in transit use and in carpooling between 1983 and 1990; for example, for nonstop worktrips the combined *shares* of ridesharing and transit use fell by 7.58 percent (from 23.21 to 15.63 percent) for one-worker households and by 5.37 percent (from 21.62 percent to 16.25 percent) for two-worker households. Transit use among two-worker households was perhaps surprisingly below that of the general workforce in both years (although two-worker households were more likely to carpool, not necessarily together).

Table 5.3. Distribution of commuting mode choices (all household members) by number of workers per household, 1983 and 1990 (nonstop worktrips only)

Number of workers in household	Private Drive-Alone		Private Drive w/Others		Public Transportation		All Others	
	1983	1990	1983	1990	1983	1990	1983	1990
1 worker	73.93	82.56	15.21	9.73	8.00	5.90	2.85	1.81
2 workers	75.36	82.48	16.49	12.45	5.13	3.80	3.02	1.28
3 workers	72.06	78.13	20.11	15.98	–	4.41	4.45	1.49
More than 3 workers	71.64	70.10	22.39	21.08	–	6.62	–	–
All	74.16	81.33	16.92	12.40	5.68	4.74	3.23	1.53

Our previous work on recent commuting trends (Gordon and Richardson, 1994) implied that the suburbanization of jobs and residences has allowed most workers to commute over longer distances, but because of higher speeds on less congested roads, they are paying only a modest cost (if any) in extra time traveled. The same result is found to be true when households are disaggregated by number of workers or when the various commuters in two- worker households (long tripmaker and short tripmaker) are studied. Table 5.5 shows that higher speeds were reported by all workers, regardless of household size and regardless of place of residence (inside or outside central cities).

Two other findings drew our attention: 1) comparing one-worker with two-worker households, total commuting time and distance for the latter are approximately double that of the former (Table 5.6) in two-worker households; but 2) the two journeys-to-work are not equal; rather there is a short and a significantly longer commute; or, as a second worker is added, marginal commuting times and distances fall (Table 5.5); lower marginal commuting costs may in many cases reflect the marginal change in earnings contributed by the second earner (Table 5.1); this result holds for inside as well as for outside central city residents. Putting these results together, we find that in one-worker households the distance (and time) to work is approximately one-half of the *average* for the two workers in the two worker household.

Most authors writing about household commuting (Madden, 1981; White, 1977) assume that the shorter commute is by the female household member. Our data show that whereas a majority of the shorter commutes in two-worker households is by females, in 1990 more than two-fifths (41 percent) of the shorter trips were by males. We cannot determine from the NPTS data set whether these short-distance commuters are the low-income earners in their respective

Table 5.4. National trends 1960–1990 factors, U.S. totals compared to metropolitan areas with over one million inhabitants

Factor	U.-S. Totals	Metropolitan area totals
Population		
1960	179,323,175	77,175,875
1990	248,709,873	123,814,261
Percent change	38.69	60.43
Workers		
1960	64,655,805	29,033,438
1990	115,070,274	59,704,401
Percent change	77.97	105.64
Male Female Worker Breakdown		
1960	67.7% / 32.3%	66.4% / 33.6%
1990	54.7% / 45.3%	54.4% / 45.6%
Persons/household		
1960	3.33	3.24
1990	2.63	2.65
Percent change	-21.02	-21.40
Workers/household		
1960	1.22	1.24
1990	1.25	1.31
Percent change	2.58	5.65
Vehicles/household		
1960	1.03	1.00
1990	1.66	1.59
Percent change	60.37	59.00
Vehicles/person		
1960	0.31	0.31
1990	0.61	0.58
Percent change	100.61	87.10
Percent Workers Traveling by POV		
1960	66.49	61.03
1990	88.02	83.44

Source: *Journey-to-Work Trends in the United States and its Major Metropolitan Areas. 1960–1990,* Publication FHWA-PL-94-012, U.S. Department of Transportation, Federal Highway Administration

households. This result throws into question the household decision model suggested by White (which predicts that two-worker households are attracted to suburban locations because they offer a shorter commute for the female worker); her model would not explain the large numbers of males taking shorter commutes.

Table 5.5. Nonstop worktrips: Comparisons between 1-worker and 2-worker households, by place of residence, 1983 vs. 1990 (drive-alone only)

Place of residence	Commuter	1983 Duration (min.)	1983 Distance (miles)	1983 Speed (mph)	1983 Male Freq	1983 Male %	1983 Female Freq	1983 Female %	1990 Duration (min.)	1990 Distance (miles)	1990 Speed (mph)	1990 Male Freq	1990 Male %	1990 Female Freq	1990 Female %
Inside Central City	1-worker household	18.8	8.2	25.5	89	62.24	54	37.76	19.6	10.5	31.2	451	60.46	295	39.54
	Long dist trip in 2-worker household	*24.8	*12.1	29.2	12	60.00	8	40.00	*26.0	*15.1	33.6	104	63.41	60	36.59
	Short dist trip in 2-worker household	15.4	6.9	23.8	10	50.00	10	50.00	*15.5	*6.8	*25.8	67	40.85	97	59.15
Outside Central City	1-worker household	19.2	9.8	28.5	122	67.78	58	32.22	22.8	13.3	33.3	572	65.60	300	34.40
	Long dist trip in 2-worker household	*25.6	*13.2	29.7	24	75.00	8	25.00	*28.4	*19.0	*37.4	142	59.92	95	40.08
	Short dist trip in 2-worker household	*13.0	*5.2	24.4	13	40.63	19	59.38	*15.1	*7.6	*29.8	98	41.35	139	58.65

* Significantly different from 1-worker household commuters at the 95% level of confidence.

Table 5.6. Nonstop worktrips: Comparisons between 1-worker and 2-worker households, 1983 vs. 1990 by choice of mode

Number of workers in household	1983 Both Drive-alone Time (min.)	1983 Both Drive-alone Dist (miles)	1983 Alone/Carpool Time (min.)	1983 Alone/Carpool Dist (miles)	1983 Both Carpool Time (min.)	1983 Both Carpool Dist (miles)	1990 Both Drive-alone Time (min.)	1990 Both Drive-alone Dist (miles)	1990 Alone/Carpool Time (min.)	1990 Alone/Carpool Dist (miles)	1990 Both Carpool Time (min.)	1990 Both Carpool Dist (miles)
1-worker household	19.1	9.1	–	–	26.3	13.0	21.3	12.0	–	–	24.9	14.5
2-worker household*	39.2	18.6	44.8	23.0	53.8	17.9	42.7	24.7	37.6	20.2	51.6	27.0

* Sums of both workers´ times and distances.

Table 5.7. Nonstop worktrips: Comparisons between 1-worker and 2-worker households, by household income group, 1983 vs. 1990 (drive-alone only)

		1983				1990			
Household income	Commuter	Duration	Distance	Speed	Freq	Duration	Distance	Speed	Freq
< $15,000	1-worker household	16.8	8.1	26.9	88	16.8	8.9	30.5	130
	Long dist trip in 2-worker household	14.9	7.2	26.8	5	18.8	12.0	41.1	6
	Short dist trip in 2 worker household	12.0	4.6	22.4	5	16.7	6.3	26.2	6
$15,000 – $24,999	1-worker household	19.6	9.4	26.4	114	18.0	9.6	30.7	221
	Long dist trip in 2-worker household	24.2	12.5	28.1	11	23.1	*16.3	35.8	24
	Short dist trip in 2-worker household	12.5	5.3	19.8	11	13.0	6.5	29.6	24
$25,000 – $39,999	1-worker household	20.1	9.2	27.8	84	21.8	12.4	32.9	392
	Long dist trip in 2-worker household	*27.3	*14.2	31.2	23	*25.4	14.7	32.9	80
	Short dist trip in 2-worker household	11.9	5.1	26.0	23	13.5	6.2	27.6	80
$40,000 – $54,999	1-worker household	20.4	9.8	28.9	22	23.7	13.2	32.1	214
	Long dist trip in 2-worker household	28.9	15.3	33.0	7	*28.1	*19.3	*39.9	79
	Short dist trip in 2-worker household	21.3	9.0	25.0	7	15.4	7.3	28.6	79
$55,000 +	1-worker household	20.4	10.8	28.7	15	24.5	14.3	33.8	237
	Long dist trip in 2-worker household	24.2	9.5	23.5	6	*31.7	*20.3	36.1	106
	Short dist trip in 2-worker household	17.2	6.8	25.8	6	16.9	8.6	29.8	106

* Significantly greater than 1-worker household commuters at the 95% level of confidence.

Table 5.8. Nonstop worktrips: 1-worker and 2-worker households, family status, 1983 vs. 1990 (drive-alone only)

Commuter		1983 Duration (min.)	1983 Distance (miles)	1983 Speed (mph)	1983 Male Freq	1983 Male %	1983 Female Freq	1983 Female %	1990 Duration (min.)	1990 Distance (miles)	1990 Speed (mph)	1990 Male Freq	1990 Male %	1990 Female Freq	1990 Female %
1-worker household	No children	17.6	8.0	25.5	100	58.14	72	41.86	20.2	11.0	31.6	530	58.05	383	41.95
	With children	*20.7	*10.3	*29.1	111	73.51	40	26.49	*22.8	*13.3	*33.2	487	70.48	204	29.52
2-worker household Longer trip	No children	22.5	11.0	28.9	14	60.87	9	39.13	26.9	16.9	35.8	127	58.26	91	41.74
	With children	27.5	14.2	29.9	22	75.86	7	24.14	28.4	18.3	36.3	116	65.17	62	34.83
2-worker household Shorter trip	No children	12.8	5.2	22.8	9	39.13	14	60.87	15.2	7.2	28.5	97	44.50	121	55.50
	With children	14.7	6.3	25.3	14	48.28	15	51.72	15.5	7.5	27.9	66	37.08	112	62.92

* Significantly greater than for workers in single-worker household families with no children at the 95% level of confidence

Also, while some researchers have asserted that multi-worker households adjust their residential locations to the female's job (Tkocz and Kristensen (1994) drew this conclusion from Danish data), it appears that explanations for the U.S. results would have to be more complex.

The same two relationships between one- and two-worker households' commuting distances more or less hold when household income is held constant (Table 5.7). On the other hand, higher income does have the effect of inducing longer distance commutes for all workers. Here, 1990 results (for all but the lowest income group) are emphasized because the 1983 sample is too small to be disaggregated into income groups.

As in previous research on the effects of gender, we found that an important explanatory variable is the presence or absence of children. Table 5.8 shows that trip lengths increase for households with children, suggesting household trade-offs that include access to schools and other children-related activities. Yet the similarity of findings between one-worker households and two-worker households in terms of total commuting again remain: times and distances are roughly double for the latter. The similarity of results persists when comparisons are extended to the behavior of the two commuters in the two-worker households: in households with children the long-trip commuter still travels about twice as far as the short-trip commuter.

5.3 Discussion

Household structure is clearly a major determinant of travel behavior (Kitamura and Kostyniuk, 1986). It affects the major locational choices that households make. In this paper we have focussed on the journey-to-work while Kitamura and Kostyniuk addressed other aspects of trip-making. Clearly, all trip behavior needs to be better understood if urban economics is to offer an improved reflection of reality. We have pointed out in previous work that the field's devotion to the monocentric model of cities made much of the literature questionable when it came to explaining how travel behavior affects the spatial structure of modern cities. The assumption of single-worker households that permeates the standard model may be equally troublesome. The presumption that it is always the females in two-worker households that have the shorter trip also needs to be reconsidered.

There have recently been several papers on "wasteful commuting", see Richardson et al. (1992) for a recent summary and Mills (1998). The pejorative "wasteful" is attached to commutes that exceed those predicted by the simple neoclassical urban model. While longer commutes can be explained as being more "efficient" than "wasteful" because households may be optimizing with respect to locational pulls other than proximity to work (such as schools, amenities, "clean" neighborhoods, etc.), locational choices involving two journeys to work also undermines the wasteful commuting hypothesis. The results reported here, specifically one long and one short commute that together double the distance to work of the single worker, suggest locational choices that are quite consistent with economic behavior (e.g. joint household welfare maximization over a wide range of attributes). Future research will have to examine the joint locational decisions of multi-worker households. That work will have to consider *inter alia* the gradual blurring of gender roles in the labor market.

Although the NPTS statistics are useful for illustrating the importance and key characteristics of commuting in multi-person households, they are deficient as a data base for transportation/land-use models because they do not provide details of the precise locations of either workplaces or residences. Accordingly, the research results reported here can only be a starting point for the inclusion of multiple-worker households in transportation network and other land use models. Future research needs to be focussed on the endpoints of worktrips, and in the United States at least this requires survey research, especially for multiple-worker households.

The growing importance of services consumption by households is one of the explanations for the recently observed rise in nonwork travel (Gordon et al., 1988). In addition, telecommuting options suggest that many worktrips have new substitutes. Together with the changing role of females, these changes suggest a much more complicated set of locational choice trade-offs in the decisions made by households. It has been suggested that conventional neoclassical approaches are inappropriate for handling these difficult problems (Richardson, 1988). On the other hand, as the quality of locational data improves, hedonic approaches can be used to study the various "prices" that result from households' complex demand for "locational packages". Progress in identifying the underlying demand functions would help us to learn more about the trade-offs made by households.

A revolution in macroeconomic theory occurred in the 1970s when economists began to study the microfoundations of macroeconomic behavior. A much better understanding of the aggregate economy was obtained when what happened in the economy at large was reconciled with the actions of individual decision makers. Analogously, we would argue that major improvements in transportation/land-use modeling will be obtained when they are founded on a deeper understanding of how households actually make decisions about where to live and work and about the number and types of trips that they take. This initial research into household commuting behavior offers a modest step in this direction.

References

Alonso, W. (1980), 'The population factor and urban structure', in Solomon, A. (ed.), *The Prospective City*, MIT Press, Cambridge, MA.

Alonso, W. (1964), *Location and Land Use*, Harvard University Press, Cambridge, MA.

Forstall, R.L. (1994), Memorandum of January 6, 1994, Bureau of the Census, Washington, DC.

Giuliano, G. (1989), 'Research policy and review: New directions for understanding transportation and land use', *Environment and Planning A*, vol. 21, pp. 145–159.

Giuliano, G. and Small, K. (1993), 'Is the journey to work explained by urban structure?', *Urban Studies*, vol. 30, pp. 1485–1500.

Gordon, P. and Moore II, J.E. (1989), 'Endogenizing the rise and fall of urban subcenters via discrete programming models', *Environment and Planning A*, vol. 21, pp. 1195–1203.

Gordon, P. and Richardson, H.W. (1994), *Geographic Factors Explaining Worktrip Length Changes,* U.S. Department of Transportation, Federal Highway Administration, Washington, DC.

Gordon, P., Richardson, H.W. and Kumar, A. (1989), 'Gender differences in metropolitan travel behavior', *Regional Studies*, vol. 23, pp. 499–510.

Gordon, P., Richardson, H.W. and Kumar, A. (1988), 'Beyond the journey to work', *Transportation Research A*, vol. 22, pp. 419–426.

Hu, P.S. and Young, J. (1993), '1990 NPTS databook: Nationwide personal transportation survey', U.S. Department of Transportation, Federal Highway Administration, Washington, DC.

Kitamura, R. and Kostyniuk, L.P. (1986), 'Maturing motorization and household travel: The case of nuclear-family households', *Transportation Research A*, vol. 20, pp. 245–260.

Liao, Y. (1993), 'Trip chaining in urban travel', Paper presented at University of Southern California, Urban Economics Group Seminar.

Madden, J. (1981), 'Why women work closer to home', *Urban Studies*, vol. 18, pp. 181–194.

Mills, E.S. (1998), 'Excess commuting in U.S. metropolitan areas', Chapter 4, this volume.

Richardson, H.W. (1988), 'Monocentric vs. polycentric models: The future of urban economics in regional science', *Annals of Regional Science*, vol. 22, pp. 1–12.

Richardson, H.W., Gordon, P. and Choi, Y. (1992), 'Tests of the standard urban model: A micro (trade-off) alternative', *Review of Urban and Regional Development Studies*, vol. 4, pp. 50–66.

Tkocz, Z. and Kristensen, G. (1994), 'Commuting distances and genders: A spatial urban model', *Geographical Analysis*, vol. 26, pp. 1–14.

White, M.J. (1977), 'A model of residential location choice and commuting by men and women workers', *Journal of Regional Science*, vol. 17, pp. 41–52.

Part II

Dynamics and Equilibria in Network Modelling: New Theoretical and Methodological Developments

6 Disequilibrium Network Design: A New Paradigm for Transportation Planning and Control

Terry L. Friesz
Departments of Systems Engineering and Operations Research & Engineering
George Mason University
Fairfax, VA 22030, USA

Samir Shah
PB Farradyne Inc
Rockville, MD 20852
USA

David Bernstein
Dept. of Civil Engineering and Operations Research
Princeton University
Princeton, NJ 08544, USA

6.1 Introduction

In a very broad sense, the network design problem (NDP) is a topic that has captured the attention of many researchers. This is mainly due to the immense importance of strategic capital investment decisions involving transportation infrastructure. For the purposes of this paper we limit the scope of the network design problem to highway systems. Design related decisions relevant to highways include a rich and wide variety of strategic (e.g., new right of way), tactical (one way street assignment, HOV assignment) and operational (traffic signalization, ramp metering) decisions that typically arise in transportation planning.

The equilibrium network design problem is to find an optimal network design in terms of additional facilities or capacity enhancements, when the network flow pattern is constrained to be an equilibrium. The much reported occurrence of Braess' paradox (Murchland, 1970) requires that design models have Wardropian user equilibrium (Wardrop, 1952) constraints. The transportation research literature includes many such equilibrium design models wherein equilibrium constraints are generally articulated as an equivalent optimization problem (Abdulaal and LeBlanc, 1979; LeBlanc,1975) or as an equivalent variational inequality problem (Friesz et al., 1990; 1992; 1993b; Marcotte, 1986). Unfortunately such equilibrium network design models presuppose a static environment and completely ignore the impacts of potential disequilibria which can arise due to perturbations in the capacity of the network infrastructure. Such a static perspective may lead to the occurrence of a temporal version of Braess' Paradox. The "temporal Braess' Paradox" does not yet enjoy a standard definition, but in the context of the present discussion can be viewed as occurring when a capacity altering action lowers (or leaves unchanged) overall delay in the present and near future but increases overall delay at some more distant time. Such outcomes are possible when the present value of disequilibrium impacts is substantially negative, as can occur when the immediate disequilibrium response is a sharp congestion increase or when the disequilibrium response is a mild congestion increase of relatively long duration. An example is provided by a highway construction project intended to enhance capacity but which produces traffic congestion for weeks or months prior to its completion; in this case, the present value of construction impacts may be sufficiently negative to eradicate all positive benefits in the post-construction period. This is particularly true in standard cost benefit kind of analyses, wherein near term benefits are more heavily discounted than long term benefits.

In light of the above remarks, we wish to present a "disequilibrium network design" paradigm. This new paradigm differs from the historical static one in the sense that it takes into account both the underlying time varying nature of the network and the disequilibrating effects that capacity enhancements to the network may produce. Below we articulate such a disequilibrium network design paradigm model for selecting optimal capacity enhancement *trajectories* for network arcs, with the time evolution of flows and minimum path costs are described by an appropriate disequilibrium adjustment process. In this control theoretic formulation we employ the concept of a network traffic disequilibrium, by which is meant a flow pattern which may fail to satisfy flow conservation (transportation market clearing) constraints and for which network users may experience clear advantages from changing their paths, but which evolves from a previously realized (dis)equilibrium in accordance with some plausible behavioral laws.

The aforementioned disequilibrium network design paradigm depends on a valid description of the disequilibrium adjustment process which guides the network of interest from one disequilibrium state to another, eventually settling down to a conventional steady state (static) equilibrium. Friesz et al. (1994) and Friesz et al. (1996) have developed models of this type based on realizable network generalizations of the traditional aspatial tatonnement models reported in the economics literature. These models consider day-to-day adjustments of flows and costs. Because of their central importance to disequilibrium design, a brief introduction to disequilibrium adjustment processes is presented below. Throughout all the subsequent sections, the notation introduced by Friesz et al. (1994) to describe dynamic traffic disequilibria is employed. This notation, which is similar to that traditionally employed for static equilibria is summarized in the Appendix.

6.2 Disequilibrium Dynamics for Urban Network

This section is adapted from the presentation in Friesz et al. (1996) of a tatonnement model used to describe disequilibrium dynamics representing highway passenger traffic.

Friesz et al. (1994) have described the time rates of change of path flows, represented by the vector h, and perceived origin-destination costs, represented by the vector u, as a tatonnement process[1] whose disequilibrium dynamics have the following form:

$$\frac{dh}{dt} = \eta F(ETC) \qquad (1)$$

$$\frac{du}{dt} = \kappa G(ETD) \qquad (2)$$

[1] A tatonnement process is a trading process among all agents (sellers and buyers) in an economy which is conducted by a super-auctioneer. Specifically, in the trading process the auctioneer calls out a set of prices and receives transaction offers from the agents. If these offers do not match (amount demanded is not equal to amount supplied), he calls another set of prices by following some rules and the process continues without any transaction being allowed to take place until transaction offers match.

$$h(0) = h^0 \tag{3}$$
$$u(0) = u^0, \tag{4}$$

where t is a scalar denoting time, η and κ are diagonal matrices, $F(\cdot)$ and $G(\cdot)$ are monotonic vector operators, ETC is a vector of excess travel costs and ETD is a vector of excess travel demands.

In order to understand the balance of our exposition, it is important to first understand the key concepts behind the disequilibrium model (1)-(4). In particular, it should be noted that this model borrows heavily from the dynamic adjustment or tatonnement models of aspatial microeconomic theory. In particular, the traffic network analogues of prices, namely perceived travel costs, are viewed in (1) as varying over time in direct response to excess travel demands. Excess travel demand is simply the difference between desired and actual travel rates for a given origin-destination pair. Thus, (2) is similar to the differential equations used to tell the traditional story of a fictitious auctioneer who clears the market in aspatial microeconomic theory by calling out prices and observing the demands of buyers and the supplies of sellers at each price, with one notable difference. That difference is that (2) describes realizable disequilibrium perceived costs for the entire time period of interest and not merely the unrealized prices put forward by the auctioneer to stimulate responses from buyers and sellers as in microeconomic theory, where only the market clearing (equilibrium) response is realized. Moreover, in (1) path flows, which represent travelers' attempts to achieve their travel goals through route and departure time choice, are viewed as varying over time in direct response to excess travel costs. Excess travel cost is simply the difference between actual travel cost and perceived travel cost for a given route/departure time alternative. As was true of the perceived cost dynamics, (1) describes realizable disequilibrium path flows. The realizable nature of the state variables in (1)-(4) owes primarily to the fact that there is no notion of inventorying or backordering trips as is necessary for conventional nonperishable commodities[2]. Also important to assuring realizability, and discussed in more detail below, is that the operators $F(\cdot)$ and $G(\cdot)$ in (1)-(4) may be constructed so that relevant social, technological, legal and policy constraints are enforced throughout the disequilibrium trajectories. Furthermore the simple, intuitive relationships stated in (1)-(4), as discussed subsequently, lead to a plausible model whose steady states and dynamic equilibria are identical to widely accepted *game theoretic models* of traffic flow on networks.

Disequilibrium models like (1)-(4) have as their foundation definitions of the time derivatives of those state variables whose trajectories are sought; namely

$$\frac{dh(t)}{dt} \equiv \lim_{\Delta t \to 0} \frac{h(t + \Delta t) - h(t)}{\Delta t} \tag{5}$$

$$\frac{du(t)}{dt} \equiv \lim_{\Delta t \to 0} \frac{u(t + \Delta t) - u(t)}{\Delta t} \tag{6}$$

A specific tatonnement adjustment process is constructed by selecting a model for the perturbed flows $h(t + \Delta t)$ and perturbed perceived costs $u(t + \Delta t)$. To this end,

[2] Varian (1984) points out that a traditional tatonnement model will have realizable trajectories if its state variables are expressed in flow terms and the commodity being analyzed cannot be inventoried/backordered.

it is useful to define

$$v(t) = Pr_\Omega\{u(t) + \alpha ETD\}, \quad \alpha \in \Re^1_+, \tag{7}$$

where the operator $Pr_\Omega\{\ \cdot\ \}$ denotes a projection to avoid infeasibilities onto the closed set of constraints Ω pertinent to the analysis and is defined by

$$Pr_\Omega\{x\} \equiv \arg\min_{y\in\Omega} \|x - y\| \tag{8}$$

for an arbitrary vector x. As such $v(t)$ may be viewed as the *instantaneous* revision of the perceived cost in accordance with continuously provided (excess) demand information. We further postulate that *future* perceptions of O-D travel costs are formed through adjustment of current cost perceptions at a rate proportional to demand induced change in perceived cost $[v(t) - u(t)]$. That is,

$$u(t + \Delta t) = u(t) + \kappa[v(t) - u(t)]\Delta t \tag{9}$$

We shall also need to define

$$g(t) = Pr_\Omega\{h(t) - \beta ETC[u(t), h(t)]\}, \quad \beta \in \Re^1_+, \tag{10}$$

which is the *instantaneous* revision of path preference in accordance with continuously provided (excess) cost information. In (10) the minus sign is chosen to ensure that the inclination to travel along a given path decreases as the experienced cost increases above perceived cost. The projection operator $Pr_\Omega\{\ \cdot\ \}$ is again used to maintain feasibility. We further postulate that *future* path flows are established through adjustment of current path flows at a rate proportional to cost induced change in path preference $[g(t) - h(t)]$. That is,

$$h(t + \Delta t) = h(t) + \eta[g(t) - h(t)]\Delta t \tag{11}$$

The preceding discussion yields the dynamic system

$$\left.\begin{aligned} \frac{dh}{dt} &= \eta[Pr_\Omega\{h - \alpha ETC\} - h] \\ \frac{du}{dt} &= \kappa[Pr_\Omega\{u + \beta ETD\} - u] \\ h(0) &= h^0 \\ u(0) &= u^0, \end{aligned}\right\} \tag{12}$$

where $\eta = \text{diag}(\eta_p \in \Re_+ : p \in P)$ and $\kappa = \text{diag}(\kappa_{ij} \in \Re_+ : i \in N_O, j \in N_D)$ are constant diagonal matrices. Dynamic system (12) gives an explicit form to the abstract right hand side operators $F(\cdot)$ and $G(\cdot)$ introduced respectively in (1) and (2).

From the point of view of traffic theory, a particularly important case is that of only nonnegativity restrictions on path flows and perceived costs: $\Omega = \{[h(t), u(t)] : h(t) \geq 0,\ u(t) \geq 0\ \forall\ t \in [0, T]\} \equiv \Omega_+$. In this case the projection operator (8) is quite easy to express in closed form for an arbitrary argument x:

$$Pr_{\Omega_+}\{x\} = \arg\min_{y\in\Omega_+} \frac{1}{2}(x - y)'(x - y) \tag{13}$$

The Kuhn-Tucker conditions for (13) are

$$-(x-y)-\rho=0,\quad \rho y=0,\quad \rho\geq 0 \tag{14}$$

Hence, because $y \in \Omega_+$, we have

$$y=x-\rho=\{x\}_+, \tag{15}$$

where $\{x\}_+ \equiv \max\{0,x\}$. The implication of (13) is that system (12) takes the following form when there are only nonnegativity constraints:

$$\left.\begin{array}{rcl} \dfrac{dh}{dt} &=& \eta[\{h-\alpha ETC\}_+ - h] \\ \dfrac{du}{dt} &=& \kappa[\{u+\beta ETD\}_+ - u] \\ h(0) &=& h^0 \\ u(0) &=& u^0 \end{array}\right\} \tag{16}$$

Dynamical system (16) is analyzed in detail in Friesz et al. (1994). Moreover, Friesz et al. (1994) establish that a day-to-day adjustment process based on (16) is asymptotically stable under plausible regularity conditions.

Smith et al. (1996) have shown that the disequilibrium trajectories depicted by (12) and (16) tend to be entirely within the feasible region Ω, and do not follow constraints boundaries. In fact Smith et al. (1996) have aptly termed (12) "*global projective dynamics*", and shown that, under quite mild regularity conditions, its trajectories remain for all time in the constraint set onto which the projection defining their right hand sides is made. It is this interior and anticipatory nature of their trajectories that allows various constraints to be placed on the disequilibrium trajectories to realistically describe different dynamic flow environments using tatonnement-type models.

It has already been noted that the manner in which the dynamic system (12), and hence the system (16), has been derived tends to cause its trajectories to traverse the interior of the feasible region of interest (see Smith et al., 1996), Ω. When Ω consists of only nonnegativity restrictions on path flows and on perceived costs, this interior point nature of trajectories realistically describes the tendency of drivers to continue to explore all potentially practical routes and departure-time choices prior to attaining an equilibrium, and reflects the physical reality that costs cannot be zero. More generally, (12) may describe anticipatory behavior on the part of drivers in avoiding any constraint boundary. Alternative derivations of adjustment dynamics which are constraint boundary following in nature are possible, but result in right hand sides which are substantially more complicated and not continuous. One such boundary following mechanism is that proposed by Dupuis and Nagurney (1993).

Some debate has arisen in the transportation and spatial economic research community regarding the relative merits of local versus global projective dynamics. In particular, Nagurney (1994) has claimed that global projective dynamics like (12) and (16) are "incorrect" and that only the variety of local projective dynamics she has studied are valid for describing behavior. Her main point in this regard has been that global projective dynamics anticipate constraint boundaries and avoid those boundaries, so that only at a steady state do any constraints bind. As Smith et al. (1996) have written, there is an entire body of literature on anticipatory systems in physics and economics with exactly this property of anticipation shown by global projective dynamics; furthermore, there is no reason to believe that all common physical and economic adjustment processes proceed by moving along constraint boundaries as is the case with local projective dynamics. Moreover, the urban network disequilibrium model discussed here and in Friesz et al. (1994) has only two constraints: nonnegativity of path flows and nonnegativity of perceived costs. That such constraints should not bind – i.e., should not cause path flows and perceived costs to be zero – is not only plausible but also observed. That is, to reiterate:

1. flows on paths actually considered viable by travellers continue to be used at some generally diminishing level until the learning process represented by the disequilibrium adjustment process is complete, and
2. perceived costs are simply never zero for any realistic travel options.

It was to rigorously establish these and other distinctions between global and local projective dynamics which are not initially apparent that the paper by Smith et al. (1996) was written; the interested reader is strongly encouraged to consult that paper.

The disequilibrium design paradigm developed in this paper, on the other hand, does not depend on the way the dynamics are derived, as long as they are derived from sound economic reasoning and yield steady states which are proper user equilibria.

In the following paragraphs, we describe alternative levels of detail in the mathematical descriptions of the excess operators in the disequilibrium dynamics introduced above. The possible time scales of interest in predicting urban network flows are day-to-day and within-day. In the simplest terms, day-to-day adjustments are slow relative to within-day adjustments which are fast. In particular, route choice and departure time choice may be viewed as the result of a relatively fast dynamic process, for travelers have complete control of these decisions and employ them virtually continuously in optimizing and re-optimizing their travel experiences. From (1) we see that, if route choice and departure time choice (which is intrinsic to the former if path flows are defined as the flow entering the first arc of a path) are fast, then they must be derived from within-day cost information contained in the excess cost operator; that is, from cost information which is timely or, in the extreme of abstraction, instantaneous. If only day-to-day (average) cost information is employed, there is no meaningful concept of departure time and one is modeling day-to-day changes in route choices and path flows, which is a slow dynamic process. These day-to-day path flows must by necessity be daily averages. So the choice of time scale for describing path flows impacts the particular cost model one employs. Highly similar observations may be made about the perceived cost dynamics (2). However, for simplicity of exposition, we consider subsequently only the case of costs evolving slowly over time relative to route choice and departure time choice; and, so, it is enough to use in (2) demand models which describe day-to-day fluctuations in travel.

The tatonnement model reported by Friesz et al. (1994) describes day-to-day disequilibrium adjustments wherein variables are daily averages and departure time choice is not explicitly considered. Within-day simultaneous departure time and route choice decisions are considered in a dynamic *equilibrium* model by Friesz et al. (1993a) through the introduction of carefully defined path cost operators which consider the impact of departure times. Specifically, they employ the operator

$c_p(t, H)$ describing the flow unit cost encountered in traversing path p when departure from the origin occurs at time t and traffic conditions $H = (h_p(t) : p \in P, t \in [0, T])$ are encountered.

The Friesz et al. $c_p(t, H)$ operators are derived from a simple deterministic model (refer to Friesz et al. (1993a)) which has as its main assumption that the delay on any network arc depends only on the number of vehicles in front of the vehicle of interest at the time of entry to the arc. This assumption leads to a relatively simple, but still computationally demanding, submodel for finding the operators $c_p(t, H)$. Other assumptions could be made and corresponding submodels for simultaneous route and departure (SRD) unit cost operators employed, including simulation models. Indeed, the SRD unit cost operators may, at least in principle, be determined empirically without recourse to any theory. Moreover, it should be clearly understood that the method used to derive or numerically simulate the operators $c_p(t, H)$ is irrelevant to the presentation which follows. Finally, we point out that all within-day dynamic traffic assignment and equilibrium models known to us either explicitly or implicitly presume that costs like $c_p(t, H)$ can be obtained. In cases where perfect hindsight and foresight is not assumed, the dependence on flows is changed to reflect only the present, near future and near past flows, so that the cost operator can be stated symbolically as $c_p(t, \tilde{H})$ where $\tilde{H} = \{h_p(t') : p \in P, t' \in [t-a, t+b]\}$, $a > 0$ is the maximum temporal retrospective used in determining costs, and $b > 0$ the maximum temporal prospective. No substantial difference in the presentation which follows would occur if these modified cost operators reflecting imperfect information were employed.

Using the aforementioned SRD unit cost operators in the dynamical system (1)-(4) results in a model which *simultaneously* describes within-day adjustments of path flows and day-to-day adjustments of perceived costs if one recognizes that these adjustments operate at different speeds and involve different variables. This is because the SRD operators capture the disutility of route and departure time choices from moment to moment within a day[3]. The tatonnement mechanism, on the other hand, involves subsystems which may be characterized as fast (within-day) or slow (day-to-day). Specifically, path choice and departure time choice are within-day decisions which are fast relative to the evolution of cost perceptions, which is considered here a slow, day-to-day process. In particular, the SRD unit cost operators together with notation from Appendix A, allow the excess functions to be written as

$$ETC = c(t, H) - W'u \tag{17}$$

$$ETD = d(u) - Wh, \tag{18}$$

so that a rather general form of the disequilibrium dynamics is

$$\frac{dh}{dt} = \eta F[c(t, H) - W'u] \tag{19}$$

$$\frac{du}{dt} = \kappa G[d(u) - Wh] \tag{20}$$

$$h(0) = h^0 \tag{21}$$

$$u(0) = u^0, \tag{22}$$

[3] We of course use the word "day" loosely as a surrogate for any time period of special interest, e.g. the rush hour or a sequence of travel periods.

where W is the path-OD incidence matrix[4], $c(t, H)$ is a vector of unit path costs, and $d(u)$ is a vector of transportation demands. For simplicity, we assume, as reflected in (18) and (20), that travel demands are autonomous functions of time. Evidently, if demands varied with time of day or otherwise depended explicitly on t, we would write them in the non-autonomous form $d(t, u)$.

A comment is also required about the relative coarseness, indeed discreteness, of a day-to-day time scale compared to a within-day time scale, which is presumably continuous and smooth. System (19)-(22) implicitly addresses this matter through our presumption about the relative speeds of (19) and (20). Specifically, in that route and departure choices are viewed here as a fast within-day process and perceived cost adjustments are a slow day-to-day process, we expect that $\eta_p \gg \kappa_{ij}$ for each $i \in N_O,\ j \in N_D$ and $p \in P_{ij}$. This allows day-to-day perceived cost adjustments to be viewed (approximated) as occurring in continuous time without resort to a mixed system of difference and differential equations.

6.3 Control Theoretic Formulation of Disequilibrium Design

In the case of dynamic disequilibrium network design problem, we wish to maximize the net present value of benefits to users of a transportation network over a fixed planning horizon $[0, T]$. Furthermore, for simplicity of exposition, we take the capacity enhancement for arc a at time t to be a continuous variable $y_a(t)$; while $f_a(t)$ will denote the flow on arc a at time t. Consequently, the unit cost function for arc a is of the form $c_a[f(t), y(t)]$, where $f(t) = (f_a(t) : a \in A)$ and $y(t) = (y_a(t) : a \in A)$. Therefore, because $f_a = \sum_p \gamma_{ap} h_p$ and $c_p = \sum_a \gamma_{ap} c_a$ where $\gamma_{ap} = 1$ if $a \in p$ and zero otherwise, the unit cost on path p is of the form $c_p(t, H, y)$ and $c(t, H, y) = \{c_p(t, H, y) : p \in P\}$.

Assuming that travel demands are elastic, the present value of the gross benefits to the users of the transportation network is the present value of consumers' surplus calculated from these travel demands. Hence, the net benefits accruing to users at time t will be consumer surplus less total congestion costs:

$$Z(u(t), h(t), y(t)) = \sum_{i \in N_O} \sum_{j \in N_D} \oint_{u^0}^{u(t)} d_{ij}[w(t)] dw_{ij} - \sum_{p \in P} c_p[h(t), y(t)] h_p(t) \qquad (23)$$

where $\oint$ denotes a line integral and w is dummy variable of integration.[5] In continuous time, the net benefits will be the present value of properly discounted $Z[\cdot]$ over time. If we use r to denote a constant discount rate, the present value of net benefits to the

[4] Note that W' is the transpose of W, $Wh = (\sum_p w_{ij}^p h_p : i \in N_O, j \in N_D) = (\sum_{p \in P_{ij}} h_p : i \in N_O, j \in N_D)$ and $W'u = (\sum_k \sum_l w_{kl}^p u_{kl} : i \in N_O, j \in N_D, p \in P_{ij}) = ((u_{ij} : p \in P_{ij}) : i \in N_O, j \in N_D)$.

[5] Note that the first term of (23) is instantaneous change in consumers' surplus while the second term is of course system-wide costs. The change in consumers' surplus is the sum of areas beneath the demand curves between the initial and present perceived costs.

users of the transportation network may be expressed as

$$J(u(t),h(t),y(t)) = \int_0^T e^{-rt}\left\{\sum_{i\in N_O}\sum_{j\in N_D}\oint_{u^0}^{u(t)} d_{ij}[w(t)]dw_{ij} - \sum_{p\in P} c_p[h(t),y(t)]h_p(t)\right\}dt$$

Also observe that since the travel cost vector depends on improvements, namely

$$ETC = ETC[t,u(t),H,y(t)], \tag{24}$$

the desired disequilibrium network design model has the form

$$\max \; J(u(t),h(t),y(t)) \tag{25}$$

subject to:

$$\dot{u}(t) = \kappa[\{u(t) + \alpha ETD[u(t),h(t)]\}_+ - u(t)] \quad \forall \quad t \in [0,T] \tag{26}$$

$$\dot{h}(t) = \eta[\{h(t) - \beta ETC[t,u(t),H,y(t)]\}_+ - h(t)] \quad \forall \quad t \in [0,T] \tag{27}$$

$$\pi(y(t)) \leq B \quad \forall \quad t \in [0,T] \tag{28}$$

$$u(0) = u^0 \tag{29}$$

$$h(0) = h^0 \tag{30}$$

$$y(t) \geq 0 \quad \forall \quad t \in [0,T] \tag{31}$$

Where B is the improvement budget and $\pi[\cdot]$ is an expenditure operator which will be a function of capacity enhancement.

This model is evidently an optimal control problem. In the terminology of optimal control theory, $u(t)$ and $h(t)$ are called *state* variables and $y(t)$ *control* variables. We seek the trajectories through time of both the state and the control variables. Equations (26) and (27) are the dynamics of the state variables u and h respectively, while (29) and 30) are their initial conditions. Constraint (28) is a budget constraint. Constraint (31) ensures that variables $y(t)$ are non-negative, which prohibits dis-investment.

The necessary conditions for a solution of a classical optimal control problem are given by Pontryagin's maximum principle (Pontryagin et al., 1962). However, the state equations (26) and (27) are nonstandard. Because of the presence of projection operators, the right hand sides are non-differentiable (non-smooth). Hence, the necessary conditions are not amenable to analysis via classical theory. The necessary conditions for such a optimal control problem can be derived using certain results from Bryson and Ho (1975). Following the analysis provided in Bryson and Ho (1975), one can partition the planning horizon $[0,T]$ for the optimal control formulation (25)-(31) in N finite subintervals, $[(t_{i-1},t_i), i = 1,\ldots N]$ where $t_N = T$, such that for each subinterval, the right hand sides of the state dynamics are smooth. One can then derive the necessary conditions for each stage of the optimal control problem and synthesize the results to cover the entire time horizon. For the sake of brevity, such details are not presented here; the reader can however refer to Shah (1996) for the detailed derivation of the necessary conditions. The necessary conditions for disequilibrium network design problem can also be analyzed using generalized differentials (Clark, 1983) or differential inclusions (Aubin and Cellina, 1984).

Shah (1996) also shows that the necessary conditions derived from the optimal control formulation are economically/behaviorally meaningful in the sense that they imply that marginal total costs equal marginal total benefits throughout the period of analysis.

6.3.1 Alternative Formulation

Model (25)-(31) can be further simplified and restated by exploiting the "price" dynamics (26). In particular it is useful to use the notation

$$G(ETD) = [\{u(t) + \alpha ETD[u(t), h(t)]\}_{+} - u(t)] \quad \forall \quad t \in [0, T] \tag{32}$$

as a shorthand for the right hand side of dynamics (26), so that

$$du = \kappa G(ETD)dt \tag{33}$$

One may unambiguously evaluate consumers' surplus, despite its original articulation as a line integral, by expressing it entirely in terms of ordinary integrals as follows:

$$\begin{aligned} &\sum_{i \in N_O} \sum_{j \in N_D} \oint_{u^0}^{u(t)} d_{ij}[w(t)] dw_{ij} = \\ &\qquad \sum_{i \in N_O} \sum_{j \in N_D} \kappa_{ij} \int_0^t d_{ij}[u(t')] G_{ij}[ETD_{ij}(u(t'), h(t'))] dt' \end{aligned} \tag{34}$$

As a result the optimal control formulation (25)-(26) can now be restated as:

$$\begin{aligned} \max \ J(u, h, y) \ = \ & \int_0^T e^{-rt} \Bigg\{ \sum_{i \in N_O} \sum_{j \in N_D} \kappa_{ij} \int_0^t T_{ij}[u(t')] G_{ij}[ETD_{ij}(u(t'), h(t'))] dt' - \\ & \sum_{p \in P} c_p[h(t), y(t)] h_p(t) \Bigg\} dt \end{aligned} \tag{35}$$

subject to:

$$\dot{u}(t) = \kappa[\{u(t) + \alpha ETD[u(t), h(t)]\}_{+} - u(t)] \quad \forall \quad t \in [0, T] \tag{36}$$

$$\dot{h}(t) = \eta[\{h(t) - \beta ETC[t, u(t), H, y(t)]\}_{+} - h(t)] \quad \forall \quad t \in [0, T] \tag{37}$$

$$\pi(y(t)) \leq B \quad \forall \quad t \in [0, T] \tag{38}$$

$$u(0) = u^0 \tag{39}$$

$$h(0) = h^0 \tag{40}$$

$$y(t) \geq 0 \quad \forall \quad t \in [0, T] \tag{41}$$

This last formulation, as discussed below, suggests a natural discrete time mathematical programming formulation for numerical calculations. Also note that (35)-(41) is not hierarchical in structure; it is therefore potentially much more numerically tractable than conventional bi-level static equilibrium network design formulation. In the limit $T \longrightarrow \infty$, formulation (35)-(41) will determine the static equilibrium network design if the dynamic adjustment process represented by (36) and (37) is stable.

6.4 An Algorithm

Formulation (35)-(41) is an optimal control problem with dynamics whose right hand sides are non-differentiable. As such this problem may be viewed as a non-smooth infinite dimensional mathematical program. Such problems are intrinsically difficult though not impossible to solve (see e.g. Clark, 1983). Discrete time optimal control problems are optimization problems in finite-dimensional spaces and, hence, require less mathematical sophistication in their treatment than continuous optimal control problems. An obvious solution approach is to discretize the disequilibrium network design model, thereby creating a finite dimensional mathematical program. It seems possible to use traditional methods of non-differentiable optimization such as classical search techniques to solve small and medium size mathematical programs created in this fashion.

Numerical experiments involving various example problems and algorithms are currently being conducted. The reader is referred to Shah (1996) for detailed results of such numerical experiments.

6.5 Conclusion and Future Research

The paradigm of "disequilibrium network design" based on "disequilibrium dynamics" derived from the tatonnement paradigm of microeconomics was used to present an optimal control formulation of the network design problem. Indeed it appears that this model, because of its ability to address disequilibrium issues, holds the potential to significantly impact the transportation planning process. Further theoretical and numerical analysis are required to assess the potential value of such a model.

References

Abdulaal, M., and LeBlanc, L. (1979), 'Continuous equilibrium network design model', *Transportation Research*, vol. 13B, pp. 19–32.

Aubin, J. and Cellina, A. (1984), *Differential Inclusions*, Springer-Verlag, Berlin.

Bryson, A. and Ho, Y.C. (1975), *Applied Optimal Control* (revised), John Wiley & Sons, New York, NY.

Clark, F. (1983), *Optimization and Nonsmooth Analysis*, Wiley-Interscience, New York, NY.

Dupuis, P. and Nagurney, A. (1993), 'Dynamical systems and variational inequalities', *Annals of Operations Research*, vol. 44, pp. 9–42.

Friesz, T. (1985), 'Transportation network equilibrium, design and aggregation: Key developments and research opportunities', *Transportation Research*, vol. 19A, pp. 413–427.

Friesz, T., Tobin, R., Cho, H. and Mehta, N. (1990), 'Sensitivity analysis based heuristic algorithms for mathematical programs with variational inequality constraints', *Mathematical Programming*, vol. 48, pp. 265–284.
Friesz, T., Cho, H., Mehta, N. and Tobin, R. (1992), 'Simulated annealing methods for network design problems with variational inequality constraints', *Transportation Science*, vol. 26, pp. 18–26.
Friesz, T., Bernstein, D., Tobin, R. and Wie, B. W. (1993a), 'A variational inequality formulation of the dynamic network user equilibrium problem', *Operations Research*, vol. 37, pp. 893–901.
Friesz, T., Tobin, R., Shah, S., Mehta, N. and Anandalingam, G. (1993b), 'The multiobjective equilibrium network design problem revisited: A simulated annealing approach', *European Journal of Operations Research*, vol. 65, 44–57.
Friesz, T., Bernstein, D., Mehta, N., Tobin, R. and Ganjalizadeh, S. (1994), 'Day-to-day dynamic network disequilibrium and idealized driver information systems', *Operations Research*, vol. 42, pp. 1120–1136.
Friesz, T., Bernstein, D. and Stough. R. (1996), 'Dynamic systems, variational inequalities and control theoretic models for predicting urban network flows', *Transportation Science*, vol. 30, pp. 14–31.
LeBlanc, L. (1975), 'An algorithm for the discrete network design problem', *Transportation Science*, vol. 9, pp. 283–287.
Marcotte, P. (1986), 'Network design problem with congestion effects: A case of bilevel math program', *Mathematical Programming*, vol. 34, pp. 142–162.
Murchland, J. (1970), 'Braess' paradox of traffic Flow', *Transportation Research*, vol. 4, pp. 391-394.
Nagurney, A. (1994), Comments during The conference on network infrastructure and the urban environment – Recent advances in land-use/transportation modeling, Sweden.
Pontryagin, L., Boltyanskii, V., Gramkrelidze, R. and Mischenko, E. (1962), *The Mathematical Theory of Optimal Processes*, John Wiley & Sons, New York, NY.
Shah, S. (1996), *Dynamic Disequilibrium Network Design Model*, Ph.D. Dissertation, George Mason University, Fairfax, VA.
Smith, T., Friesz, T., Bernstein, D. and Suo, Z. (1996), 'A comparative analysis of two minimum-norm projective dynamics and their relationship to variational inequalities', forthcoming in *SIAM Journal of Optimization.*
Varian, H. (1984), *Microeconomic Analysis*, W. W. Norton and Company, New York, NY.
Wardrop, J. (1952), 'Some theoretical aspects of road traffic research', *Proceedings of Institute of Civil Engineers*, vol. 1, part II, pp. 325–378.

Appendix: Summary of Notation

a	is an index denoting an arc
$\alpha, \beta \in \Re_+$	are tatonnement parameters
p	is an index denoting a path
$m(p)$	is the number of arcs on path p
A	is the set of arcs; $\|A\| = m$
N	is the set of nodes; $\|N\| = n$
$N_O \subseteq N$	is the set of nodes which are trip origins
$N_D \subseteq N$	is the set of nodes which are trip destinations
$P = \bigcup_{i \in N_O,\, j \in N_D} P_{ij}$	is the complete set of network paths; $\| P \| = \rho$
P_{ij}	is the set of paths connecting origin i to destination j
$Pr_\Omega\{v\}$	is the projection of the vector v onto the set Ω
$\{v\}_+$	$= \max(0, v)$
γ_{ap}	$= \begin{cases} 1 & \text{if } a \in p \\ 0 & \text{otherwise} \end{cases}$

$\Gamma = (\gamma_{ap})$	is the arc-path incidence matrix
w_{ij}^p	is an element of the path-OD incidence matrix; specifically, $w_{ij}^p = 1$ if path p connects origin-destination (OD) pair (i, j) and $w_{ij}^p = 0$ otherwise
$W = (w_{ij}^p)$	is the path-OD incidence matrix
$h_p(t)$	is the departure rate from the origin or flow on path p at time t
$h(t) = (h_p(t) : p \in P)$	is the full vector of path flows at time t
$H = (h(t) : t \in [0, T])$	is the full history of path flows
$\kappa_{ij}, \eta_p \in \Re$	are tatonnement parameters
$\eta = diag(\eta_p \in \Re_+ : p \in P)$	is a matrix of path flow tatonnement parameters
$\kappa = diag(\kappa_{ij} \in \Re_+ : i \in N_O, j \in N_D)$	is a matrix of perceived cost tatonnement parameters
$\tau_{a_i}^p(t)$	is the time of exiting the ith arc of path p provided that departure from the origin occurs at time t
$\theta_{a_i}^p(\beta)$	is the inverse exit time function giving the time of departure from the origin node of path p when exit from arc a_i occurs at time β
$c_p(t, H)$	is the unit generalized travel cost for path p given that departure from the origin occurs at time t and traffic conditions H are encountered
$u_{ij}(t)$	is the perceived travel cost between origin i and destination j at time t
$u(t) = (u_{ij}(t) : i \in N_O, j \in N_D)$	is the full vector of perceived travel costs at time t
$U = (u(t) \; : \; t \in [0, T])$	is the full history of perceived costs
$d_{ij}[u(t)]$	is the travel demand between i and j at time t
$d[u(t)] = (d_{ij}[u(t)] : i \in N_O, j \in N_D)$	is the full vector of travel demands
$[0, T]$	is the closed time interval of interest or planning horizon
$y_a(t)$	is the magnitude of the improvement to link a, i.e., the increase in the effective capacity of link a at time t
B	is the fixed exogenous budget available for improvement
$\pi(y_a(t))$	is the cost of making an improvement of magnitude y_a on link a at time t
r	is the constant discount rate

7 Infinite Dimensional Formulations of Some Dynamic Traffic Assignment Models

David Bernstein
Department of Civil Engineering and Operations Research
Princeton University
Princeton, NJ 08544, USA

Terry L. Friesz
Departments of Systems Engineering and Operations Research & Engineering
George Mason University
Fairfax, VA 22030, USA

7.1 Introduction

Traffic assignment models attempt to determine the usage of each route and/or link in a transportation network, given information about the number of trips being taken between various locations, the characteristics of the network, and the characteristics of the vehicles on the network. Though the term "assignment" seems to connote a prescriptive process in which vehicles are assigned to particular routes, there are both descriptive/positive and prescriptive/normative traffic assignment models.

Obviously, these models are of practical significance because of *congestion*. If transportation networks were not congested, there would be little practical interest in traffic assignment models. As it turns out, this is also what makes these models interesting from a theoretical standpoint. Indeed, as observed by Wardrop (1952), the route choices of drivers both cause and are influenced by congestion. This feedback process is illustrated in Fig. 7.1 and has been modeled in two ways. One approach is to ignore the adjustment mechanism and focus exclusively on the resulting equilibrium (see, for example, Beckmann et al., 1956; Smith 1979; Dafermos, 1980). The other approach is to explicitly model the day-to-day adjustment process (see, for example, Smith, 1979; Cascetta, 1991; Dupuis and Nagurney, 1993; Friesz et al., 1994).

In spite of the importance of congestion in traffic assignment, traditional traffic assignment techniques use very simple models of congestion. In fact, they completely ignore the inherently dynamic nature of congestion. That is, they employ link performance functions which describe the relationship between the average number of vehicles using a link during a period, and the average travel time experienced by those vehicles during that period (see, for example, Branston, 1976, or Boardman and Lave, 1977). In addition, in keeping with their use of aggregate performance functions, they ignore departure-time choices and the relationship between those choices and travel-times. Finally, they ignore how the route and departure-time choices of travelers change in the short-term (say, in response to congestion, information, or guidance).

One way of incorporating the dynamics of congestion is illustrated in Fig. 7.2. Unlike the traditional paradigm, this new paradigm distinguishes between day-to-day adjustment behavior and within-day adjustment behavior. Perhaps more importantly, this new paradigm explicitly considers the timing of trips (i.e., the trajectory of route choices) and how/when different vehicles interact. That is, the dynamics of

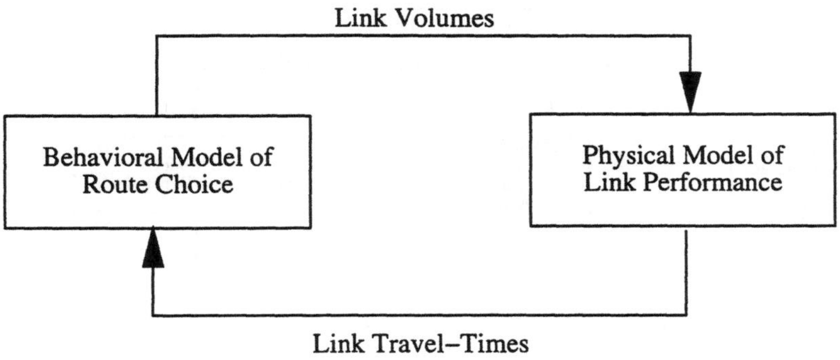

Fig. 7.1. The classical traffic assignment paradigm

congestion formation are considered as are the impacts of time-varying travel costs on behavior.

Though no complete model of this kind exists, some of the pieces have been considered. First, a variety of different models of the (hopefully) resulting equilibrium have been developed, some of which include only route choices and some of which include both route and departure-time choices (see, for example, Smith and Ghali, 1990; Drissi-Kaitouni, 1990; Cascetta, 1991; Janson (1991); Friesz et al., 1993; Ran et al., 1993; and Bernstein et al., 1993). They are perhaps best viewed as models of *default* (or usual) behavior and are most appropriate for transportation planning applications. In addition, there have been several attempts to to model the adjustments that drivers make to their routes (and departure-times) in response to changing conditions (see, for example, Chang and Mahmassani, 1988; and Ben-Akiva et al., 1991). They are perhaps best viewed as *pre-trip or en-route adjustment models* and are most appropriate for traffic control applications.

Where the developers of these dynamic traffic assignment models seem to have had the most difficulty is in the development of the physical model of traffic flow/dynamics.[1] While it is relatively easy to specify individual link performance functions, network effects complicate things dramatically. For example, consider the deterministic queueing approach introduced by Vickrey (1969). Though it is quite easy to model the travel time for a vehicle entering a link at a specific time as a function of the number of other vehicles in the queue at the tail of the link at that time, it is very difficult to model the same travel time as a function of the departure patterns over the entire network (i.e., incorporating multiple origin-destination pairs, paths, and departure-times).

Two broad approaches have been suggested to replace the closed-form perform-

[1] Indeed, early dynamic traffic assignment models took an entirely different approach, using link exit-rate functions (see, for example, Merchant and Nemhauser, 1978a;1978b; Carey and Srinivasan, 1988; and Friesz et al., 1989). However, this approach, in addition to being somewhat less intuitive, has some problems which make it unworkable in many cases.

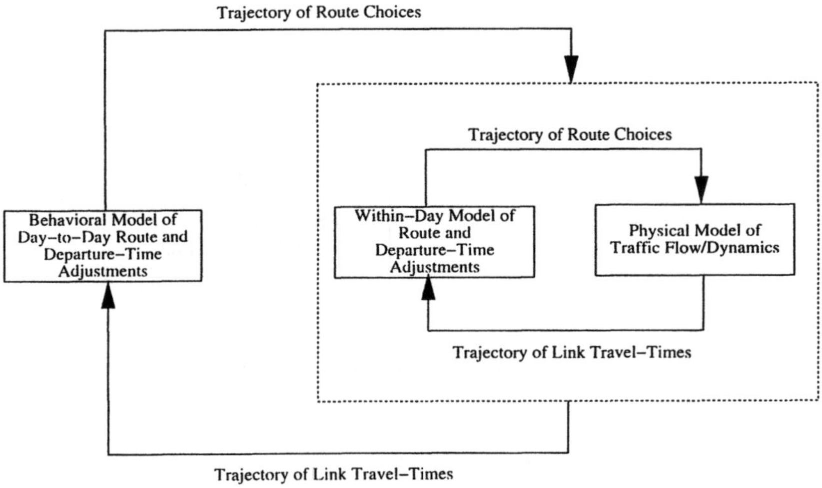

Fig. 7.2. A dynamic traffic assignment paradigm

ance functions used in static traffic assignment models. One is to use time-based simulation (see, for example, Mahmassani and Peeta, 1992; and Bernstein et al., 1992). In this approach, at distinct points in time, vehicles are essentially "moved" through the network interacting with other vehicles as appropriate. Unfortunately, while this does enable the calculation of link travel-times, it provides no information about the properties of the traffic flow process (e.g., monotonicity, continuity). Hence, this approach makes it impossible to understand the properties of the larger model (e.g., the existence of equilibria, the convergence of algorithms). The alternative is to use a system of differential and/or integral equations (see, for example, Friesz et al., 1993; and Bernstein et al., 1993). In this approach, the different links are "tied together" in such a way that flow conservation is maintained. The difficulty with this is that the resulting system of equations is very difficult to solve, even on small networks.

In this paper we (informally) present a third approach to modeling traffic dynamics. Specifically, we develop an infinite dimensional fixed-point formulation which is motivated using an event-based simulation. We believe that this approach may enable us to both easily obtain solutions and determine their properties. We then show how this approach can be incorporated into two types of dynamic traffic assignment models. The first is an equilibrium model of route and departure-time choice, and the second is a disequilibrium adjustment model. Finally, we conclude with a discussion of future research.

7.2 A Model of Traffic Dynamics

As discussed above, one of the most attractive features of static traffic assignment models is that they admit closed-form link performance functions which can be expressed in terms of path choices. In particular, consider a network consisting of a finite set of *nodes*, N, and a finite set of *arcs* (or *links*), A. Then, a *path* (or *route*), p, is simply an acyclic ordered set of arcs, $< a_1^p, \ldots, a_{m(p)}^p >$, that connects an *origin* and a *destination*. We denote the finite set of all paths in the network by P. So, letting $\Re$, $\Re_+$, and $\Re_{++}$ denote the *real numbers*, *nonnegative reals*, and *positive reals* respectively, the static traffic assignment problem attempts to determine the number of vehicles using each path during a particular period, which we denote by $h = (h_1, \ldots, h_{|P|}) \in \Re_+^{|P|}$.[2] Given the *arc-path incidence matrix*, $\Delta = (\Delta_{ap} : a \in A, p \in P) \in \{0,1\}^{|A|\times|P|}$, where $\Delta_{ap} = 1$ if arc a is an element of path p and 0 otherwise, the number of vehicles using each arc during a particular period, $x = (x_1, \ldots, x_{|A|}) \in \Re_+^{|A|}$, is given by $x = \Delta h$. Hence, if the performance (i.e., travel-time) of a link, $c_a : \Re_+^{|A|} \to \Re_{++}$, is a function of the number of vehicles on that link (or, in fact, every link) during the period, then it is easy to see that c_a can be written explicitly as a function of h.

On the other hand, it turns out to be quite difficult to incorporate the dynamics of traffic flow into a traffic assignment model. In particular, even if a link's performance (i.e., travel-time) at a particular time is a function of the number of vehicles on that link at that time, it is very difficult to model link performance as a function of departure-time and route choices. This is because the number of vehicles on a link at a particular time is itself determined by the travel-time on (potentially) all of the links in the network.

As mentioned in the introduction, two approaches have been proposed to model these interactions. One involves time-based simulation and the other involves systems of integral and/or differential equations. The approach we take here is to instead develop a fixed-point formulation of the problem.

7.2.1 An Event-Based Simulation

Perhaps the best way to motivate this formulation is to consider an *event-based simulation* of the dynamics of vehicles traveling on a network. First, let $Q_p > 0$ denote the known (integer) number of vehicles that are going to use path p, and $L_{p,n}$ denote the known time that the nth vehicle using path p departs from the origin. Now, let $\delta_a^{p,n}$ denotes the *travel-time* on arc a experienced by the nth vehicle using path p. We assume that $\delta_a^{p,n}$ is a function only of the number of vehicles on link a when this vehicle enters link a. That is, letting $f_a^{p,n}$ denote the number of vehicles on arc a when the nth vehicle on path p enters a, we write:

$$\delta_a^{p,n} = g_a(f_a^{p,n}) \tag{1}$$

for all $p \in P$, $a \in p$, and $n = 1, \ldots, Q_p$. The *travel delay* function, g_a, is assumed to be *FIFO-preserving* for all $a \in A$. That is, the travel delay functions are assumed to enforce a first-in-first-out discipline (see, for example, Friesz et al., 1993, Theorem 1).

Now, the time that the nth vehicle using path p *enters* link a is denoted by $e_a^{p,n}$ and the time it *exits* link a is denoted by $o_a^{p,n}$. So:

$$o_a^{p,n} = e_a^{p,n} + \delta_a^{p,n} \tag{2}$$

[2] Some people would argue that the objective is actually to determine the number of vehicles using each arc during the period. This distinction is unimportant for our present purposes.

for all $p \in P$, $a \in p$, and $n = 1, \ldots, Q_p$, where:

$$e_{a_1}^{p,n} = L_{p,n} \tag{3}$$

and

$$e_{a_i}^{p,n} = L_{p,n} + \sum_{j=1}^{i-1} \delta_{a_j}^{p,n} \quad i = 2, \ldots, m(p) \tag{4}$$

for all $p \in P$ and $n = 1, \ldots, Q_p$. We adopt the convention that $e_a^{p,n} = \infty$ when $a \notin p$.

As it turns out, given $L = (L_{p,n} : p \in P, n = 1, \ldots, Q_p)$ it is possible to sequentially solve for $\delta = (\delta_a^{p,n} : p \in P, n = 1, \ldots, Q_p, a \in p)$ using event-based simulation. In this case, an *event* is simply a vehicle entering a link. The *event queue*, $\mathcal{Q}$, contains a list of all vehicles (and their paths), the next arc that they will enter, and the time that the next arc is entered, $[n, p, a_i, e_{a_i}^{p,n}]$. It is sorted based on the *event time* (i.e., the time that each vehicle will enter the next arc). Initially, the event queue contains each vehicle and the first arc in its path, and is sorted based on each vehicle's departure-time from the origin.

Each event is processed in three steps. First, the number of vehicles on the link about to be entered is calculated. Specifically, $f_a^{p,n}$ is the number of vehicles that entered link a before vehicle p, n and exited link a after vehicle p, n entered. Second, the travel-time, exit-time and next link's entry-time are calculated using (1), (2) and (4) respectively. This also yields the next event time for this vehicle (i.e., the time it enters the next arc). Third, this new event is inserted in the queue.

As an example, consider the start of the simulation. The first event processed will be the departure of the first vehicle from its origin. In the case of the first vehicle, no other vehicles are yet on the network. Hence, the travel-time experienced by this first vehicle will be the free-flow travel time and the exit-time can be calculated accordingly. The next event time for this vehicles is the time that it enters the second arc in its path (i.e., which is simply the time that it exits the first arc in its path). This event is then inserted into the event queue and processing continues. In this case, the second event will either be a new vehicle departing from the origin (either on this or another path), or the first vehicle entering the second arc along its path. In either case, the simulation proceeds as described above.

The complete process can be summarized as follows:

Step 0. Create the initial event queue, $\mathcal{Q}$. It will contain $[n, p, a_i, L_{p,n}]$ for all p and $n = 1, \ldots, Q_p$. Sort $\mathcal{Q}$ by the departure-times, $L_{p,n}$.

Step 1. Select the next event in the queue. Let a denote the arc about to be entered, and (n, p) denote the vehicle and path. Calculate $f_a^{p,n}$.

Step 2. Calculate $\delta_a^{p,n}$ using (1). Set $o_a^{p,n} = e_a^{p,n} + \delta_a^{p,n}$. If a is not the last arc in path p, set $e_b^{p,n} = o_a^{p,n}$ where b denotes the successor of arc a in path p.

Step 3. If a is not the last arc in path p insert $[n, p, b, e_b^{p,n}]$ into $\mathcal{Q}$ and re-sort $\mathcal{Q}$ based on the event-times.

Step 4. If the event queue is not empty, GOTO Step 1.

This sequential process works because of the assumed FIFO-preserving properties of the travel delay functions. Because of its simplicity it can be executed quite rapidly even on large networks with large numbers of vehicles.

7.2.2 A Fixed-Point Formulation Using Departure-Times

With this event-based simulation as motivation, we can now develop a fixed-point formulation of the traffic dynamics problem that can be used for dynamic traffic assignment. However, given that most dynamic traffic assignment models make use of departure rates in continuous time, we must now treat n as if it is real-valued rather than integer-valued (see Bernstein, 1994, for a discussion of a simultaneous route and departure-time choice equilibrium model with discrete vehicles and continuous time). This has two important consequences.

First, with n real-valued the sequential process described above is no longer appropriate. In particular, with $n \in [0, Q_p]$ there are an infinite number of decision variables. Hence, instead of the sequential approach described above, we will now solve for all travel-times simultaneously.

Second, we must re-examine how f_a^p can be calculated. In particular, since traffic is real-valued we can no longer "count" vehicles.

We begin by modifying our notation somewhat in order to both illustrate the functional dependencies more carefully and to capture the infinite dimensional nature of the problem. To that end, we now let $Q_p \in \Re_{++}$ denote the known number of p-vehicles (i.e., vehicles that are going to use path p, and let $L_p : [0, Q_p] \to \Re_+$ denote the departure-time function for p-vehicles. Similarly, we let $\delta_a^p : [0, Q_p] \to \Re_+$ denote the travel-time function on arc a for p-vehicles. Now, given $L = (L_1, \ldots, L_p)$ and $\delta = (\delta_a^p : p \in P, a \in p)$, the entry- and exit-times on any arc $a \in p$ for the nth p-vehicle are denoted by $e_a^p(n|L, \delta)$ and $o_a^p(n|L, \delta)$. As before:

$$e_{a_1}^p(n|L, \delta) = L_p(n|L, \delta) \quad p \in P, n \in [0, Q_p] \tag{5}$$

$$e_{a_i}^p(n|L, \delta) = L_p(n|L, \delta) + \sum_{j=1}^{i-1} \delta_{a_j}^p(n) \quad p \in P, i = 2, \ldots, m(p), n \in [0, Q_p] \tag{6}$$

$$o_a^p(n|L, \delta) = e_a^p(n|L, \delta) + \delta_a^p(n) \quad p \in P, a \in p, n \in [0, Q_p]. \tag{7}$$

Now, if the inverse functions, $[e_a^p]^{-1}(t|L, \delta)$ and $[o_a^p]^{-1}(t|L, \delta)$ exist for all a and p, then:

$$f_a^p(n|L, \delta) = \sum_{q:a \in q} ([e_a^q]^{-1}[e_a^p(n|L, \delta)|L, \delta] - [o_a^q]^{-1}[e_a^p(n|L, \delta)|L, \delta] \tag{8}$$

for all $p \in P$, $a \in p$, and $n \in [0, Q_p]$. Hence, given L and FIFO-preserving travel functions, $g_a : \Re_+ \to \Re_{++}, a \in A$, the problem is to find a δ such that:

$$\delta_a^p(n) = g_a[f_a^p(n|L, \delta)] \tag{9}$$

for all $p \in P$, $a \in p$, and $n \in [0, Q_p]$. This is clearly an infinite dimensional fixed-point problem, the properties of which are well-known (see, for example, Istratescu, 1981).

This fixed-point approach to modeling traffic dynamics has one important advantage over the time-based simulation approach. Specifically, it enables us to consider the properties of the travel-time functions. As will become clear below, this is particularly useful when trying to study dynamic traffic assignment models.

This fixed-point approach also has one important advantage over the integral/differential equation system approach. Specifically, it enables us to easily add constraints to the model. For example, it would be very easy to add a link capacity constraint to the above formulation. That is, letting K_a denote the instantaneous capacity of arc a, one could require that δ satisfy $f_a^p(n|L,\delta) \le K_a$ for all a, p containing a, and $n \in [0, Q_p]$.

7.2.3 A Fixed-Point Formulation Using Departure-Rates

The fixed-point formulation developed above, while easy to understand, cannot be directly incorporated into most existing dynamic traffic assignment (DTA) models. This is because most existing DTA models make use of departure-rates rather than departure-times [see Jauffred and Bernstein (1996) for a discussion of a simultaneous route and departure-time model that does make use of departure times]. Hence, we now "invert" the above formulation so that it is more amenable to use in other models. That is, rather than expressing the various functions in terms of vehicle n, we express them in terms of the (corresponding) departure-time t. This is possible because, given a *departure rate* function, $h_p : [0,T] \to \Re_+$, for each path, $p \in P$, the number of p-vehicles that have departed before time t is given by $I_p(t) = \int_0^t h_p(\omega)d\nu(\omega)$, where ν denotes a Lebesgue measure. Hence, under reasonable conditions, it is possible to establish a correspondence between vehicles and their departure-times and transform the above fixed-point problem.

To do so, we let the complete pattern of departures be denoted by $h = (h_p : p \in P)$ and we let the travel-time on arc a of path p for a vehicle that departed from the origin at time t be denoted by $\delta_a^p(t)$. Similarly, the entry- and exit-times on any arc $a \in p$ for the p-vehicle that departs at time t are denoted by $e_a^p(t|h,\delta)$ and $o_a^p(t|h,\delta)$. Then:

$$e_{a_1}^p(t|h,\delta) = t \quad p \in P \tag{10}$$

$$e_{a_i}^p(t|h,\delta) = t + \sum_{j=1}^{i-1} \delta_{a_j}^p(t) \quad p \in P, i = 2, \ldots, m(p) \tag{11}$$

and

$$o_{a_i}^p(t|h,\delta) = e_{a_i}^p(t|h,\delta) + \delta_{a_i}^p(t) \quad p \in P, a \in p \tag{12}$$

$$= t + \sum_{j=1}^{i} \delta_{a_j}^p(t) \quad p \in P, i = 1, \ldots, m(p). \tag{13}$$

So, assuming the inverse functions, $[e_a^p]^{-1}(t|h,\delta)$ and $[o_a^p]^{-1}(t|h,\delta)$ exist for all a and p:

$$f_a^p(t|h,\delta) = \sum_{q:a\in q} \int_0^{[e_a^p]^{-1}(t|h,\delta)} h_q(\omega)d\nu(\omega) - \int_0^{[o_a^p]^{-1}(t|h,\delta)} h_q(\omega)d\nu(\omega). \tag{14}$$

for all $p \in P$, $a \in p$, and $n \in [0, Q_p]$.

The resulting fixed-point problem is to find a δ such that:

$$\delta_a^p(t) = g_a[f_a^p(t|h,\delta)] \tag{15}$$

for all $p \in P$, $a \in p$, and t.

7.3 A Simultaneous Route and Departure-Time Choice Equilibrium Model

We now consider how the above model of traffic dynamics can be incorporated into the simultaneous route and departure-time choice (SRD) equilibrium model of Friesz et al. (1993). Specifically, we consider a model of commuter behavior in which each commuter chooses both a route, $p \in P$, and a departure-time, $t \in [0, T]$. The set of origin-destination pairs (O-D pairs), denoted by W, is assumed to be finite and partitions the path set, P, into sets of w-paths, P_w, among which commuters traveling between O-D pair w must choose.

The number of commuters between any O-D pair, $w \in W$, is denoted by $Q_w \in \Re_{++}$, and the vector $Q = (Q_w : w \in W) \in \Re_{++}^{|W|}$ is designated as the (fixed) *travel demand vector* for the system. For each path, $p \in P$, the *departure rate* function is denoted by $h_p : [0, T] \to \Re_+$. That is, the departure rate on path p at time t is denoted by $h_p(t) \in \Re_+$. Each complete pattern of departures is described by the vector, $h = (h_p : p \in P)$. Thus, letting ν denote a Lebesgue measure on $[0, T]$, the set of *feasible path flow vectors* under travel demand Q is given by

$$H_Q = \left\{ h : \sum_{p \in P_w} \int_0^T h_p(t) d\nu(t) = Q_w \quad \forall\, w \in W \right\}. \tag{16}$$

In general, each feasible departure pattern gives rise to a cost pattern in which the cost on path p for departures at time t given a departure pattern h is denoted as $c_p(t|h)$.

Loosely speaking, we assume, as in Friesz et al. (1993), that the only flow patterns which can persist are those in which all used path and departure-time choices have minimum cost. More formally, if for any measurable set, $S \subseteq [0, T]$, with $\nu(S) > 0$ and any measurable function, $F : S \to \Re$, the *essential infimum* of F on S is given by

$$\text{ess inf}\{F(s) : s \in S\} = \sup\{x \in \Re : \nu\{s \in S : F(s) < x\} = 0\} \tag{17}$$

and

$$\mu_p(h) = \text{ess inf}\{c_p(t|h) \;:\; t \in [0, T]\} \tag{18}$$

then the relevant lower bound on achievable costs for a w-commuter is given by:

$$\mu_w(h) = \min\{\mu_p(h) : p \in P_w\}. \tag{19}$$

So, letting $\forall_\nu(t)$ denote the phrase "for ν-almost all $t \in [0, T]$", we can define an equilibrium as follows:

Definition (SRD Equilibrium). *A departure rate pattern, $h \in H_Q$ is said to be a* simultaneous route and departure-time choice equilibrium *(SRD equilibrium) for Q if and only if h satisfies the following condition for all $w \in W$, and $p \in P_w$:*

$$h_p(t) > 0 \Longrightarrow c_p(t|h) = \mu_w(h)\forall_\nu(t). \tag{20}$$

To develop a specific SRD equilibrium model, we now assume that the cost of a trip is a function of both the travel-time and the *schedule cost* (which accounts for travelers arriving at their destination at undesirable times). That is, letting $D_p(t)$ denote the travel-time on path p for a departure at time t, the schedule delay cost is given by:

$$\Phi_p[t + D_p(t)] = \begin{bmatrix} \alpha[(T^* - \Delta) - (t + D_p(t))] & \text{if} \quad (T^* - \Delta) > [t + D_p(t)] \\ 0 & \text{if} \quad (T^* - \Delta) \leq [t + D_p(t)] \leq (T^* + \Delta) \\ \beta[(t + D_p(t)) - (T^* + \Delta)] & \text{if} \quad (T^* + \Delta) < [t + D_p(t)], \end{bmatrix} \tag{21}$$

where $\alpha \in \Re_+$ denotes the dollar penalty associated with early arrival, $\beta \in \Re_+$ denotes the dollar penalty associated with late arrival, $\Delta \in \Re_+$ denotes half the arrival time window, and $[T^* - \Delta, T^* + \Delta]$ denotes the set of "equally acceptable" arrival times.

The only remaining task in this development of a specific SRD equilibrium model is the specification of the path travel-time functions and this is where the fixed-point formulation of the traffic dynamics problem comes into play. In particular, given a link travel-time function, δ the path travel-time functions are given by:

$$D_p(t) = \sum_{a \in p} \delta_a^p(t) \tag{22}$$

for all $p \in P$ and $t \in [0, T]$.

Thus, we may now define the generalized travel cost operators as follows:

$$c_p(t|h, \delta) = \gamma \cdot \sum_{a \in p} \delta_a^p(t) + \Phi_p\left(t + \sum_{a \in p} \delta_a^p(t)\right) \quad \forall\, p \in P \tag{23}$$

where $\gamma \in \Re_+$ is the value of time.

It now follows from Friesz et al. (1993, Theorem 2) and the discussion above that our objective is to find a departure rate pattern, $\hat{h} \in H_Q$, and travel-time pattern, $\hat{\delta}$, such that:

$$\sum_{p \in P} \int_0^T c_p(t|\hat{h}, \hat{\delta})[h_p(t) - \hat{h}_p(t)]d\nu(t) \geq 0 \tag{24}$$

$$\hat{\delta}_a^p(t) = g_a[f_a^p(t|\hat{h}, \hat{\delta})] \quad p \in P, a \in p, t \in [0, T] \tag{25}$$

for all $h \in H_Q$. Of course, this can easily be written as a single fixed-point problem rather than a mixed variational inequality, fixed-point problem.

7.4 A Dynamic Disequilibrium Adjustment Model

We now consider how the model of traffic dynamics developed above can be incorporated into a model of adjustment behavior rather than a model of default behavior. To begin, we consider a model of disequilibrium path-adjustment behavior developed by Friesz et al. (1992) which has the following form:

$$\frac{dh}{dt} = \eta F(ETC) \tag{26}$$
$$\frac{du}{dt} = \kappa G(ETD) \tag{27}$$
$$h(0) = h^0 \tag{28}$$
$$u(0) = u^0, \tag{29}$$

where $h \in \Re_+^{|P|}$ is a vector of route flows, $u \in \Re_{++}^{|W|}$ is a vector of O-D costs, $\eta \in \Re^{|P|\times|P|}$ and $\kappa \in \Re^{|W|\times|W|}$ are diagonal matrices, $F : \Re^{|P|} \rightarrow \Re^{|P|}$ and $G : \Re^{|P|} \rightarrow \Re^{|P|}$ are monotonic vector operators, $ETC \in \Re^{|P|}$ is a vector of excess travel costs and $ETD \in \Re^{|W|}$ is a vector of excess travel demands. As in traditional static traffic assignment models, this general path-adjustment process ignores departure-time choices and traffic dynamics. One particular model is:

$$ETC = c(h) - W'u \tag{30}$$
$$ETD = d(u) - Wh, \tag{31}$$

where W is the path-OD incidence matrix[3], and $d(u) \in R_{++}^{|W|}$ is a vector of transportation demands. Thus, the complete model takes the form:

$$\frac{dh}{dt} = \eta F[c(h) - W'u] \tag{32}$$
$$\frac{du}{dt} = \kappa G[d(u) - Wh] \tag{33}$$
$$h(0) = h^0 \tag{34}$$
$$u(0) = u^0. \tag{35}$$

The basic premise behind this model is that people have information about the cost on the path that they are going to take (e.g., from radio/television reports), $c_p(h)$, and they have some perception of what the cost "should be" from their origin to their destination, u_w. As time progresses, the number of people choosing each path changes in response to the information they have about the difference between the cost of using that path and what the cost "should be". In addition, as time progresses

[3] Note that W' is the transpose of W, $Wh = (\sum_p w_{ij}^p h_p : i \in N_O, j \in N_D) = (\sum_{p \in P_{ij}} h_p : i \in N_O, j \in N_D)$ and $W'u = (\sum_k \sum_l w_{kl}^p u_{kl} : i \in N_O, j \in N_D, p \in P_{ij}) = ((u_{ij} : p \in P_{ij}) : i \in N_O, j \in N_D)$.

they also adjust their perception of what the O-D cost should be in response to the difference between the number of people that "should be" traveling and the number that actually are. It is interesting to note that the steady states of this model (if they exist) are traditional static equilibria. It is also interesting to note that one can imagine situations in which d^{-1} exists and is known, in which case this system can be simplified significantly.

Of course, the two adjustment mechanisms in this model can be characterized as being either fast or slow. For example, one can imagine a situation in which people change their path choices frequently and quickly but change their beliefs about the true O-D cost much more slowly. In this case, $\eta_p \gg \kappa_{ij}$ for each $i \in N_O,\ j \in N_D$ and $p \in P_{ij}$. Alternatively, one could assume that these adjustment processes move at the same speed.

As Friesz et al. (1992) point out, the relative speeds of the adjustment processes become much more important and interesting when departure-time choices and traffic dynamics are incorporated into this model. If we now consider that case, then $h_p : [0,T] \to \Re_+$ again denotes the departure rate function for path p, $h = (h_p : p \in P)$ again denotes the complete pattern of departures, $u_w : [0,T] \to \Re_{++}$ denotes the O-D cost function for O-D pair w, and $u = (u_w : w \in W)$ denotes the complete O-D cost pattern. Further, making use of the fixed-point formulation of the traffic dynamics model, $c_p(t|h,\delta)$ again denotes the generalized travel cost on path p for departures at time t given h and δ. Hence, to solve the complete disequilibrium adjustment model we must find h, u, and δ such that:

$$\frac{dh}{dt} = \eta F[c(t|h,\delta) - W'u] \tag{36}$$

$$\frac{du}{dt} = \kappa G[d(u) - Wh] \tag{37}$$

$$h(0) = h^0 \tag{38}$$

$$u(0) = u^0, \tag{39}$$

$$\delta_a^p(t) = g_a[f_a^p(t|h,\delta)] \quad p \in P, a \in p, t \in [0,T] \tag{40}$$

In principle, this system can be readily solved. However, several observations about this system deserve some additional comment.

7.5 Conclusion and Directions for Future Research

In this paper we have attempted to describe a new approach for modeling traffic dynamics and for incorporating traffic dynamics into existing equilibrium and disequilibrium models of path and departure-time choice. Though we believe that this approach has promise, a considerable amount of work still needs to be done before dynamic traffic assignment models can become a standard part of the transportation planner's and traffic engineer's toolboxes.

The first step is to consider the fixed-point formulation of the traffic dynamics problem more fully. This will involve developing conditions which ensure the existence and uniqueness of solutions to the fixed-point problem. It will also involve implementing convergent algorithms for solving the fixed-point problem in a reasonable amount of time.[4] The key here will be to impose requirements on δ that are consistent with the invertability assumptions used above that do not make the feasible set non-compact.

The second step is to consider the formulation of the SRD equilibrium problem presented here more fully. This will involve developing conditions which ensure that such equilibria exist, as well as the development of convergent algorithms for the SRD equilibrium problem.

The final step is to consider the formulation of the disequilibrium adjustment model presented here. This will involve work regarding the existence and stability of steady-states as well as other properties of the trajectories. It will also involve developing algorithms for solving the system of differential equations.

References

Beckmann, M.,McGuire, C. and Winsten, C. (1956), *Studies in the Economics of Transportation*, Yale University Press, New Haven, CT.

Ben-Akiva, M., de Palma, A. and Kaysi,I. (1991) 'Dynamic network models and driver information systems', *Transportation Research*, vol. 25A, pp. 251–266.

Bernstein, D., Friesz, T.L., Tobin, R.L., Shenoi, R. and Wie, B.W. (1992), 'Solving a variational inequality formulation of the simultaneous route and departure-time choice equilibrium problem', presented at the 39th North American Meetings of the Regional Science Association International.

Bernstein, D., Friesz, T.L., Tobin, R.L. and Wie, B.-W. (1993), 'Variational control models of dynamic network traffic equilibrium', *Proceedings of the 12th International Symposium on Transportation and Traffic Theory*, Elsevier, Amsterdam, pp. 107–126.

Bernstein, D. (1994) 'Non-existence of nash equilibria to the departure-time choice problem', presented at the Joint National Meetings of ORSA/TIMS.

Boardman, A.E. and Lave, L.B. (1977), 'Highway congestion and congestion tolls', *Journal of Urban Economics*, vol. 4, pp. 340–359.

Branston, D. (1976), 'Link capacity functions: A review', *Transportation Research*, vol. 10, pp. 223–236.

Carey, M. and Srinivasan, A. (1988), 'Congested network flows: Time varying demands and start-time policies', *European Journal of Operational Research*, vol. 36, pp. 227–240.

Cascetta, E. (1991), 'A day-to-day and within-day dynamic stochastic assignment model', *Transportation Research*, vol. 25A, pp. 277–291.

Chang, G-L. and Mahmassani, H.S. (1988), 'Travel time prediction and departure time adjustment behavior dynamics in a congested traffic system', *Transportation Research*, vol. 22B, pp. 217–232.

Dafermos, S.C. (1980), 'Traffic equilibria and variational inequalities', *Transportation Science*, vol. 14, pp. 42–54.

Dupuis, P. and Nagurney, A. (1993), 'Dynamical systems and variational inequalities', *Annals of Operations Research*, vol. 44, pp. 9-42.

Drissi-Kaitouni, O. (1990), 'A model for the dynamic traffic assignment problem', Centre de Recherche sur les Transports, Publication No. 702, Universite de Montreal.

[4] Of course, the definition of "reasonable" will be quite different in planning applications and real-time control applications. This may necessitate the development of different algorithms/heuristics for different purposes.

Friesz, T.L., Luque, J., Tobin, R.L. and Wie, B.W. (1989), 'Dynamic network traffic assignment considered as a continuous time optimal control problem,' *Operations Research*, vol. 37, pp. 893–901.

Friesz, T.L., Bernstein, D. and Stough, R. (1992), 'Variational inequalities, dynamical systems, and control theoretic models for predicting time-varying urban network flows', presented at the North American Meetings of the Regional Science Association International.

Friesz, T.L., Bernstein, D., Smith, T.E., Tobin, R.L. and Wie, B.-W. (1993), 'A variational inequality formulation of the dynamic network user equilibrium problem', *Operations Research*, vol. 41, 1993, pp. 179–191.

Friesz, T.L., Mehta, N., Tobin, R.L., Bernstein, D. and Ganjalizadeh, S. (1994), 'Day-to-day adjustments in the presence of an imperfect driver information system', *Operations Research*, forthcoming .

Istratescu, V.I. (1981) *Fixed Point Theory*, Riedel Publishing, Dordrecht.

Janson, B.N. (1991) 'Dynamic traffic assignment for urban road networks', *Transportation Research*, vol. 25B, pp. 143–161.

Jauffred, F. and Bernstein, D. (1996), 'An alternative formulation of the simultaneous route and departure-time choice equilibrium problem',*Transportation Research C*, vol. 4, pp. 339–357.

Mahmassani, H.S. and Peeta, S. (1992), 'Dynamic assignment for route guidance decisions', presented at the 39th North American Meetings of the Regional Science Association International.

Merchant, D.K. and Nemhauser, G.L. (1978a) 'A Model and an Algorithm for the Dynamic Traffic Assignment Problem', *Transportation Science*, vol. 12, pp. 183–199.

Merchant, D.K. and Nemhauser, G.L. (1978b) 'Optimality Conditions for a Dynamic Traffic Assignment Model', *Transportation Science*, vol. 12, pp. 200–207.

Ran, B., Boyce, D.E. and LeBlanc, L.J. (1993), 'Towards a new class of instantaneous dynamic user-optimal traffic assignment models', *Operations Research*, vol. 41, pp. 192–202.

Smith, M.J. (1979) 'Existence, uniqueness, and stability of traffic equilibria', *Transportation Research*, vol. 13B, pp. 295–304.

Smith, M.J. and Ghali, M.O. (1990) 'The dynamics of traffic assignment and traffic control: A theoretical study', *Transportation Research*, vol. 24B, pp. 409–422.

Vickrey, W.S. (1969) 'Congestion theory and transport investment', *Papers and Proceedings of the American Economic Review*, vol. 59, pp. 251–261.

Wardrop, J. (1952), 'Some theoretical aspects of road traffic research', *Proceedings of Institute of Civil Engineers*, vol. 1, part II, pp. 325–378.

8 Introduction to Projected Dynamical Systems for Traffic Network Equilibrium Problems

Anna Nagurney
School of Management
University of Massachusetts
Amherst, MA 01003, USA

Ding Zhang
School of Business
State University of New York at Oswego
Oswego, NY 13126, USA

8.1 Introduction

It is now well-known that a plethora of equilibrium problems, notably, network equilibrium problems, can be uniformly formulated and studied as finite-dimensional variational inequality problems. Indeed, it was precisely the traffic network equilibrium problem, as stated by Smith (1979), and identified by Dafermos (1980) to be a variational inequality problem, that gave birth to the ensuing research activity in variational inequality theory and applications in transportation science, regional science, operations research, and, more recently, in economics.

Typically, using this methodology, one first formulates the governing equilibrium conditions as a variational inequality problem. In the case of traffic equilibrium, the conditions date to Wardrop (1952) and Beckmann et al. (1956). Qualitative properties of existence and uniqueness of solutions to a variational inequality problem can then be studied using the standard theory (cf. Kinderlehrer and Stampacchia (1980)) or by exploiting problem structure (cf. Nagurney (1993)). Finally, a variety of algorithms for the computation of solutions to finite-dimensional variational inequality problems are now available (see, e.g., Dafermos (1983), Bertsekas and Tsitsiklis (1989), Nagurney (1993), and Florian and Hearn (1995), and the references therein).

Finite-dimensional variational inequality theory by itself, however, provides no framework for the study of the dynamics of competitive systems. Rather, it captures the system at its equilibrium state and, hence, the focus of this tool is static in nature.

Dupuis and Nagurney (1993) proved that, given a variational inequality problem, there is a naturally associated dynamical system, the stationary points of which correspond precisely to the solutions of the variational inequality problem. This association was first noted by Dupuis and Ishii (1991). This dynamical system, first referred to as a *projected dynamical system* by Zhang and Nagurney (1995), is non-classical in that its right-hand side, which is a projection operator, is discontinuous. The discontinuities arise because of the constraints underlying the variational inequality problem, modeling the application in question. Hence, classical dynamical systems theory (cf. Coddington and Levinson (1955), Lefschetz (1957), Hartman (1964), and Perko (1991)) is no longer applicable.

Nevertheless, as demonstrated rigorously in Dupuis and Nagurney (1993), a projected dynamical system may be studied through the use of the Skorokhod Problem (Skorokhod, 1961), a tool originally introduced for the study of stochastic differential equations with a reflecting boundary condition. Existence and uniqueness of a solution path, which is essential for the dynamical system to provide a reasonable model, were also established therein.

This paper surveys and synthesizes the recent research in the formulation, stability analysis, and computation of dynamic traffic network problems in the scenarios of elastic travel demand and fixed travel demand, using the approach of projected dynamic systems. For a survey of dynamic traffic models, see the book by Ran and Boyce (1994). The advantages of the projected dynamical systems approach are several-fold, and are highlighted below.

First, projected dynamical systems theory provides insights into the travelers' dynamic behavior in making their trip decisions (in the elastic demand scenario) and in adjusting their route choices (in both the elastic and the fixed demand scenarios).

Second, it provides a powerful theory for stability analysis of the trip-route choice adjustment processes. For example, the stability analysis addresses such questions as:

1. Whether and how the travelers' dynamic behavior in adjusting their route choices to avoid congestion will lead to a traffic equilibrium?

2. Given that the current traffic flow pattern is near an equilibrium, will the travelers' adjustment process stay close to this equilibrium flow pattern throughout time?

3. Given an observed or computed traffic equilibrium flow pattern, will it resume after some perturbation caused by either planned events or random incidents?

These questions are of concern not only to transportation planners and analysts but also form a critical basis upon which the concept of an equilibrium flow pattern rests. Indeed, if stability cannot be justified, then an equilibrium is not likely to prevail.

We note that such stability issues could not be addressed in a static framework using, for example, a finite-dimensional variational inequality formulation of the traffic network equilibrium conditions. Consequently, stability analysis provided by projected dynamical systems theory allows the examination and evaluation of traffic network equilibria in a more realistic dynamic setting, while, at the same time, capturing and retaining its static properties.

In fact, stability concerns surrounding traffic equilibria have motivated considerable research. Among others, Smith (1984) proposed a distinct route choice adjustment process, and established its global stability assuming that the path costs are monotone. In the case of symmetric path cost functions, Bernstein (1990) considered the stability of individual equilibria using classical techniques that do not require monotonicity. Horowitz (1984) started to explore the stability of stochastic equilibrium on a simple network. In addition, Mahmassani (1990) empirically studied stability issues by investigating the behavior of commuters in adjusting their route choices on a simulated traffic network.

Third, projected dynamical systems theory provides alternative algorithmic schemes for the computation of not only traffic equilibria but, also, the approximate trajectory of the travelers' dynamic behavior. In fact, these iterative schemes can be interpreted as time discretizations of the trip-route choice adjustment process in the projected dynamical systems formulation, and, therefore, they are referred to as discrete time algorithms. The proof of their convergence crucially depends on the stability of the continuous time adjustment processes.

To-date, two discrete time algorithms, based on the projected dynamical systems methodology, namely, the Euler method and the Heun method, have been employed in the computation of traffic equilibria. Both algorithms are special cases of the general iterative scheme of Dupuis and Nagurney (1993) and have variable step sizes. In particular, the Euler method can be thought of as a projection method with a variable step size for solving the governing variational inequality problem. The convergence of both algorithms has been established under strict monotonicity conditions of the link user cost functions in the case of fixed demand traffic network problems (cf. Nagurney and Zhang (1997a)), and of disutility functions (cf. Nagurney and Zhang (1998; 1997b)) and demand functions (cf. Zhang and Nagurney (1997)), respectively, for elastic demand traffic network problems. We note that various step-size rules have been used, together with other assumptions on the function that enters the variational inequality problem, in order to establish the convergence of projection methods with variable step size for variational inequality problems (see, e.g., Bakusinskii and Polyak (1974), Alber (1979), Fukushima (1986), and Flam (1992)). However, among those, the one presented in this chapter is the weakest one, and, until now, it has usually been incorporated with strong monotonicity conditions to guarantee convergence. Other research on the study of step-size rules for the gradient projection method (for optimization problems) includes that of Goldstein (1964), Armijo (1966), and Bertsekas (1976).

Finally, we note that the Euler method and the Heun method are suitable for exploitation of massively parallel computer architectures and have been implemented on such architectures for the massively parallel computation of dynamic traffic network problems with elastic demands by Nagurney and Zhang (1998; 1997c).

Our goal in this paper is to survey and to synthesize the recently derived results on projected dynamical systems and dynamic traffic modeling, analysis, and computation. Additional references can be found in the citations.

The organization of the paper is as follows. Section 8.2 summarizes the basic elements of projected dynamical system theory, including its one-to-one correspondence with the associated variational inequality problem. It also recalls the fundamental stability analysis results, and the general iterative scheme for computational purposes, with the induced Euler method and Heun method.

Section 8.3 reviews the projected dynamical system methodology for the formulation, analysis, and computation of traffic network problems with elastic demand in the case of known travel disutility functions. The trip-route choice adjustment process for this model was first proposed in Dupuis and Nagurney (1993). Zhang and Nagurney (1996) studied the stability issues. The convergence of the discrete time algorithms was established in Nagurney and Zhang (1998; 1997b), where numerical results were also provided.

Section 8.4 focuses on the transportation network model with elastic demand and with known travel demand functions. The trip-route choice adjustment process, its stability analysis, and the discrete time algorithms for this model were originally presented in Zhang and Nagurney (1997).

Section 8.5 summarizes the research on the use of projected dynamical systems for fixed demand transportation network problems. The route choice adjustment process and its behavioral intuition were proposed in Nagurney and Zhang (1997a). In that paper, the stability issues were studied and a discrete time algorithm for the computation of the traffic equilibrium pattern was presented, along with convergence results.

The paper concludes with a summary and discussion in Section 8.6.

8.2 Projected Dynamical Systems

In this section we review some recent results in the development of a new tool for the study of network equilibrium problems in a dynamic setting, which we have termed *projected dynamical systems* theory (cf. Zhang and Nagurney (1995)). One of the notable features of this tool, whose rigorous theoretical foundations were laid by Dupuis and Nagurney (1993), is its relationship to the variational inequality problem. Projected dynamical systems theory, however, goes further than finite-dimensional variational inequality theory in that it extends the static study of equilibrium states by introducing an additional time dimension in order to allow for the analysis of disequilibrium behavior that precedes the equilibrium.

In particular, we associate with a given variational inequality problem, a non-classical dynamical system, called a projected dynamical system. The projected dynamical system is interesting both as a dynamical model for the system whose equilibrium behavior is described by the variational inequality, and, also, because its set of stationary points coincides with the set of solutions to a variational inequality problem. In this framework, the feasibility constraints in the variational inequality problem correspond to discontinuities in the right-hand side of the differential equation, which is a projection operator. Consequently, as noted in the introduction, the projected dynamical system is not amenable to analysis via the classical theory of dynamical systems.

Clearly, since network equilibrium problems are characterized by conservation of flow constraints, the classical theory of dynamical systems, which cannot handle constraints, is no longer applicable. Projected dynamical systems theory, in contrast, provides natural dynamical analogues of network equilibrium problems which have been studied primarily at their equilibrium states and, usually, as variational inequality problems. The proofs of the theorems, etc., in this section are omitted for the sake of brevity, with the reader referred to the appropriate citations in the literature.

This section is organized as follows. In Section 8.2.1 the variational inequality problem is reviewed. We then present the definition of a projected dynamical system, which evolves within a constraint set K. Its stationary points are identified with the solutions to the corresponding variational inequality problem with the same constraint set. For theoretical soundness, we state in a theorem the fundamental properties of such a projected dynamical system in regards to the existence and uniqueness of solution paths to the governing ordinary differential equation. We then provide an interpretation of the ordinary differential equation that defines the projected dynamical system, along with a description of how the solutions may be expected to behave.

In Section 8.2.2 we present some general stability results using the monotonicity condition. Section 8.2.3 then recalls the general iterative scheme and gives conditions for convergence. For additional background, discussion, and proofs, see the books by Nagurney (1993) and Nagurney and Zhang (1997b).

8.2.1 The Variational Inequality Problem and A Projected Dynamical System

We now present the definition of a variational inequality problem (VI) and that of a projected dynamical system (PDS).

Definition 1 (The Variational Inequality Problem). *For a closed convex set $K \subset R^k$ and vector function $F : K \mapsto R^k$, the variational inequality problem,* VI(F, K), *is to determine a vector $x^* \in K$, such that*

$$\langle F(x^*)^T, x - x^* \rangle \geq 0, \quad \forall x \in K, \tag{1}$$

where $\langle \cdot, \cdot \rangle$ denotes the inner product in R^k.

As is well-known, the variational inequality has been used to formulate a plethora of network equilibrium problems ranging from traffic network equilibrium problems to spatial oligopolistic market equilibrium problems (cf. Nagurney (1993), Florian and Hearn (1995), and the references therein).

Finite-dimensional variational inequality theory, however, provides no framework for studying the underlying dynamics of systems, since it considers only equilibrium solutions in its formulation. Hence, in a sense, it provides a static representation of a system at its "steady state".

One would, therefore, like a theoretical framework that permits one to study a system not only at its equilibrium point, but also in a dynamical setting. In this way one can gain insights into the behavior of the system throughout time. Towards this end, we now recall projected dynamical systems.

The definition of a projected dynamical system (PDS) is given with respect to a closed convex set K, which is usually the constraint set underlying a particular application, such as, for example, network equilibrium problems, and a vector field F whose domain contains K. As noted in Dupuis and Nagurney (1993), it is expected that such projected dynamical systems will provide mathematically convenient approximations to more "realistic" dynamical models that might be used to describe non-static behavior. The relationship between a projected dynamical system and its associated variational inequality problem with the same constraint set is then highlighted.

For completeness, we also recall the fundamental properties of existence and uniqueness of the solution to the ordinary differential equation (ODE) that defines such a projected dynamical system.

Let $K \subset R^k$ be closed and convex. Denote by ∂K and K^0 the boundary and interior of K, respectively. Given $x \in \partial K$, define the set of inward normals to K at x by

$$n(x) = \{\gamma : \|\gamma\| = 1, \text{ and } \langle \gamma, x - y \rangle \leq 0, \forall y \in K\}. \tag{2}$$

For notational convenience, we define $n(x)$ to be $\{\gamma : \|\gamma\| = 1\}$ for x in the interior of K.

When K is a convex polyhedron, K takes the form $\cap_{i=1}^N K_i$, where each K_i is a closed half-space with inward normal n_i. Let P be the norm projection, defined as

$$P_K(x) = \text{argmin}_{z \in K} \|z - x\|. \tag{3}$$

Then P projects onto K "along n", in that if $y \in K$, then $P(y) = y$, and if $y \notin K$, then $P(y) \in \partial K$, and $P(y) - y = \alpha\gamma$ for some $\alpha > 0$ and $\gamma \in n(P(y))$.

Given $x \in K$ and $v \in R^k$, define the projection of the vector v at x (with respect to K) by

$$\Pi_K(x, v) = \lim_{\delta \to 0} \frac{(P_K(x + \delta v) - x)}{\delta}. \tag{4}$$

The class of pertinent ordinary differential equations takes the form:

$$\dot{x} = \Pi_K(x, -F(x)), \quad x(0) = x_0 \in K, \tag{5}$$

where K is a closed convex set, corresponding to the constraint set in a particular application, and $F(x)$ is a vector field defined on K. We refer to $\dot{x} = \Pi_K(x, -F(x))$ as ODE(F, K).

Note that the right-hand side of the ordinary differential equation (5) is associated with a projection operator and is, hence, discontinuous on the boundary of K. Therefore, one needs to articulate what one means by a solution to an ODE with a discontinuous right-hand side.

Definition 2 (A Projected Dynamical System). *Define the projected dynamical system* (PDS) $x_0(t) : K \times R \mapsto K$ *as the family of solutions to the Initial Value Problem* (IVP) *(5) for all* $x_0 \in K$.

It is clear from the definition that $x_0(0) = x_0$. For convenience, we will sometimes write $x_0 \cdot t$ for $x_0(t)$ and say interchangeably that x^* is an equilibrium or stationary point of the PDS(F, K).

The relationship between the ODE(F, K) and the VI(F, K) is revealed in the following theorem. This association was first noted by Dupuis and Ishii (1991) but remained unexplored for some time.

Theorem 1 (Dupuis and Nagurney (1993)). *Assume that K is a convex polyhedron. Then the stationary points of the* ODE(F, K) *coincide with the solutions of* VI(F, K).

Before stating the fundamental theorem of projected dynamical systems, we introduce the following assumption needed for the theorem.

Assumption 1 (Linear Growth Condition). *There exists a $B < \infty$ such that the vector field $-F : R^k \mapsto R^k$ satisfies the linear growth condition: $\|F(x)\| \leq B(1 + \|x\|)$ for $x \in K$, and also*

$$\langle (-F(x) + F(y))^T, x - y \rangle \leq B\|x - y\|^2, \quad \forall x, y \in K. \tag{6}$$

Theorem 2 (Existence, Uniqueness, and Continuous Dependence) (Dupuis and Nagurney (1993)). *Assume Assumption 1. Then*

(i) For any $x_0 \in K$, there exists a unique solution $x_0(t)$ to the initial value problem (5);

(ii) If $x_n \to x_0$ as $n \to \infty$, then $x_n(t)$ converges to $x_0(t)$ uniformly on every compact set of $[0, \infty)$.

The second statement of Theorem 2 is sometimes called the continuous dependence of the solution path to ODE(F, K) on the initial value. By virtue of Theorem 2, the PDS(F, K) is well-defined and inhabits K whenever Assumption 1 holds.

Lipschitz continuity is a condition that plays an important role in the study of variational inequality problems. It also is a critical concept in the classical study of dynamical systems. We recall it here for the convenience of the reader.

Definition 3 (Lipschitz Continuity). *$F : K \mapsto R^k$ is locally Lipschitz continuous if for every $x \in K$ there is a neighborhood $N(x)$ and a positive number $L(x) > 0$ such that*

$$\|F(x') - F(x'')\| \leq L(x)\|x' - x''\|, \quad \forall x', x'' \in N(x). \tag{7}$$

When (7) holds uniformly on K for some constant $L > 0$, i.e.,

$$\|F(x') - F(x'')\| \leq L\|x' - x''\|, \quad \forall x', x'' \in K, \tag{8}$$

then F is said to be Lipschitz continuous on K.

Lipschitz continuity implies Assumption 1 and is, therefore, a sufficient condition for the fundamental properties of projected dynamical systems stated in Theorem 2.

8.2.2 Stability Results

Here we review some recently obtained stability results using monotonicity conditions. Additional stability results, along with accompanying proofs, can be found in Zhang and Nagurney (1995) and in Nagurney and Zhang (1997b).

Stability analysis is crucial to the understanding of the behavior of dynamical models. In particular, one would like to know answers to such questions as:

1. If a system starts near an equilibrium, will it stay close to it forever?
2. Given the current state of the competitive system, will it asymptotically approach an equilibrium?

We now recall some of the stability definitions from Zhang and Nagurney (1995) that are relevant to the applications addressed in this chapter.

Definition 4 (Global Monotone Attractor). *An equilibrium point x^* is a global monotone attractor if*

$$d(x,t) = \|x \cdot t - x^*\| \tag{9}$$

is a nonincreasing function of t. x^* is a strictly global monotone attactor if $d(x,t)$ is monotonically decreasing to zero in t for all $x \in K$.

Definition 5 (Globally Exponentially Stable Equilibrium Point). *An equilibrium point x^* is globally exponentially stable if there exists constants $b > 0$ and $\mu > 0$ such that*

$$\|x_0 \cdot t - x^*\| \leq b\|x_0 - x^*\| \exp(-\mu t), \quad \forall t \geq 0, \ \forall x_0 \in K. \tag{10}$$

We now recall stability results under various monotonicity conditions. The following results proved in Zhang and Nagurney (1995) show that monotonicity implies a global monotone attractor, strict monotonicity implies a strictly global monotone attractor at x^*, and strong monotonicity implies exponential stablility. We begin with the following:

Theorem 3 (Zhang and Nagurney (1995)). *Suppose that x^* solves* VI(F, K). *If $F(x)$ is monotone at x^*, then x^* is a global monotone attractor.*

Theorem 4 (Zhang and Nagurney (1995)). *Suppose that x^* solves* VI(F, K). *If $F(x)$ is strictly monotone at x^*, then x^* is a strictly global monotone attractor.*

Under the strong monotonicity, we have

Theorem 5 (Zhang and Nagurney (1995)). *Suppose that x^* solves* VI(F, K). *If $F(x)$ is strongly monotone at x^*, then x^* is globally exponentially stable.*

8.2.3 The General Iterative Scheme

In this subsection we recall a general iterative scheme devised by Dupuis and Nagurney (1993) that induces a variety of numerical procedures. These procedures are designed to estimate stationary points of the projected dynamical system (5); equivalently, to determine solutions to the variational inequality

$$\langle F(x^*)^T, x - x^* \rangle \geq 0, \quad \forall x \in K. \tag{11}$$

The algorithms for obtaining a solution to (11) all take the form

$$X_{\tau+1} = P_K(X_\tau - a_\tau F_\tau(X_\tau)), \tag{12}$$

where $\{a_\tau, \tau \in N\}$ is a sequence of positive scalars and the sequence of vector fields $\{F_\tau(\cdot), \tau \in N\}$ are "approximations" to $F(\cdot)$.

We first give the precise conditions for the convergence theorem and a general discussion of the conditions. Subsequently, several examples of the functions $\{F_\tau(\cdot), \tau \in N\}$ are given.

The following notation is needed for the statement of Assumption 2.

For each $x \in R^k$, let the set-valued function $\bar{F}(x)$ be defined as

$$\bar{F}(x) = \bigcap_{\epsilon > 0} \text{cov}\left(\overline{\{F(y) : \|x - y\| \leq \epsilon\}}\right)$$

where the overline indicates the closure and $\text{cov}(A)$ denotes the convex hull of the set A. Then $\bar{F}(x)$ is convex and upper semicontinuous, and, in particular, $\bar{F}(x) = F(x)$, when F is continuous at x.

For any $z \in R^k, A \subset R^k$, let

$$d(z, A) := \inf_{y \in A} \|z - y\|$$

denote the distance between z and A. Then

$$d(z, A) = \|z - P_A(z)\|,$$

when A is closed and convex.

The conditions for the convergence theorem are now in order.

Assumption 2. *Suppose that we fix an initial condition $X_0 \in K$ and we define the sequence $\{X_\tau, \tau \in N\}$ by (12). We assume the following conditions:*

(i) $\sum_{\tau=0}^{\infty} a_\tau = \infty, \quad a_\tau > 0, \quad a_\tau \to 0, \quad as \quad \tau \to \infty.$

(ii) $d(F_\tau(x), \bar{F}(x)) \to 0$ *uniformly on compact subsets of* K *as* $\tau \to \infty$.

(iii) *Define* $y(\cdot)$ *to be the unique solution to* $\dot{x} = \Pi_K(x, -F(x))$ *that satisfies* $y(0) = y \in K$. *The* $w-$*limit set*

$$w(K) = \cup_{y \in K} \cap_{t \geq 0} \overline{\cup_{s \geq t} \{y(s)\}}$$

is contained in the set of stationary points of $\dot{x} = \Pi_K(x, -F(x))$.

(iv) *The sequence* $\{X_\tau, \tau \in N\}$ *is bounded.*

(v) *The solutions to* $\dot{x} = \Pi_K(x, -F(x))$ *are stable in the sense that given any compact set* K_1 *there exists a compact set* K_2 *such that* $\cup_{y \in K \cap K_1} \cup_{t \geq 0} \{y(t)\} \subset K_2$.

Examples. We now give examples for the vector field $F_\tau(x)$. The simplest and most obvious example is when $F_\tau(x) = F(x)$ for all $\tau \in N$ and $x \in K$. This would correspond to the basic *Euler scheme* in the numerical approximation of standard ODEs. Another example is given by

$$F_\tau(x) = \frac{1}{2}\left[F(x) + F(P(x - a_\tau F(x)))\right].$$

This can be interpreted as a *Heun-type* scheme. If the function $F(\cdot)$ is defined in a sufficiently large neighborhood of K, an *alternative Heun-type scheme* is given by

$$F_\tau(x) = \frac{1}{2}\left[F(x) + F(x - a_\tau F(x))\right].$$

One can proceed in the obvious manner, and generate algorithms that correspond to all of the standard methods of approximating solutions to ODEs, such as Runge-Kutta type algorithms and algorithms with controlled step size.

The convergence result is now stated.

Theorem 6 (Dupuis and Nagurney (1993)). *Let* S *denote the solutions to the variational inequality (11), and assume Assumption 1 and Assumption 2. Suppose* $\{X_\tau, \tau \in N\}$ *is the scheme generated by (12). Then* $d(X_\tau, S) \to 0$ *as* $\tau \to \infty$.

Corollary 1 (Dupuis and Nagurney (1993)). *Assume the conditions of Theorem 6, and also that* S *consists of a finite set of points. Then* $\lim_{\tau \to \infty} X_\tau$ *exists and equals a solution to the variational inequality (11).*

8.3 Elastic Demand Traffic Network Problems with Known Travel Disutility Functions

In this section we summarize the application of projected dynamical system theory to elastic demand traffic network problems in the case of known travel disutility functions.

We first recall, in Section 8.3.1, the projected dynamical system formulation of a travel-route choice adjustment process, proposed by Dupuis and Nagurney (1993).

The adjustment process models the travelers' day-to-day dynamic behavior of making trip decisions and route choices associated with a travel disutility perspective. Then, in Section 8.3.2, the stability results of this travel-route choice adjustment process obtained by Zhang and Nagurney (1996) are reviewed, which address whether and how the travelers' dynamic behavior in attempting to avoid congestion leads to a traffic equilibrium pattern. Finally, Section 8.3.3 recalls the discrete time algorithms devised for the computation of traffic network equilibria with elastic demands and with known travel disutility functions. The convergence of these discrete time algorithms was established by Nagurney and Zhang (1998; 1997b).

8.3.1 A Trip-Route Choice Adjustment Process

We consider a network $[N, L]$ consisting of nodes $[N]$ and directed links $[L]$. Let a denote a link of the network connecting a pair of nodes, and let p denote a path (assumed to be acyclic) consisting of a sequence of links connecting an origin/destination (O/D) pair w. P_w denotes the set of paths connecting the O/D pair w with n_{P_w} paths. We let W denote the set of O/D pairs and P the set of paths in the network.

Let x_p represent the flow on path p and let f_a denote the load on link a. The following conservation of flow equation must hold for each link a:

$$f_a = \sum_p x_p \delta_{ap}, \tag{13}$$

where $\delta_{ap} = 1$, if link a is contained in path p, and 0, otherwise. Expression (13) states that the load on a link a is equal to the sum of all the path flows on paths that contain the link a.

Moreover, if we let d_w denote the demand associated with an O/D pair w, then we must have that for each O/D pair w

$$d_w = \sum_{p \in P_w} x_p, \tag{14}$$

where $x_p \geq 0$, for all p, that is, the sum of all the path flows on paths connecting the O/D pair w must be equal to the demand d_w. Let x denote the column vector of path flows with dimension n_P.

Let c_a denote the user cost associated with traversing link a, and let C_p the user cost associated with traversing path p. Then

$$C_p = \sum_a c_a \delta_{ap}. \tag{15}$$

In other words, the cost of a path is equal to the sum of the costs on the links comprising that path. We group the link costs into the column vector c with n_A components, and the path costs into the column vector C with n_P components. We also assume that we are given a travel disutility function λ_w for each O/D pair w. We group the travel disutilities into the column vector λ with J components.

We assume that, in general, the cost associated with a link may depend upon the entire link load pattern, that is,

$$c_a = c_a(f) \tag{16}$$

and that the travel disutility associated with an O/D pair may depend upon the entire demand pattern, that is,

$$\lambda_w = \lambda_w(d), \tag{17}$$

where f is the n_A-dimensional column vector of link loads and d is the J-dimensional column vector of travel demands.

Definition 6 (Traffic Network Equilibrium) (Beckmann et al. (1956), Dafermos (1982)). *A vector $x^* \in R_+^{n_P}$, which induces a vector d^* through (14), is a traffic network equilibrium if for each path $p \in P_w$ and every O/D pair w:*

$$C_p(x^*) \begin{cases} = \lambda_w(d^*), & \text{if} \quad x_p^* > 0 \\ \geq \lambda_w(d^*), & \text{if} \quad x_p^* = 0. \end{cases} \tag{18}$$

In equilibrium, only those paths connecting an O/D pair that have minimal user costs are used, and their costs are equal to the travel disutility associated with traveling between the O/D pair.

The equilibrium conditions (18) have been formulated as a variational inequality problem by Dafermos (1982). In particular, we have:

Theorem 7 (Dafermos (1982)). *$(x^*, d^*) \in K^1$ is a traffic network equilibrium pattern, that is, satisfies equilibrium conditions (18) if and only if it satisfies the variational inequality problem:*

(Path Flow Formulation)

$$\langle C(x^*)^T, x - x^* \rangle - \langle \lambda(d^*)^T, d - d^* \rangle \geq 0, \quad \forall\, (x, d) \in K^1, \tag{19}$$

where $K^1 \equiv \{(x, d) : x \geq 0;$ and (14) holds$\}$, or, equivalently, $(f^*, d^*) \in K^2$ satisfies the variational inequality problem:

(Link Flow Formulation)

$$\langle c(f^*)^T, f - f^* \rangle - \langle \lambda(d^*)^T, d - d^* \rangle \geq 0, \quad \forall\, (f, d) \in K^2, \tag{20}$$

where $K^2 \equiv \{(f, d) : x \geq 0;$ and (13) and (14) hold$\}$.

Since users of a network select paths to reach their destinations from their origins, we consider variational inequality (19) as the basic one for obtaining the dynamical system equivalence. Specifically, note that, in view of constraint (14), one may define $\hat{\lambda}(x) \equiv \lambda(d)$, in which case, one may rewrite variational inequality (19) in the path flow variables x only, that is, we seek to determine $x^* \in R_+^{n_P}$, such that

$$\langle (C(x^*) - \bar{\lambda}(x^*))^T, x - x^* \rangle \geq 0, \quad \forall x \in R_+^{n_P}, \tag{21}$$

where $\bar{\lambda}(x)$ is the $n_{P_{w_1}} \times n_{P_{w_2}} \times \dots n_{P_{w_J}}$-dimensional column vector with components:

$$(\hat{\lambda}_{w_1}(x), \dots, \hat{\lambda}_{w_1}(x), \dots, \hat{\lambda}_{w_J}(x), \dots, \hat{\lambda}_{w_J}(x)),$$

where J is the number of O/D pairs. If we now let $F(x) \equiv (C(x) - \bar{\lambda}(x))$ and $K \equiv \{x : x \in R_+^{n_P}\}$, then, clearly, (21) can be placed into standard form given by (1).

The dynamical system, first presented in Dupuis and Nagurney (1993), whose stationary points correspond to solutions of (21), is given by:

$$\dot{x} = \Pi_K(x, \bar{\lambda}(x) - C(x)), \quad x(0) = x_0 \in K. \tag{22}$$

The adjustment process interpretation of (22), as discussed in Dupuis and Nagurney (1993), is as follows: Users of a transportation network select at the greatest rate those paths whose differences between the travel disutilities (demand prices) and path costs are maximal; in other words, those paths whose costs are minimal relative to the travel disutilities. If the travel cost on a path exceeds the travel disutility associated with the O/D pair, then the flow on that path will decrease; if the travel disutility exceeds the cost on a path, then the flow on that path will increase. If the difference between the travel disutility and the path cost drives the path flow to be negative, then the projection operator guarantees that the path flow will be zero. The process continues until there is no change in path flows, that is, until all used paths have path costs equal to the travel disutilities; unused paths will have costs which exceed the disutilities. Namely, according to (22), the travelers adjust their route choices until an equilibrium is reached.

The following example, given in a certain discrete time realization, shows how the dynamic mechanism of the trip-route choice adjustment (22) would reallocate the traffic flow among the paths and would react to changes in the travel disutilities.

Example 1. Consider a simple transportation network consisting of two nodes, with a single O/D pair w, and two links a and b representing the two disjoint paths connecting the O/D pair. Suppose that the link costs are

$$c_a(f_a) = f_a + 2, \quad c_b(f_b) = 2f_b,$$

and the travel disutility function is given by

$$\lambda_w(d_w) = -d_w + 5.$$

Note that here a path consists of a single link and, hence, we can use x and f interchangeably. Suppose that, at time $t = 0$, the flow on link a is 0.7, the flow on link b is 1.5; hence, the demand is 2.2, and the travel disutility is 2.8, that is,

$$x_a(0) = 0.7, \quad x_b(0) = 1.5, \quad d_w(0) = 2.2, \quad \lambda_w(0) = 2.8,$$

which yields travel costs: $c_a(0) = 2.7$ and $c_b(0) = 3.0$.

According to the trip-route choice adjustment process (22), the flow changing rates at time $t = 0$ are:

$$\dot{x}_a(0) = \lambda_w(0) - c_a(0) = 0.1, \quad \dot{x}_b(0) = \lambda_w(0) - c_b(0) = -0.2.$$

If a time increment of 0.5 is used, then at the next moment $t = 0.5$, the flows on link a and link b are:

$$x_a(0.5) = x_a(0) + 0.5\dot{x}_a(0) = 0.7 + 0.5 \times 0.1 = 0.75,$$

$$x_b(0.5) = x_b(0) + 0.5\dot{x}_b(0) = 1.5 - 0.5 \times 0.2 = 1.4,$$

which yields travel costs: $c_a(0.5) = 2.75$ and $c_b(0.5) = 2.8$, a travel demand $d_w(0.5) = 2.15$, and a travel disutility $\lambda_w(0.5) = 2.85$. Now, the flow changing rates are given by:

$$\dot{x}_a(0.5) = \lambda_w(0.5) - c_a(0.5) = 2.85 - 2.75 = 0.1,$$

$$\dot{x}_b(0.5) = \lambda_w(0.5) - c_b(0.5) = 2.85 - 2.8 = 0.05.$$

The flows on link a and link b at time $t = 1.0$ would, hence, then be:

$$x_a(1.0) = x_a(0.5) + 0.5\dot{x}_a(0.5) = 0.75 + 0.5 \times 0.1 = 0.80,$$

$$x_b(1.0) = x_b(0.5) + 0.5\dot{x}_b(0.5) = 1.4 + 0.5 \times 0.05 = 1.425,$$

which yields travel costs: $c_a(1.0) = 2.80$ and $c_b(1.0) = 2.85$, a travel demand $d_w(1.0) = 2.225$, and a travel disutility $\lambda_w(1.0) = 2.775$. Now, the flow changing rates are given by:

$$\dot{x}_a(1.0) = \lambda_w(1.0) - c_a(1.0) = 2.775 - 2.800 = -0.025,$$

$$\dot{x}_b(1.0) = \lambda_w(1.0) - c_b(1.0) = 2.775 - 2.850 = -0.075.$$

The flows on link a and link b at time $t = 1.5$ would be:

$$x_a(1.5) = x_a(1.0) + 0.5\dot{x}_a(1.0) = 0.8 - 0.5 \times 0.025 = 0.7875,$$

$$x_b(1.5) = x_b(1.0) + 0.5\dot{x}_b(1.0) = 1.425 - 0.5 \times 0.075 = 1.3875,$$

which yields travel costs: $c_a(1.5) = 2.7875$ and $c_b(1.5) = 2.775$, a travel demand $d_w(1.5) = 2.175$, and a travel disutility $\lambda_w(1.0) = 2.82$.

In this example, we see that, as time elapses, the trip-route choice adjustment process (22) adjusts the flow volume on the two links so that the difference between the travel costs of link a and link b is being reduced, from 0.3, to 0.05, and, finally, to 0.0125; and, the difference between the disutility and the travel costs on the used links is also being reduced from 0.2, to 0.1, and to 0.045. In fact, the traffic equilibrium (cf. Definition 6), with: $x_a^* = 0.8$ and $x_b^* = 1.4$, which induces the demand $d_w^* = 2.2$, is almost attained in only 1.5 time units.

8.3.2 Stability Analysis

We now present the stability results of the route choice adjustment process (22). The questions that motivate transportation planners and analysts to study the stability of a transportation system include: Will any initial flow pattern be driven to an equilibrium by the adjustment process? Also, will a flow pattern near an equilibrium always stay close to it? These concerns of system stability are important in traffic assignment and form, indeed, a critical base for the very concept of an equilibrium flow pattern.

For the specific application of transportation network problems, the following definitions of global stability of the transportation system and the local stability of an equilibrium are adapted from the general stability concepts of projected dynamical systems presented in Section 8.2 (cf. Zhang and Nagurney (1996)).

Definition 7 (Stability at an Equilibrium). *An equilibrium flow pattern x^* is stable if it is a global monotone attractor for the corresponding route choice adjustment process.*

Definition 8 (Asymptotical Stability at an Equilibrium). *An equilibrium flow pattern x^* is asymptotically stable if it is a strictly global monotone attractor for the corresponding route choice adjustment process.*

Definition 9 (Stability of the System). *A route choice adjustment process is stable if all its equilibrium flow patterns are stable.*

Definition 10 (Asymptotical Stability of the System). *A route choice adjustment process is asymptotically stable if all its equilibrium flow patterns are asymptotically stable.*

The above definitions are stated in a way that are also adaptable for the stability analysis in the subsequent two sections, namely, in the context of elastic demand transportation problems with known demand functions and fixed demand transportation problems.

We now present the stability results in Zhang and Nagurney (1996) for the trip-route choice adjustment process (22).

Theorem 8 (Zhang and Nagurney (1996)). *Suppose that the link cost functions c are monotone increasing in the link load pattern f and that the travel disutility functions λ are monotone decreasing in the travel demand d. Then the trip-route choice adjustment process (22) is stable.*

Theorem 9 (Zhang and Nagurney (1996)). *Assume that there exists some equilibrium path flow pattern. Suppose that the link cost functions c and negative disutility functions $-\lambda$ are strictly monotone in the link load f and the travel demand d, respectively. Then, the trip-route choice adjustment process (22) is asymptotically stable.*

Theorem 8 states that, provided that monotonicity of the link cost functions and the travel disutility functions holds true, then any flow pattern near an equilibrium will stay close to it forever. Under the stict monotonicity assumption, on the other hand, Theorem 9 can be interpreted as saying that any initial flow pattern will eventually be driven to an equilibrium by the route choice adjustment process.

However, in most situations, it may be unrealistic for us to acquire the entire structure of the link travel cost and the travel disutility functions. Instead, we may only have access to observed data, which can be thought of as local data, and we may have to rely on and make reasonable assumptions as to the management of these local data. Does there exist some local stability theory that only requires that the local data satisfy the local stability of an observed or computed equilibrium flow pattern? Specifically, will a certain equilibrium flow pattern resume after some local perturbation caused by either certain planned events or random incidents?

For instance, examples of such events or incidents include: the construction of road sections, the installation of certain facilities, a social or political event in the region that may attract additional traffic demand, or some unexpected destruction to the traffic network caused by a natural disaster or an accident.

The local stability analysis follows to address precisely such questions. We first present some necessary definitions.

Definition 11 (A Regular Traffic Network Equilibrium). *A path flow pattern x^* is called a regular (traffic network) equilibrium if all the paths that are not used have higher travel costs and those that are used have the same costs. Mathematically, this can be expressed as, for every O/D pair $w \in W$:*

$$x_p^* = 0,\ p \in P_w \iff C_p(x^*) > \min_{q \in P_w} C_q(x^*),\ p \in P_w. \tag{23}$$

A traffic network equilibrium is irregular if it is not regular.

One sees immediately that a regular equilibrium flow pattern is always a Wardropian equilibrium, and that an irregular Wardropian equilibrium must have a pair of paths joining an O/D pair with the same travel costs where one path is used, whereas the other is not at equilibrium. This suggests that irregular equilibrium flow patterns are not likely to emerge in reality and we can focus our attention on regular equilibrium flow patterns for practical considerations. Interestingly, we will show that regular equilibria do have privilege in stability analysis.

In the context of the elastic demand traffic network problem, Definition 11 is equivalent to:

Definition 12 (cf. Definition 6). *A path flow pattern x^*, which induces a demand pattern $d^* = d(x^*)$ through (1), is a regular traffic network equilibrium, if, for every O/D pair $w \in W$ and every path $p \in P_w$, the following holds:*

$$C_p(x^*) \begin{cases} = \lambda_w(d^*), & \text{if} \quad x_p^* > 0 \\ > \lambda_w(d^*), & \text{if} \quad x_p^* = 0. \end{cases} \tag{24}$$

Let x^* be any fixed equilibrium flow pattern. We say that a path p is an equilibrium active path if $x_p^* > 0$ and say that a path p is an equilibrium inactive path if $x_p^* = 0$.

Let

$$U(x^*) = \{p \in P :\ x_p^* > 0\} \tag{25}$$

denote the set of all equilibrium active paths at x^*, and let

$$V(x^*) = \{p \in P :\ x_p^* = 0\} \tag{26}$$

denote the set of all equilibrium inactive paths at x^*.

We may write U and V, respectively, for $U(x^*)$ and $V(x^*)$ when there is no chance for confusion, and denote their cardinality by u and v. With notice to the earlier conventions, one has that

$$u + v = m, \quad \text{and} \quad U \cup V = P. \tag{27}$$

Corresponding to the above notation, any path flow pattern x can be decomposed into equilibrium active path flows

$$x^U = (x_p)_{p \in U} \in R_+^u \tag{28}$$

and

$$x^V = (x_p)_{p \in V} \in R_+^v. \tag{29}$$

Let x^* be a regular equilibrium flow pattern. Denote the $u \times u$ sub-Jacobian matrix of $\hat{\lambda}$ with respect to the equilibrium active path flows $x_p^*, p \in U(x^*)$, by $\nabla_U \hat{\lambda}(x^*)$, that is,

$$\nabla_U \hat{\lambda}(x^*) = \left(\frac{\partial \left[\hat{\lambda}(x^*) \right]_q}{\partial x_p} \right)_{p,q \in U(x^*)}, \tag{30}$$

where $\left[\hat{\lambda}(x^*) \right]_q = \hat{\lambda}_w(x^*) = \hat{\lambda}_w(d^*)$, for $q \in P_w$.

Similarly, denote the $u \times u$ sub-Jacobian matrix of path costs C with respect to x_p^*, $p \in U(x^*)$, by $\nabla_U C(x^*)$. Namely,

$$\nabla_U C(x^*) = \left(\frac{\partial C_q(x^*)}{\partial x_p} \right)_{p,q \in U(x^*)}. \tag{31}$$

Using relationships (14) and (15) one can write $\nabla_U \hat{\lambda}(x^*)$ and $\nabla_U C(x^*)$ more explicitly. Suppose that there are J O/D pairs $w = \{w_1, \cdots, w_J\}$ and corresponding to the equilibrium flow x^*, one has

$$U(x^*) = \bigcup_{j=1}^{J} U_j(x^*), \tag{32}$$

where $U_j(x^*) = U(x^*) \cap P_{w_j}$. If the cardinality of $U_j(x^*)$ is denoted by u_j, then we have

$$u = \sum_{j=1}^{J} u_j.$$

For any two O/D pairs w_i, w_j, denote the partial derivative $\partial \lambda_{w_i}(d^*)/\partial d_{w_j}$ by λ_{ij}, and denote by E_{ij} the $u_i \times u_j$ matrix all of whose entries are 1. Then we can write

$$\nabla_U \hat{\lambda}(x^*) = \begin{pmatrix} \lambda_{11} E_{11} & \cdots & \lambda_{1J} E_{1J} \\ \cdot & \cdots & \cdot \\ \lambda_{J1} E_{J1} & \cdots & \lambda_{JJ} E_{JJ} \end{pmatrix}, \tag{33}$$

where it is understood that $\lambda_{ij} E_{ij}$ does not appear if either u_i or u_j is zero.

On the other hand, according to (15), one has

$$\frac{\partial C_q(x^*)}{\partial x_p} = \frac{\partial}{\partial x_p} \sum_{a \in L} \delta_{aq} c_a(x^*) = \sum_{a \in L} \delta_{aq} \sum_{b \in L} \frac{\partial c_a(f^*)}{\partial f_b} \frac{\partial f_b(x^*)}{\partial x_p}$$

$$= \sum_{a,b \in L} \delta_{aq} \delta_{bp} \frac{\partial c_a(f^*)}{\partial f_b} = \Delta_q^T \nabla c(f^*) \Delta_p, \tag{34}$$

where $\Delta_p = (\delta_{ap})_{a \in L}$ and $\Delta_q = (\delta_{aq})_{a \in L}$ are column vectors, and $\nabla c(f^*)$ is the Jacobian matrix of the link costs with respect to the link flows. With this notation, $\Delta_U = (\Delta_p)_{p \in U}$, and (34) can now be written as

$$\nabla_U C(x^*) = (\Delta_U)^T \nabla c(f^*) (\Delta_U). \tag{35}$$

The next theorem of Zhang and Nagurney (1996) presents a criterion to verify the local asymptotical stability of the route choice adjustment process at a regular equilibrium flow pattern. This criterion does not require monotonicity assumptions on the costs and the disutilities.

Theorem 10 (Zhang and Nagurney (1996)). *Let x^* be a regular equilibrium flow pattern. Then, x^* is asymptotically stable if all the eigenvalues of*

$$J_U = \nabla_U \hat{\lambda}(x^*) - \nabla_U C(x^*) = \nabla_U \hat{\lambda}(x^*) - (\Delta_U)^T \nabla c(f^*)(\Delta_U) \tag{36}$$

have negative real parts.

In order to explore the asymptotical stability of the route choice adjustment process at a regular equilibrium flow, Theorem 10 states that we only need to calculate the eigenvalues of a matrix of dimensions $u \times u$. One sees that this may be computationally easier than verifying the monotonicity of the travel disutilities and link user costs.

8.3.3 Discrete Time Algorithms

The Euler method and the Heun method were employed by Nagurney and Zhang (1998; 1997b) for the computation of solutions to dynamic elastic demand traffic network problems with known travel disutility functions, and their convergence was also established therein. We refer the reader to these references for numerical results, including traffic network examples that are solved on a massively parallel computer architecture.

Recall from Section 8.2.3 that, at iteration τ, the Euler method computes

$$x^{\tau+1} = P_K(x^\tau - a_\tau F(x^\tau)), \tag{37}$$

whereas, according to the Heun method, at iteration τ one computes

$$x^{\tau+1} = P_K(x^\tau - a_\tau \frac{1}{2}\left[F(x^\tau) + F(P(x^\tau - a_\tau F(x^\tau)))\right]). \tag{38}$$

In the case that the sequence $\{a_\tau\}$ in the Euler method (cf. (37)) is fixed, say, $\{a_\tau\} = \rho$, for all iterations τ, then the Euler method collapses to a projection method.

In the context of the dynamic traffic network problem with known travel disutility functions, the projection operation in the above discrete time algorithms can be evaluated explicitly and in closed form. Indeed, at each iteration τ of Euler method, (37) takes the form: For each path $p \in P$ in the transportation network, compute the path flow $x_p^{\tau+1}$ according to:

$$x_p^{\tau+1} = \max\{0, x_p^\tau + a_\tau(\lambda_w(d^\tau) - C_p(x^\tau))\}. \tag{39}$$

Each iteration of the Heun method (cf. (38)), in turn, consists of two steps. First, at iteration τ one computes the approximate path flows:

$$\bar{x}_p^\tau = \max\{0, x_p^\tau + a_\tau(\lambda_w(d^\tau) - C_p(x^\tau))\}, \forall p \in P, \tag{40}$$

and updates the approximate travel demands:

$$\bar{d}_w^\tau = \sum_{p \in P_w} \bar{x}_p^\tau, \forall w \in W. \tag{41}$$

Let

$$\bar{x}^\tau = \{\bar{x}_p^\tau, p \in P\} \tag{42}$$

and

$$\bar{d}^\tau = \{\bar{d}_w^\tau, w \in W\}. \tag{43}$$

Then, for each path $p \in P$ in the transportation network one computes the updated path flows $x_p^{\tau+1}$ according to:

$$x_p^{\tau+1} = \max\{0, x_p^\tau + \frac{a_\tau}{2}\left[\lambda_w(d^\tau) - C_p(x^\tau) + \lambda_w(\bar{d}^\tau) - C_p(\bar{x}^\tau)\right]\}, \forall p \in P, \tag{44}$$

and updates the travel demands $d_w^{\tau+1}$ according to:

$$d_w^{\tau+1} = \sum_{p \in P_w} x_p^{\tau+1}, \forall w \in W.$$

It is worth noting that both the Euler method and the Heun method at each iteration yield subproblems in the path flow variables, each of which can be solved not only in closed form, but also, simultaneously. Hence, these algorithms in the context of this model can be interpreted as massively parallel algorithms and can be implemented on massively parallel architectures. Indeed, this has been done so by Nagurney and Zhang (1998).

In order to establish the convergence of the Euler method and the Heun method, it is convenient to regularize the link cost structures.

Definition 13 (A Regular Cost Function). *The link cost function c is called regular if, for every link $a \in L$,*

$$c_a(f) \longrightarrow \infty, \text{ as } f_a \longrightarrow \infty, \tag{45}$$

holds uniformly true for all link flow patterns.

We note that the above regularity condition on the link cost functions is natural from a practical point of view and it does not impose any substantial restrictions. In reality, any link has an upper bound in the form of a capacity. Therefore, letting $f_a \longrightarrow \infty$ is an artificial device under which one can reasonably deduce that $c_a(f) \longrightarrow \infty$, due to the congestion effect. Consequently, any practical link cost structure can be theoretically extended to a regular link cost structure to allow for an infinite load.

Theorem 11 below shows that both the Euler method (39) and the Heun method (40)-(44) converge to the traffic network equilibrium under reasonable assumptions.

Theorem 11 (Nagurney and Zhang (1998)). *Suppose that the link cost function c is regular and strictly monotone increasing, and that the travel disutility function λ is strictly monotone decreasing. Let $\{a_\tau\}$ be a sequence of positive real numbers that satisfies*

$$\lim_{\tau \to \infty} a_\tau = 0 \tag{46}$$

and

$$\sum_{\tau=0}^{\infty} a_\tau = \infty. \tag{47}$$

Then both the Euler method (39) and the Heun method (40)-(44) produce sequences $\{x^\tau\}$ that converge to some traffic network equilibrium path flow pattern.

8.4 Elastic Demand Traffic Network Problems with Known Travel Demand Functions

In this section we consider elastic demand traffic network problems in which the travel demand functions rather than the travel disutility functions are assumed to be given. This section is organized as follows. Section 8.4.1 presents the projected dynamical system formulation of the route adjustment process. Section 8.4.2 then presents the stability analysis of such an adjustment process. Finally, Section 8.4.3 discusses discrete time algorithms for the computation of the elastic demand traffic equilibrium. The main results in this section are quoted from Zhang and Nagurney (1997).

8.4.1 A Trip-Route Choice Adjustment Process

Consider, as in Section 8.3, a transportation network. We retain the notation of Section 8.3 except for the following changes. We assume now that the demand d_w, associated with traveling between O/D pair w, is now a function, in general, of the travel disutilities associated with traveling between all the O/D pairs, that is,

$$d_w = d_w(\lambda). \tag{48}$$

Note that the expression (13) relating the link loads to the path flows is still valid, as is the nonnegativity assumption on the path flows. In addition, the link cost and path cost functions are defined according to (15)-(16).

The traffic network equilibrium conditions take on here the following form (cf. Beckmann et al. (1956) and Dafermos and Nagurney (1984)):

Definition 14 (Traffic Network Equilibrium). *A path flow pattern x^* and a travel disutility pattern λ^* is a traffic network equilibrium pattern if, for every O/D pair w and each path $p \in P_w$, the following equalities and inequalities hold:*

$$C_p(x^*) \begin{cases} = \lambda^*_w, & \text{if } x^*_p > 0 \\ \geq \lambda^*_w, & \text{if } x^*_p = 0, \end{cases} \tag{49}$$

and

$$d_w(\lambda^*) \begin{cases} = \sum_{p \in P_w} x^*_p, & \text{if } \lambda^*_w > 0 \\ \leq \sum_{p \in P_w} x^*_p, & \text{if } \lambda^*_w = 0. \end{cases} \tag{50}$$

Condition (49) is analogous to condition (18) where now the equilibrium travel disutilities λ^* are to be determined, rather than the equilibrium travel demand d^*.

Condition (50), in turn, has the following interpretation: if the travel disutility (or price) associated with traveling between an O/D pair w is positive, then the "market" clears for that O/D pair, that is, the sum of the path flows on paths connecting that O/D pair are equal to the demand associated with that O/D pair; if the travel disutility (or price) is zero, then the sum of the path flows can exceed the demand.

Here we can immediately write down the governing variational inequality formulation in path flow and travel disutility variables (see, also, e.g., Dafermos and Nagurney (1984) and Nagurney (1993)).

Theorem 12 (Variational Inequality Formulation). $(x^*, \lambda^*) \in R_+^{n_P+J}$ *is a traffic network equilibrium if and only if it satisfies the variational inequality problem:*

$$\sum_w \sum_{p \in P_w} [C_p(x^*) - \lambda_w^*] \times \left[x_p - x_p^*\right] - \sum_w \left[d_w(\lambda^*) - \sum_{p \in P_w} x_p^*\right] \times [\lambda_w - \lambda_w^*] \geq 0,$$

$$\forall (x, \lambda) \in R_+^{n_P+J}, \tag{51}$$

or, in vector form:

$$\langle (C(x^*) - \tilde{B}^T \lambda^*)^T, x - x^* \rangle - \langle (d(\lambda^*) - \tilde{B} x^*)^T, \lambda - \lambda^* \rangle \geq 0,$$

$$\forall (x, \lambda) \in R_+^{n_P+J}, \tag{52}$$

where $\tilde{B}$ *is the* $J \times n_P$*-dimensional matrix with element* $(w, p) = 1$*, if* $p \in P_w$*, and 0, otherwise.*

For the above elastic demand traffic assignment problem, we now recall the following travel route choice adjustment process due to Zhang and Nagurney (1997). As in the elastic demand model of Section 8.3.1, suppose that the traffic flow on each path p connecting the O/D pair w changes at a rate that is equal to the difference between the travel disutility λ_w and the path cost C_p. When the disutility exceeds the path cost, that is, $\lambda_w > C_p$, the path flow will increase; when $\lambda_w < C_p$, the path flow x_p will decrease. Hence, the travelers continuously switch from more costly routes to less costly routes relative to the travel disutilities at the rate $\lambda_w - C_p$. Moreover, in order to maintain feasibility, one must ensure that the path flows remain nonnegative.

The travel disutility λ_w associated with traveling between O/D pair w, on the other hand, changes at a rate equal to the difference between the demand d_w and the sum of the path flows on paths connecting O/D pair w. If the sum exceeds the demand, the travel disutility (or price) will decrease; if the demand exceeds the sum, the travel disutility (or price) will increase. In addition, nonnegativity of the travel disutilities must be maintained.

This process continues until there is no change in path flows and travel disutilities. Based on the above assumption, Zhang and Nagurney (1997) proposed a trip-route choice adjustment process, formulated as the following projected dynamical system.

For all $w \in W$ and all $p \in P_w$:

$$\dot{x}_p = \begin{cases} \lambda_w - C_p(x), & \text{if } \ x_p > 0 \\ \max\{0, \lambda_w - C_p(x)\}, & \text{if } \ x_p = 0, \end{cases} \tag{53}$$

and

$$\dot{\lambda}_w = \begin{cases} d_w(\lambda) - \sum_{p \in P_w} x_p, & \text{if } \ \lambda_w > 0 \\ \max\{0, d_w(\lambda) - \sum_{p \in P_w} x_p\}, & \text{if } \ \lambda_w = 0. \end{cases} \tag{54}$$

Letting $X \equiv (x, \lambda)$; $F(X) \equiv \left\{ \begin{array}{c} C(x) - \tilde{B}^T \lambda \\ -d(\lambda) + \tilde{B} x \end{array} \right\}$, (53) and (54) may be expressed in vector form as:

$$\dot{X} = \Pi_K(X, -F(X)). \tag{55}$$

In light of Theorem 1, the traffic network equilibria coincide with the set of stationary points of the trip-route choice adjustment process (53) and (54).

We now examine through the following example how the traffic flows and the the travel disutilities change over time according to the trip-route choice adjustment (53)-(54).

Example 2. Consider a transportation network with the same network topology as Example 1. Suppose that the link costs are identical to those in Example 1 but now, instead of a given disutility function, the travel demand function is now given, where

$$d_w(\lambda_w) = -\lambda_w + 5.$$

Note that the demand in this example is the "inverse" of the travel disutility in Example 1. Suppose now that at time $t = 0$, the flow on link a is 0.7, the flow on link b is 1.5, and the travel disutility is 2.8, that is,

$$x_a(0) = 0.7, \quad x_b(0) = 1.5, \quad \lambda_w(0) = 2.8,$$

which yields travel costs: $c_a(0) = 2.7$ and $c_b(0) = 3$, and the travel demand $d(0) = 2.2$.

According to the trip-route choice adjustment process (53)-(54), the flow changing rates and the disutility changing rate at time $t = 0$ are:

$$\dot{x}_a(0) = \lambda_w(0) - c_a(0) = 0.1, \quad \dot{x}_b(0) = \lambda_w(0) - c_b(0) = -0.2,$$

$$\dot{\lambda}_w(0) = d(0) - x_a(0) - x_b(0) = 0.$$

If a time increment of 0.5 is used again, then at the next moment $t = 0.5$, the flows on link a and link b are

$$x_a(0.5) = x_a(0) + 0.5\dot{x}_a(0) = 0.75, \quad x_b(0.5) = x_b(0) + 0.5\dot{x}_b(0) = 1.4,$$

and the travel disutility is

$$\lambda_w(0.5) = \lambda_w(0) + 0.5 \times \dot{\lambda}_w(0) = 2.8.$$

Therefore, the travel costs are: $c_a(0.5) = 2.75$ and $c_b(0.5) = 2.8$, and the elastic demand is $d_w(0.5) = -\lambda_w(0.5) + 5 = 2.2$.

Hence, the flow changing rates at $t = 0.5$ are:

$$\dot{x}_a(0.5) = \lambda_w(0.5) - c_a(0.5) = 0.05, \quad \dot{x}_b(0.5) = \lambda_w(0.5) - c_b(0.5) = 0,$$

and the disutility changing rate is

$$\dot{\lambda}_w(0.5) = d_w(0.5) - x_a(0.5) - x_b(0.5) = 0.05.$$

At the next moment $t = 1.0$, the flows on link a and link b are

$$x_a(1.0) = x_a(0.5) + 0.5\dot{x}_a(0.5) = 0.775,$$

$$x_b(1.0) = x_b(0.5) + 0.5\dot{x}_b(0.5) = 1.4,$$

and the travel disutility is

$$\lambda_w(1.0) = \lambda_w(0.5) + 0.5\dot{\lambda}_w(0.5) = 2.825.$$

Therefore, the travel costs are: $c_a(1.0) = 2.775$ and $c_b(0.5) = 2.8$, and the demand is $d_w(1.0) = 2.175$.

Hence, the flow changing rates at $t = 1.0$ are

$$\dot{x}_a(1.0) = \lambda_w(1.0) - c_a(1.0) = 0.05, \quad \dot{x}_b(1.0) = \lambda_w(1.0) - c_b(1.0) = 0.025,$$

and the disutility changing rate is

$$\dot{\lambda}_w(1.0) = d_w(1.0) - x_a(1.0) - x_b(1.0) = 0.$$

At the next moment $t = 1.5$, the flows on link a and link b are

$$x_a(1.5) = x_a(1.0) + 0.5\dot{x}_a(1.0) = 0.8,$$

$$x_b(1.5) = x_b(1.0) + 0.5\dot{x}_b(1.0) = 1.4125,$$

and the travel disutility is

$$\lambda_w(1.5) = \lambda_w(1.0) + 0.5\dot{\lambda}_w(1.0) = 2.825.$$

This example illustrates how an initial flow pattern given by: $x_a = 0.7$,$x_b = 1.5$, and $\lambda_w = 2.8$, evolves in time and would reach an equilibrium. Here, the equilibrium is: $x_a^* = 0.8$, $x_b^* = 1.4$, and $\lambda_w^* = 2.8$.

8.4.2 Stability Analysis

In this subsection we present the stability analysis for the route choice adjustment process described by (53)-(54). We provide the global stability results under monotonicity assumptions analogous to Theorems 8 and 9. Besides the significance of these stability properties by themselves, as motivated earlier in Section 8.3, the results are utilized in the establishment of the convergence of discrete time algorithms.

The following two theorems of Zhang and Nagurney (1997) present the global stability properties of the route choice adjustment process (53)-(54).

Theorem 13 (cf. Theorem 8) (Zhang and Nagurney (1997)). *Suppose that the link cost functions c are monotone increasing in the link load pattern f and that the demand travel functions d are monotone decreasing in the travel disutilities λ. Then the route choice adjustment process (53)-(54) is stable.*

Theorem 14 (cf. Theorem 9) (Zhang and Nagurney (1997)). *Assume that there exists some equilibrium path flow and disutility pattern. Suppose that the link cost functions c are strictly monotone increasing in the link load f and that the demand functions d are strictly monotone decreasing in travel disutilities λ. Then, the route choice adjustment process (53)-(54) is asymptotically stable.*

8.4.3 Discrete Time Algorithms

The two discrete time algorithms, the Euler method and the Heun method, were considered by Zhang and Nagurney (1997) for the computation of traffic network equilibria with elastic demands and in the case of known travel demand functions. The convergence was established therein and several numerical examples were also reported in that paper. For a massively parallel implementation of the Euler method for this model, along with additional numerical results, see Nagurney and Zhang (1997c).

In the context of the elastic demand traffic network problem with demand functions, the Euler method becomes: At each iteration τ, compute for each path $p \in P$ in the transportation network the path flow $x_p^{\tau+1}$ according to:

$$x_p^{\tau+1} = \max\{0, x_p^\tau + a_\tau(\lambda_w^\tau - C_p(x^\tau))\}, \tag{56}$$

and for each O/D pair $w \in W$ in the transportation network compute the travel disutility $\lambda_w^{\tau+1}$ according to:

$$\lambda_w^{\tau+1} = \max\{0, \lambda_w^\tau + a_\tau(d_w(\lambda^\tau) - \sum_{p\in P_w} x_p^\tau)\}. \tag{57}$$

Each iteration of the Heun method, in turn, consists of two steps. First, at iteration τ, compute:

$$\bar{x}_p^\tau = \max\{0, x_p^\tau + a_\tau(\lambda_w^\tau - C_p(x^\tau))\}, \quad \forall p \in P, \tag{58}$$

and

$$\bar{\lambda}_w^\tau = \max\{0, \lambda_w^\tau + a_\tau(d_w(\lambda^\tau) - \sum_{p\in P_w} x_p^\tau)\}, \quad \forall w \in W. \tag{59}$$

Let

$$\bar{x}^\tau = \{\bar{x}_p^\tau, p \in P\} \quad \text{and} \quad \bar{\lambda}^\tau = \{\bar{\lambda}_w^\tau, w \in W\}. \tag{60}$$

Then, for each path $p \in P$ in the transportation network compute the updated path flows $x_p^{\tau+1}$ according to:

$$x_p^{\tau+1} = \max\{0, x_p^\tau + \frac{a_\tau}{2}\left[\lambda_w^\tau - C_p(x^\tau) + \bar{\lambda}_w^\tau - C_p(\bar{x}^\tau)\right]\}, \tag{61}$$

and the updated travel disutilities for each O/D pair $w \in W$ according to:

$$\lambda_w^{\tau+1} = \max\{0, \lambda_w^\tau + \frac{a_\tau}{2}\left[d_w(\lambda^\tau) - \sum_{p\in P_w} x_p^\tau + d_w(\bar{\lambda}^\tau) - \sum_{p\in P_w} \bar{x}_p^\tau\right]\}. \tag{62}$$

It is worth noting that, for both the Euler and the Heun methods applied to this elastic demand problem, the encountered subproblems in path flows and in travel disutility variables can be computed not only in closed form, but, also, simultaneously. Hence, these traffic network equilibrium problems are also amenable to solution via massively parallel architectures using our proposed methodology of projected dynamical systems. This approach may be particularly promising in light of the fact that urban transportation networks are very large in practice.

In an endeavor to establish the convergence of the Euler method (56)-(57) and the Heun method (58)-(62), Zhang and Nagurney (1997) proposed some additional assumptions. In a practical realization of a traffic network equilibrium problem with elastic demands as functions of the travel disutilities, the following conditions are expected to prevail:

(i) For every O/D pair w, the travel demand function $d_w(\lambda)$ has an upper bound;
(ii) For every O/D pair w and every path p connecting w, one has $\lambda_w \leq C_p(x)$, for sufficiently large x_p;
(iii) For every O/D pair w, one has $d_w(\lambda) \leq \sum_{p \in P_w} x_p$, for sufficiently large λ_w.

These conditions can be interpreted as follows. If the traffic flow on path p is extremely high, then the travel cost on this path will exceed the travel disutility λ_w. Also, the sum of the path flows between O/D pair w will be abundant, attracted by the sufficiently high disutility associated with this O/D pair, and, hence, adequate for the travel demand $d_w(\lambda)$, which should be lower because $d_w(\lambda)$ is usually a decreasing function of λ_w. Mathematically, these three conditions can be expressed as:

Assumption 3 (Zhang and Nagurney (1997)). *Assume that there exist sufficiently large constants M_d, M_x, and M_λ, such that*

$$d_w(\lambda) \leq M_d, \quad \forall \lambda \in R^J_+ \tag{63}$$

$$\lambda_w \leq C_p(x), \quad \textit{if} \quad x_p \geq M_x \tag{64}$$

$$d_w(\lambda) \leq \sum_{p \in P_w} x_p, \quad \textit{if} \quad \lambda_w \geq M_\lambda, \tag{65}$$

for any w and p.

The convergence of the Euler method and the Heun method for the computation of traffic network equilibrium with elastic demands and with known travel demand functions is now stated in the following theorem.

Theorem 15 (Zhang and Nagurney (1997)). *Suppose that the link costs c are strictly monotone increasing in link loads f and that the travel demands d are strictly monotone decreasing in travel disutilities λ. In addition, assume that Assumption 3 holds true. Let $\{a_\tau\}$ be a sequence of positive real numbers that satisfies*

$$\lim_{\tau \to \infty} a_\tau = 0 \tag{66}$$

and

$$\sum_{\tau=0}^{\infty} a_\tau = \infty. \tag{67}$$

Then, both the Euler method (56)-(57) and the Heun method (58)-(62) produce sequences $\{X^\tau\} = \{(x^\tau, \lambda^\tau)\}$ that converge to some equilibrium flow and disutility pattern.

Zhang and Nagurney (1997) and Nagurney and Zhang (1997b; 1997c) tested the Euler and Heun methods on various elastic demand traffic network examples with known demand functions.

8.5 Fixed Demand Traffic Network Problems

In this section, we address the dynamic fixed demand traffic network problem. Here we are interested in dynamic models of the day-to-day behavior of travelers. Namely, we provide a route choice adjustment process in Section 8.5.1, its stability analysis in Section 8.5.2, and discuss a discrete time algorithm for computational purposes in Section 8.5.3. In particular, we first recall the path flow and link load variational inequality formulations of the traffic network equilibrium conditions in the case of fixed travel demands, introduced in Smith (1979) and Dafermos (1980). The route choice adjustment process, first proposed in Nagurney and Zhang (1997a), is then presented, along with a discussion of the travelers' behavioral intuition in switching from more costly routes to less costly routes. The given projected dynamical system that formulates this route choice adjustment process may serve as a theoretical foundation for studying the behavior of the travelers from day-to-day in avoiding congestion.

We present in Section 8.5.2 the global stability properties of the proposed route choice adjustment process that address the questions as to whether the travelers' adjustment of their route choices will eventually drive any travel pattern to some traffic equilibrium, and as to whether a flow pattern that is close to a equilibrium flow pattern will remain close. Smith (1984) proposed a different adjustment process to model the day-to-day adjustment behavior of the users of a transportation network with fixed demands, and provided its stability properties. In Smith's model, it is assumed that the users compare the travel costs on all routes connecting the same origin/destination pair, and switch to every less costly route according to a certain ratio, whereas in the projected dynamical system model presented here, it is assumed that the users directly respond to the travel costs that they are currently incurring and the behavioral intuition of avoiding congestion results in a reallocation of the path flows within the same O/D pair through a projection. The stability issues have also been empirically studied by Mahmassani (1990), who investigated the behavior of commuters in adjusting their route choices on a simulated traffic network.

Finally, we propose a discrete time algorithm in Section 8.5.3, the Euler method, for the computation of solutions to the dynamic traffic network problem with fixed travel demands. In contrast to the elastic demand models of Sections 8.3 and 8.4, the feasible set is no longer the nonnegative orthant. Nevertheless, each iteration of the Euler method can be decomposed into simple quadratic programming subproblems with special network structure. In turn, these subproblems can be solved exactly and in closed form, using the exact equilibration method of Dafermos and Sparrow (1969). The convergence of the algorithm has been established exploiting the stability results. The major results in this section are cited from Nagurney and Zhang (1997a).

8.5.1 A Route Choice Adjustment Process

We retain the notation of Sections 8.3 and 8.4. However, in contrast to the previous two sections, it is assumed now that there is a fixed and known travel demand associated with traveling between each O/D pair in the network. Let d_w denote the traffic demand between O/D pair w, which is assumed to be known and fixed. The demand must satisfy, for each $w \in W$,

$$d_w = \sum_{p \in P_w} x_p, \tag{68}$$

where $x_p \geq 0, \forall p$, that is, the sum of the path flows between an O/D pair w must be equal to the demand d_w.

As previously, we have the conservation of flow equation for each $a \in L$,

$$f_a = \sum_p \delta_{ap} x_p, \tag{69}$$

which states that the traffic load on a link a is equal to the sum of all the path flows on paths p that traverse link a.

A user traveling on path p again incurs a (path) travel cost C_p satisfying

$$C_p = \sum_a \delta_{ap} c_a. \tag{70}$$

Following Wardrop (1952) and Beckmann et al. (1956), the traffic network equilibrium conditions are given as follows.

Definition 15 (Fixed Demand Traffic Network Equilibrium). *A path flow pattern x^*, which satisfies the demand through (68), is a traffic network equilibrium, if, for every O/D pair w and each path $p \in P_w$, the following equalities and inequalities hold:*

$$C_p(x^*) \begin{cases} = \lambda_w, & \text{if} \quad x_p^* > 0 \\ \geq \lambda_w, & \text{if} \quad x_p^* = 0, \end{cases} \tag{71}$$

where λ_w is the travel disutility incurred in equilibrium.

Again, as in the elastic demand models of Sections 8.3 and 8.4, in equilibrium, only those paths connecting an O/D pair that have minimal user travel costs are used, and those paths that are not used have costs that are higher than or equal to these minimal travel costs. However, here the demands and travel disutilities are no longer functions.

The equilibrium conditions (71) have been formulated as a variational inequality problem by Smith (1979) and Dafermos (1980). In particular, we present two formulations, in path flows and link loads, respectively:

Theorem 16 (Variational Inequality Formulation in Path Flows). $x^* \in K$ *is a traffic network equilibrium in path flows if and only if it solves the following variational inequality problem:*

$$\langle C(x^*)^T, x - x^* \rangle \geq 0, \quad \forall x \in K, \tag{72}$$

where $K \equiv \{x \in R_+^{n_P} :$ and (68) holds $\}$.

Theorem 17 (Variational Inequality Formulation in Link Loads). $f^* \in K^1$ *is a traffic network equilibrium in link loads if and only if it satisfies the following variational inequality problem:*

$$\langle c(f^*)^T, f - f^* \rangle \geq 0 \quad \forall f \in K^1, \tag{73}$$

where $K^1 \equiv \{f : \exists x \geq 0 : (68)\, and\, (69)\, hold\}$.

For the above fixed demand traffic assignment problem, Nagurney and Zhang (1997a) proposed the following travel route choice adjustment process. Suppose that the travelers between a certain O/D pair switch from more costly routes to less costly routes connecting this O/D pair. Thus, it is proposed that the travelers' incentive to avoid more costly routes will dynamically adjust the allocation of the path flows among each O/D pair w so that the sum of the flows equals the travel demand d_w. Moreover, each path flow $x_p, p \in P_w$, will remain nonnegative, while, at the same time, being affected by its own path cost $C_p, p \in P_w$. Therefore, the model proposes that the traffic flows on the paths change at a rate that is equal to the projection of minus the path costs onto the constraint set K given in (72), i.e., $\Pi_K(x, -C(x))$. In fact, this represents the "best" feasible direction to reallocate the traffic flows in the expectation of reducing the users' travel costs.

This process continues until there is no change in path flows, namely, until all used paths have equal travel costs and unused paths have higher (or equal) travel costs.

Mathematically, the route choice adjustment process is defined as:

$$\dot{x} = \Pi_K(x, -C(x)). \tag{74}$$

A direct application of Theorem 1 indicates that the set of the stationary points of the above travel route choice adjustment process coincides with the set of traffic network equilibria with fixed demands.

Theorem 18 *A path flow pattern $x^* \in K$ satisfies the variational inequality problem (72) if and only if it is a stationary point for the ordinary differential equation (74), that is,*

$$0 = \Pi_K(x^*, -C(x^*)). \tag{75}$$

The behavioral assumption underlying the route choice adjustment process for the fixed demand traffic network is analogous to those for the elastic demand traffic networks. However, here the travelers are required to make the trips between their origin and destination pairs and, therefore, are only concerned with which route to select in order to avoid congestion. The dynamic model of the route choice adjustment process recalled here is based on the following arguments. Suppose that the travelers between a certain O/D pair will switch from more costly routes to less costly routes and that the traffic flows on the paths connecting this O/D pair change at a rate that is proportional to minus the costs on the paths, while the sum of these traffic flows is maintained unchanged. Thus, the travelers' incentive to avoid more costly routes will dynamically adjust the allocation of the path flows among the paths in each O/D pair w so that the sum of the flows is equal to the travel demand d_w.

For illustrative purposes, we present a small example to show how the travelers would adjust their route choices in order to avoid congestion in a fixed demand transportation network according to the dynamics of the proposed route choice adjustment process (74).

Example 3. Consider the same transportation network as in Examples 1 and 2. Suppose now that the travel demand is fixed to be $d_w = 3$ and the link cost functions are as in Examples 1 and 2.

Let the initial flow pattern at time $t = 0$ be: $x_a(0) = 1$ and $x_b(0) = 2$. The travel cost vector is then $C(0) = (3, 4)$. According to (74), the traffic flows change at the rate $\Pi_K(x(0), -C(0)) = (0.5, -0.5)$. If a time increment of 0.5 is used, then at the next moment $t = 0.5$, the flows on link a and link b are:

$$x_a(0.5) = x_a(0) + 0.5 \times 0.5 = 1.25, \quad x_b(0.5) = x_b(0) - 0.5 \times 0.5 = 1.75,$$

which yields travel costs: $c_a(0.5) = 3.25$ and $c_b(0) = 3.5$.

The flow changing rate at $t = 0.5$ can be calculated by

$$\Pi_K(x(0.5), -C(0.5)) = (0.125, -0.125).$$

Therefore, the route choice adjustment process at time $t = 1.0$ has the link flows

$$\begin{aligned} x_a(1.0) &= x_a(0.5) + 0.5 \times 0.125 = 1.3125, \\ x_b(1.0) &= x_b(0.5) + 0.5 \times (-0.125) = 1.6875, \end{aligned}$$

which yields the travel costs: $c_a(1.0) = 3.3125$ and $c_b(1.0) = 3.375$.

The difference between the travel costs on the two links causes the reallocation of the traffic flow at time $t = 1.5$, according to $\Pi_K(x(1.0), -C(1.0)) = (0.03125, -0.03125)$. Hence, at $t = 1.5$ the flows will be

$$\begin{aligned} x_a(1.5) &= x_a(1.0) + 0.5 \times 0.03125 = 1.328125, \\ x_b(1.5) &= x_b(1.0) + 0.5 \times (-0.03125) = 1.671875, \end{aligned}$$

and the travel costs are: $c_a(1.5) = 3.328125$ and $c_b(1.5) = 3.34375$.

In this example, we see that, the route choice adjustment process (74) reallocates the flows on the two links so that the difference of the travel costs is being reduced from 0.25 to 0.0625, and to 0.015625. In fact, it almost attains the equilibrium: $x_a^* = 1.333$ and $x_b^* = 1.666$ at time $t = 1.5$.

8.5.2 Stability Analysis

We now present the stability analysis for the route choice adjustment process described by (74), in the context of fixed demand traffic network problem. Besides the significance of these stability properties by themselves, the results are utilized in the establishment of the convergence of discrete time algorithms.

The following two stability results under the monotonicity assumptions are cited from Nagurney and Zhang (1997a).

Theorem 19 (Nagurney and Zhang (1997a)). *Suppose that the link costs c are monotone increasing in the link load f. Then the route choice adjustment process (74) is stable.*

The next theorem illustrates that the route travel choice adjustment process (74) eventually drives any traffic flow pattern to an equilibrium path flow pattern, provided that the link costs are strictly monotone.

Theorem 20 (Nagurney and Zhang (1997a)). *Suppose that the link costs c are continuous and strictly monotone increasing in the link loads f. Then, the route choice adjustment process (74) is asymptotically stable.*

8.5.3 A Discrete Time Algorithm

A special case of the general iterative scheme, the Euler method, was employed by Nagurney and Zhang (1997a; 1997b) for the computation of solutions to the dynamic traffic network problem with fixed travel demands. Recall that, according to the Euler method, at iteration τ, one must compute

$$x^{\tau+1} = P_K(x^\tau - a_\tau F(x^\tau)). \tag{76}$$

In the case that the sequence $\{a_\tau\}$ in the Euler method is fixed, say, $\{a_\tau\} = \rho$, for all iterations τ, then the Euler method collapses to a projection method. In the context of the traffic network equilibrium problem with fixed travel demands, the projection operation can no longer be computed in closed form, but, rather, is the solution to a quadratic programming problem. Nevertheless, because of the special structure of the resulting subproblems, each of them can be computed using an "exact" equilibration algorithm proposed by Dafermos and Sparrow (1969).

In particular, at each iteration τ of the Euler method, (76) takes the form: Compute the vector of path flows $x^{\tau+1}$ according to:

$$x^{\tau+1} = \min_{x \in K} \frac{1}{2} x^T \cdot x - (x^\tau - a_\tau F(x^\tau))^T \cdot x. \tag{77}$$

In view of the feasible set K, subproblem (77), in turn, can be decomposed into J subproblems, one for each O/D pair w. Each such subproblem is a quadratic programming problem with special structure that can be solved exactly and in closed form using exact equilibration. In particular, subproblem (77) is equivalent to the solution of: For each O/D pair w, compute:

$$\min \frac{1}{2} \sum_{p \in P_w} x_p^2 + \sum_{p \in P_w} h_p^\tau x_p \tag{78}$$

subject to:

$$\sum_{p \in P_w} x_p = d_w \tag{79}$$

and

$$x_p \geq 0, \quad \forall p \in P_w, \tag{80}$$

where

$$h_p^\tau = a_\tau C_p(x^\tau) - x_p^\tau. \tag{81}$$

Note that in the above subproblem (78)-(81), the network subproblem that is actually solved takes place on a network, in view of constraints (79) and (80), that has disjoint paths, that is, the paths connecting each O/D pair have no links in common. Moreover, each path consists of a single link. This is a remarkable feature of the Euler method in path flow variables.

To conclude this section, the following theorem provides the convergence of the proposed Euler method in the context of the fixed demand traffic network problem.

Theorem 21 (Nagurney and Zhang (1997a)). *Suppose that the link costs c are strictly monotone increasing. Let $\{a_\tau\}$ be a sequence of positive real numbers that satisfies (66) and (67). Then the Euler method given by*

$$x^{\tau+1} = P_K(x^\tau - a_\tau C(x^\tau)), \tag{82}$$

where $C(x)$ is the vector of path travel costs and $K = \{x \in R_+^{n_P} :$ and (68) holds $\}$ converges to some traffic network equilibrium path flow pattern.

8.6 Summary and Conclusions

In this chapter we have provided an introduction to the recently developed tool of projected dynamical systems in the context of traffic network equilibrium problems. The chapter provides both the qualitative analysis of such a non-classical dynamical system, in terms of existence and uniqueness of a solution path, stability analysis, as well as computational procedures, along with convergence results.

Three distinct dynamic traffic models that are modeled as projected dynamical systems are presented and reviewed. In particular, we consider two elastic demand models. In the first one studied in this chapter it is assumed that the travel disutility functions are given whereas in the second it is assumed that the travel demand functions are given. We then turn to the fixed demand problem. The route choice adjustment processes are given for all of these models, stability analysis is conducted, and the Euler method and the Heun method is discussed for computational purposes.

This chapter provides a synthesis of this methodology for day-to-day dynamic traffic network problems and can be used as a starting point for pursuing additional work in this field.

Acknowledgments

This chapter was written, in part, while the first author was a Distinguished Guest Professor at KTH in Stockholm, Sweden. The hospitality and cordiality of the Divisions of Regional Planning and Optimization and System Theory are much appreciated. The research reported herein was supported, in part, by the National Science Foundation, under the Faculty Awards for Women Program, Grant No. DMS-9024071.

References

Alber, Y.I. (1979), 'On the solution of equations and variational inequalities with maximal monotone operators', *Soviet Math. Doklady*, vol. 20, pp. 871–876.

Armijo, L. (1966), 'Minimizaton of functions having continuous partial derivatives', *Pacific Journal of Mathematics*, vol. 16, pp. 1-3.

Bakusinskii, A.B. and Polyak, B.T. (1974), 'On the solution of variational inequalities', *Soviet Math. Doklady*, vol. 15, pp. 1705–1710.

Beckmann, M., McGuire, C.B. and Winsten, C.B. (1956), *Studies in the Economics of Transportation*, Yale University Press, New Haven, CT.

Bernstein, D. (1990), 'Programmability of continuous and discrete network equilibria', Ph.D. Dissertation, University of Pennsylvania, Philadelphia, PA.

Bertsekas, D.P. (1976), 'On the Goldstein-Levitin-Polyak gradient projection method', *IEEE Transactions on Automatic Control*, vol. 21, pp. 174–184.

Bertsekas, D.P. and Tsitsiklis, J.N. (1989), *Parallel and Distributed Computation*, Prentice–Hall, Englewood Cliffs, NJ.

Coddington, E.A. and Levinson, N. (1955), *Theory of Ordinary Differential Equations*, McGraw-Hill Book Company, New York, NY.

Dafermos, S. (1980), 'Traffic equilibrium and variational inequalities', *Transportation Science*, vol. 14, pp. 42–54.

Dafermos, S. (1982), 'The general multimodal traffic equilibrium problem with elastic demand', *Networks*, vol. 12, pp. 57–72.

Dafermos, S. (1983), 'An iterative scheme for variational inequalities', *Mathematical Programming*, vol. 26, pp. 40–47.

Dafermos, S. and Nagurney, A. (1984), 'Stability and sensitivity analysis for the general network equilibrium-travel choice model', in Volmuller, J. and Hamerslag R. (eds), *Proceedings of the 9th International Symposium on Transportation and Traffic Theory*, VNU Science Press, Utrecht, pp. 217–234.

Dafermos, S.C. and Sparrow, F.T. (1969), 'The traffic assignment problem for a general network', *Journal of Research of the National Bureau of Standards*, vol. 73B, pp. 91–118.

Dupuis, P. and Ishii, H. (1991), 'On Lipschitz continuity of the solution mapping to the Skorokhod problem, with applications', *Stochastics and Stochastic Reports*, vol. 35, pp. 31–62.

Dupuis, P. and Nagurney, A. (1993), 'Dynamical systems and variational inequalities', *Annals of Operations Research*, vol. 44, pp. 9–42.

Flam, S.D. (1992), 'On finite convergence and constraint identification of subgradient projection methods', *Mathematical Programming*, vol. 57, pp. 427–437.

Florian, M. and Hearn, D. (1995), 'Network equilibrium models and algorithms', in Ball, M.O., Magnanti, T.L., Monma, C.L. and Nemhauser, G.L. (eds), *Network Routing*, Handbooks in Operations Research and Management Science, vol. 8, Elsevier Science B. V., Amsterdam, pp. 485–550.

Fukushima, M. (1986), 'A relaxed projection method for variational inequalities', *Mathematical Programming*, vol. 35, pp. 58–70.

Goldstein, A.A. (1964), 'Convex programming in Hilbert space', *Bulletin of the American Mathematical Society*, vol. 70, pp. 709–710.

Hartman, P. (1964), *Ordinary Differential Equations*, John Wiley & Sons, New York, NY.

Horowitz, J.L. (1984),'The stability of stochastic equilibrium in a two-link transportation network', *Transportation Research*, vol. 18B, pp. 13–28.

Kinderlehrer, D. and Stampacchia, G. (1980), *An Introduction to Variational Inequalities and Their Applications*, Academic Press, New York, NY.

Lefschetz, S. (1957), *Differential Equations. Geometric Theory*, Interscience, New York, NY.

Mahmassani, H. (1990), 'Dynamic models of commuter behavior: Experimental investigation and application to the analysis of planned traffic disruptions', *Transportation Research*, vol. 24A, pp. 465-.484.

Nagurney, A. (1993), *Network Economics: A Variational Inequality Approach*, Kluwer Academic Publishers, Boston, MA.

Nagurney, A. and Zhang, D. (1997a),'Projected dynamical systems in the formulation, stability analysis, and computation of fixed demand traffic network equilibria', *Transportation Science*, vol. 31, pp. 147–158.

Nagurney, A. and Zhang, D. (1997b), *Projected Dynamical Systems and Variational Inequalities with Applications*, Kluwer Academic Publishers, Boston, MA.

Nagurney, A. and Zhang, D. (1997c), 'Massively parallel computation of dynamic traffic networks modeled as projected dynamical systems', in Pardalos, P.M., Hearn, D.W. and Haper, W.W. (eds), *Network Optimization*, Lecture Notes in Economics and Mathematical Systems, vol. 450, Springer-Verlag, Berlin, pp. 374–396.

Nagurney, A. and Zhang, D. (1998), 'A massively parallel implementation of a discrete time algorithm for the computation of dynamic elastic demand traffic problems modeled as projected dynamical systems', *Journal of Economic Dynamics and Control*, in press.
Perko, L. (1991), *Differential Equations and Dynamical Systems*, Springer-Verlag, New York, NY.
Ran, B. and Boyce, D.E. (1994), *Dynamic Urban Transportation Network Models*, Lecture Notes in Economics and Mathematical Systems, vol. 417, Springer-Verlag, Berlin.
Skorokhod, A.V. (1961), 'Stochastic equations for diffusions in a bounded region', *Theory of Probability and its Applications*, vol. 6, pp. 264–274.
Smith, M.J. (1979), 'Existence, uniqueness, and stability of traffic equilibria', *Transportation Research*, vol. 13B, pp. 295–304.
Smith, M.J. (1984), 'The stability of a dynamic model of traffic assignment : An application of a method of Lyapunov', *Transportation Science*, vol. 18, pp. 245–252.
Wardrop, J.G. (1952), 'Some theoretical aspects of road traffic research', in *Proceedings of the Institute of Civil Engineers*, Part II, pp. 325–278.
Zhang, D. and Nagurney, A. (1995), 'On the stability of projected dynamical systems', *Journal of Optimization Theory and Applications*, vol. 85, pp. 97–124.
Zhang, D. and Nagurney, A. (1996), 'On the local and global stability of a travel route choice adjustment process', *Transportation Research*, vol. 30B, pp. 245–262.
Zhang, D. and Nagurney, A. (1997), 'Formulation, stability, and computation of traffic network equilibria as projected dynamical systems', *Journal of Optimization Theory and Applications*, vol. 93, pp. 417–444.

9 Worker and Workplace Heterogeneity and Residential Location: A Historical Perspective on Stockholm

Björn Hårsman
Inregia AB, P.O. Box 12519
S-102 29 Stockholm, Sweden and
Department of Infrastructure and Planning
Royal Institute of Technology
S-100 44 Stockholm, Sweden

John M. Quigley
The Goldman School,
and Department of Economics
University of California
Berkeley, CA 94720, USA

9.1 Introduction

Most studies of the role of transportation in residential location assume that an individual's human capital and labor force experience are of no relevance to the location of his or her worksite and to the tradeoff between commuting and housing costs. Recent models often assume a polycentric metropolitan area and do recognize non-central workplace concentrations, but concentrations are undifferentiated by industry or occupation. Traditional traffic models rely upon the number of jobs and workers in different subareas, distance costs, income and sometimes age and family type to explain an observed commuting pattern. It is implicitly assumed that all workers are equally attracted to all kinds of jobs and that, moreover, all workers have the same chance of getting any job.

This paper presents an analysis of the spatial differentiation of industry and occupation within a large metropolitan area. We analyze the spatial distribution of workplaces, differentiated by industry and occupation, over time. Using data on the spatial distribution of employment by industry and occupation for the Stockholm metropolitan area, we investigate trends in the spatial pattern of employment from 1970–1990. In a parallel fashion, we also investigate changes in the spatial distribution of the residential sites of workers, differentiated by occupation and industry, and also by education level during the same period.

These data also permit a cross-sectional comparison of the joint residential and workplace distribution of workers in metropolitan Stockholm using more detailed data for 1990. This latter analysis indicates the importance of spatial differentiation by industrial types and human capital in understanding commuting patterns.

The paper also analyzes data linking the residential and workplace location of the various subgroups to commuting and transport behavior.

The objective of the paper is largely empirical and descriptive. Through a variety of decomposition techniques, we establish the extent to which disaggregated origin-destination data support the common, but implicit assumption that the diversity of job requirements and differences in human capital do not influence commuting patterns. The basic tool for this analysis is the computation of entropy measures and the decomposition of these measures along categorical lines. The classic methodological references are to Theil (1967; 1972), and all the measures we employ are easily derived from the structure introduced and popularized by Theil. It is, perhaps, more surprising that these systematic techniques for decomposing joint frequency distributions in intuitively meaningful ways have not received more widespread attention in applied fields such as urban transport. In the following sections, we also develop and illustrate alternative descriptive measures of regional balance or regional homogeneity. Currently the most widely used measure of intra regional "balance" is the ratio of jobs to population. We illustrate the use of equally simple, but more meaningful, alternative measures.

9.2 The Stockholm Metropolitan Region

The Stockholm Metropolitan Region is built on a series of islands emanating from a medieval center dating from the twelfth century. Fig. 9.1 indicates the major arterials providing access to the historic center of the region. The transport system depicted in Fig. 9.1 is consistent with a single concentrated worksite located in the inner city. This is the historical pattern of the city′s development. Both the number of jobs and the number of workers living in the region have increased during the last decades. As is shown in Table 9.1, there are more jobs than workers, and net commuting into the region more than doubled in two decades.

Table 9.1. Number of jobs and workers (thousands) in the Stockholm Region, 1970–1990

	1970	1980	1990
Jobs	685	804	951
Workers	665	781	909
Net commuting	20	23	42

The increasing employment has gone hand in hand with a restructuring of employment by industry, occupation and educational level. The largest changes in industrial composition are the decline of manufacturing (including heavy and light industry) jobs and the increase in employment in health care and social care.

The occupational structure has shifted from administrative work (managers, accountants, cashiers, etc.) and goods handling work (farmers, craftsmen and assembly, etc.) towards knowledge work (technical scientific work and other knowledge-based work such as legal services and journalism) and other service work (nurses, policemen, barbers, etc).

The educational level of workers has increased substantially. The fraction of workers with low education (11 years or less) has decreased from 74 to 55 percent, and the fraction with high education (at least 15 years) has almost doubled, from 8 to 15 percent.

The expansion and restructuring of the labor market has been accompanied by a significant dispersion of both residents and jobs.

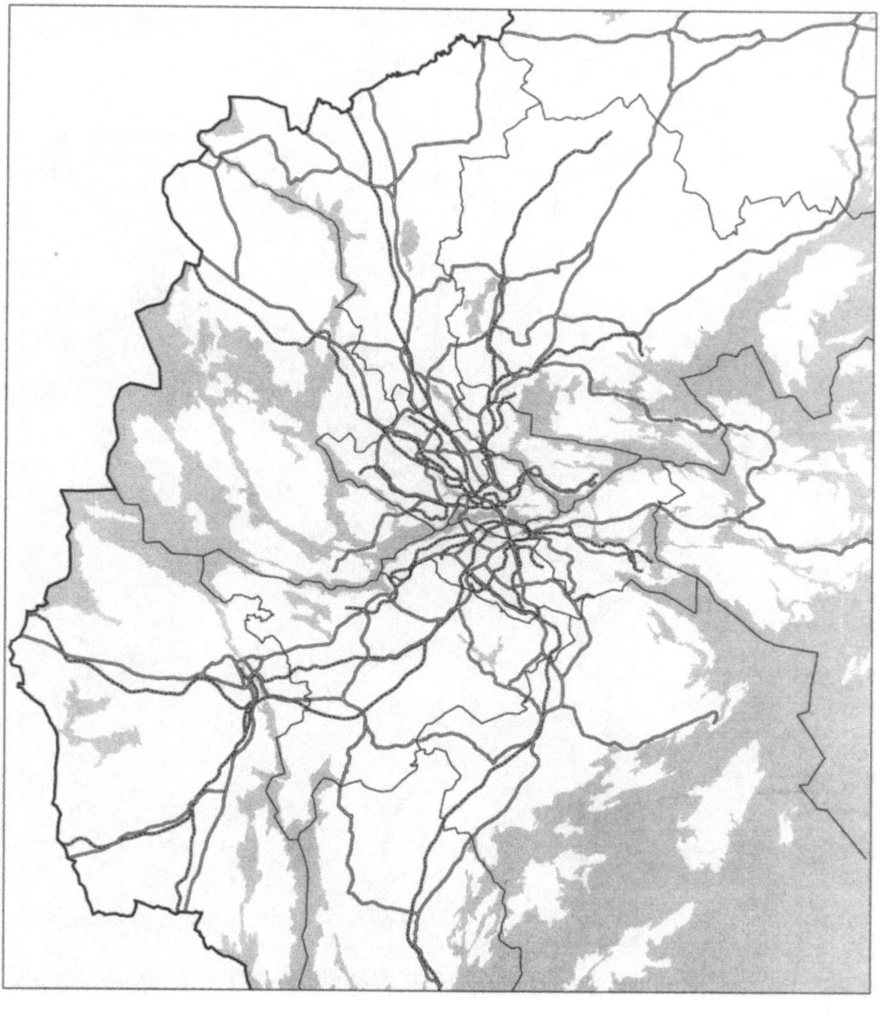

Fig. 9.1. The Stockholm Region: main roads and tracks 1994

Fig. 9.2 provides an overview of the distribution of jobs over subareas during the past quarter century. The region is divided into fourteen zones representing subareas used by the Regional Planning Authorities.

As the figure indicates, 46 percent of regional employment was concentrated in the inner city of Stockholm in 1970. By 1990, this concentration had been reduced to 34 percent, and there had been systematic increases in employment shares at non central locations.

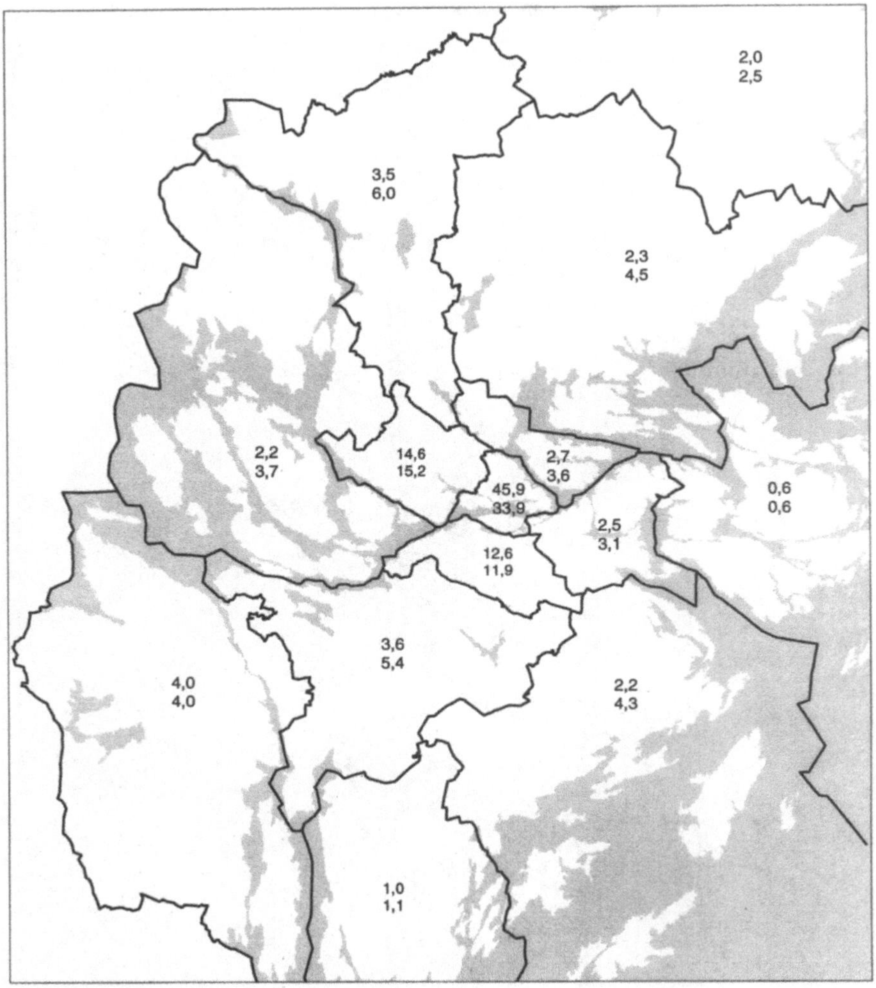

Fig. 9.2. Employment in the Stockholm Region by worksites. Percent 1970 (upper) and 1990 (lower)

Fig. 9.3 indicates the distribution of worker residences during the same period. As expected, the residential distribution is considerably less centralized. Nevertheless, the same pattern of decentralization is apparent in the data. The population share of the inner city share declined from 18 to 15 percent during the 1970–1990 period, and the population share in the southern part of the city declined from 24 to 15 percent. The population shares of the inner suburbs remained roughly constant, while the shares of the outer suburbs increased substantially – but from a relatively small base.

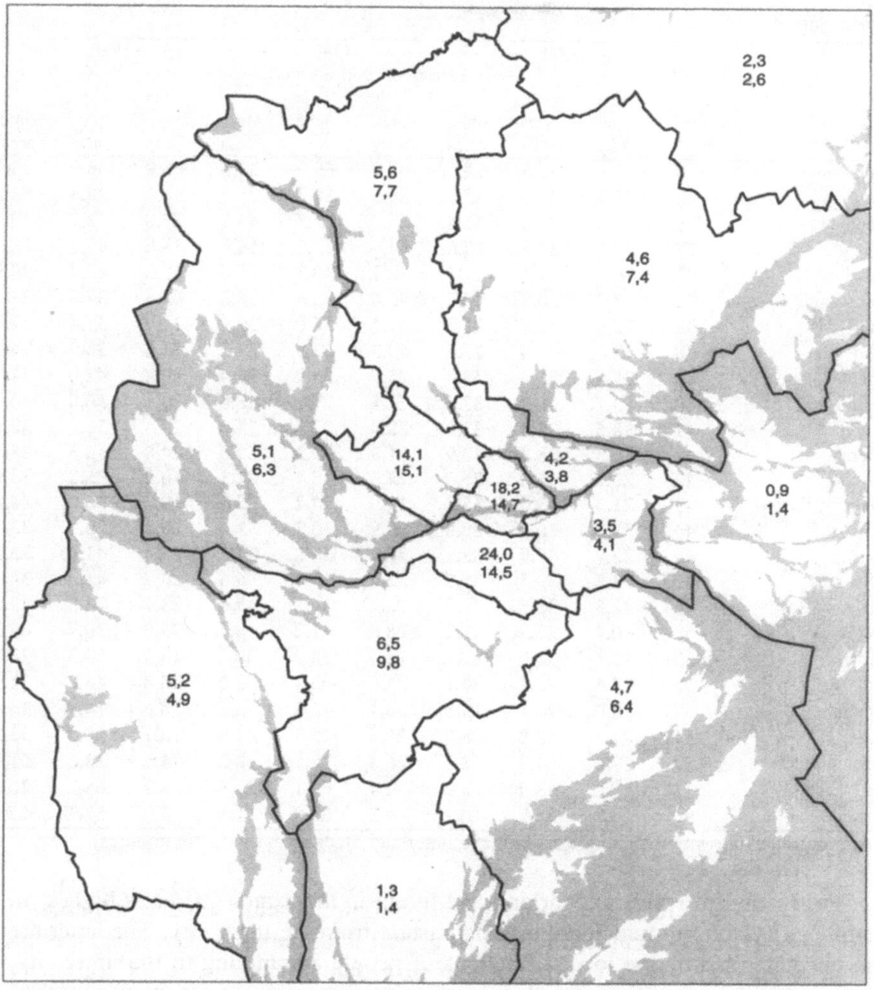

Fig. 9.3. Employment in the Stockholm Region by residential areas. Percent 1970 (upper) and 1990 (lower)

Clearly the spatial patterns of both workplaces and residence places have become less differentiated over time. Worksites are less concentrated in particular central locations, and residence sites are less segregated from worksites.

The dispersion of jobs and workers has changed the commuting pattern. Between 1971 and 1987 the share of all commuting directed towards the inner city decreased from 47 to 39 per cent. Table 9.2 shows the pattern of worktrips summarized by destination into three categories: worktrips to the same zone as that of residence; worktrips to the inner city; worktrips to other destinations. Table 9.3 reports variations by industry, education, occupation and sex.

Table 9.2. Pattern of worktrips by destination for different residential areas in Stockholm, 1975–1990

		1975			1980			1990		
	Travel				Percent of residents working in					
Residential area	time to inner city*	the inner city	the same area	other areas	the inner city	the same area	other areas	the inner city	the same area	other areas
Sthlm-inner	0	71.5	71.5	28.5	69.2	69.2	30.8	66.9	66.9	33.1
Solna	22	36.7	37.6	25.7	36.5	36.2	27.4	38.4	28.4	33.2
Sundbyberg	26	27.0	27.6	45.4	28.4	26.8	44.8	27.8	32.7	39.5
Lidingö	28	44.1	38.8	17.0	44.6	37.2	18.1	38.6	43.2	18.3
Nacka	31	39.6	41.1	19.3	38.1	40.3	21.6	35.3	44.4	20.3
Sthlm-south	32	47.7	36.0	16.2	45.8	35.4	18.8	42.4	37.5	20.1
Sthlm-west	32	38.0	36.4	25.6	37.1	36.0	26.9	34.9	37.4	27.7
Danderyd	38	42.0	31.4	26.6	40.3	32.5	27.2	34.8	38.9	26.3
Järfälla	40	25.6	35.0	39.4	25.7	35.9	38.4	20.2	42.0	37.7
Huddinge	40	33.7	28.7	37.7	32.8	29.2	38.0	29.1	30.0	40.9
Tyresö	41	38.5	28.5	33.0	38.4	27.7	34.0	32.4	35.1	32.6
Haninge	42	30.3	38.8	30.9	31.3	39.2	29.6	24.8	46.6	28.6
Värmdö	42	20.7	54.3	25.0	24.3	48.4	27.3	30.9	35.2	33.9
Upplands Väsby	43	21.3	43.6	35.2	19.9	44.7	35.3	19.2	36.0	44.8
Täby	44	38.3	30.9	30.8	36.9	30.7	32.4	30.2	38.7	31.1
Ekerö	46	28.9	37.9	33.2	28.7	36.8	34.5	25.4	41.4	33.2
Sollentuna	46	33.8	30.5	35.7	32.0	31.4	36.6	24.7	41.2	34.1
Upplands-Bro	46	21.5	35.6	42.9	20.2	34.2	45.6	23.7	28.8	47.5
Botkyrka	52	30.4	27.9	41.7	28.6	31.2	40.2	25.0	30.7	44.3
Sigtuna	53	13.3	59.0	27.8	12.3	58.2	29.5	13.5	59.3	27.3
Österåker	55	27.4	43.8	28.9	25.6	45.3	29.0	20.8	48.0	31.2
Södertälje	57	8.0	76.1	15.8	8.1	75.8	16.2	8.5	74.9	16.6
Vaxholm	57	27.4	43.8	28.9	25.6	45.3	29.0	18.1	50.7	31.2
Vallentuna	57	27.0	34.9	38.1	26.3	34.5	39.2	24.5	23.6	51.9
Nynäshamn	72	7.2	68.9	23.9	7.4	69.1	23.6	8.2	65.2	26.6
Norrtälje	90	3.3	85.7	10.9	4.3	83.9	11.8	5.3	82.1	12.6

* Approximate one-way rush hour travel time to inner city Stockholm by car (in minutes).

The fraction of residents working and living in the same subarea is highest in the inner city and in municipalities most distant from the inner city. The latter are also characterized by the lowest fractions of people commuting to the inner city. As expected, commuting to the inner city is most common in closeby municipalities. There is also a clear downward trend for the fraction commuting to the inner city.

The commuting pattern doesn't vary as much among different categories of workers as it does among subareas. However, there are some clear differences. By way of example, the commuting distance tends to be shorter for those employed in the healthcare industry than in public administration and also for those with lower levels of education. Women are also more likely than men to work within their residential areas.

We will begin our analysis by asking a series of questions about the internal homogeneity of worksites and residence places. For these and for all other calculations, we rely upon a partition of the metropolitan region into 26 zones or sub areas. 23 of these correspond to the suburban towns surrounding the central city[1]. The central city itself is divided into three parts: the Central Business District (CBD) or inner city, the western portion, and the southern part. Table 9.2 shows the rush hour travel time by car from each suburban town to the inner city.

We have detailed information about the occupational and industrial structure of jobs in each of these 26 zones, as well as information on the sex and educational level of employees. We have similar information of the characteristics of residents of each of these areas.

Table 9.3. Pattern of worktrips by destination for different categories for Stockholm, 1990

		Percent of residents working in			
		same residential area	inner city	other areas	total
Industry	Manufacturing	45.8	14.5	39.7	100
	Local services	50.1	23.7	26.2	100
	Nat/reg services	35.3	38.1	26.6	100
	Public adm	37.9	38.0	24.1	100
	Education/research	51.9	19.4	28.7	100
	Health/social care	58.0	13.2	28.7	100
	Other	47.3	19.4	33.3	100
Education	Low	49.1	22.8	28.1	100
	Medium	42.0	26.6	31.4	100
	High	40.3	27.8	32.1	100
Occupation	Knowledge	42.2	24.5	33.2	100
	Administration	37.6	38.7	23.7	100
	Service	49.7	21.6	28.7	100
	Goods handling	49.3	15.8	35.0	100
Sex	Men	41.3	24.3	34.4	100
	Women	49.8	25.0	25.1	100
Total		**45.6**	**24.7**	**29.7**	**100**

Note: Residential areas are defined in Table 9.2.

9.3 Quantitative Measures of Heterogeneity

Entropy measures are natural quantitative indicators of the heterogeneity or homogeneity of populations. These measures can be applied to describe the residential segregation of different types of workers, for example, or the clustering of employment by industry within a geographical area. Depending upon the specific focus of interest, the entropy measure can be formulated in a variety of ways.

A most basic measure is the extent to which different geographical areas – be they residence places or workplaces – are homogeneous with respect to some characteristic or set of categories. An index of heterogeneity based on entropy

[1] The municipalities of Botkyrka and Salem have been grouped together to form a single zone. The other suburban zones correspond to municipal boundaries.

concepts can be computed from counts of individuals (e.g. residents or workers) by category in each of the geographical areas. For example, let N_{ij} denote the number of individuals in category i associated with subarea j. $N._j$ is the total number of individuals in subarea j, and $N..$ is the grand total. M is the number of subareas, j=1, 2,..., M. Consider the following index of spatial heterogeneity.

$$I_j = \sum_i \frac{N_{ij}}{N_{.j}} \ln\left(\frac{N_{ij}}{N_{.j}} / \frac{N_{i.}}{N..}\right) \tag{1}$$

Equation (1) indicates how the mix with respect to category i in area j differs from the mix at the regional level. If the mix in area j is identical to the mix for the region as a whole, then the expression in brackets in equation (1) is equal to 1, and I_j=0.

Let $\bar{I}$ be the weighted average of the index of heterogeneity of the individual subareas

$$\bar{I} = \sum_j \frac{N_{.j}}{N_{..}} I_j \tag{2}$$

The value of $\bar{I}$ equals zero when the mix of categories within each subarea is the same and is equal to the mix of categories at the regional level.

Inserting (1) into (2) yields, after some manipulation,

$$\bar{I} = \sum_i \frac{N_{i.}}{N_{..}} \ln\left(1 / \frac{N_{i.}}{N_{..}}\right) - \sum_j \frac{N_{.j}}{N_{..}} \left\{\left(\sum_i \frac{N_{ij}}{N_{.j}} \ln\left(1 / \frac{N_{ij}}{N_{.j}}\right)\right)\right\} \tag{3}$$

$$= I_i - \sum_j \frac{N._j}{N..} I_j$$

$$= I_i - \bar{I}_j$$

The first term on the right hand side is the entropy of the distribution over categories at the regional level. The second term is a weighted average of the entropies of each of the subareas in the region. Clearly, the value of $\bar{I}$ equals zero if the entropy of each subarea is equal to the entropy of the region. The maximum value of $\bar{I}$ is obtained when the average entropy attains its minimum value. This suggests that a relative measure of heterogeneity or segregation (S) can be formed by dividing $\bar{I}$ by the first term of (3)

$$S_i = \bar{I} / I_i = 1 - \bar{I_j} / I_i \tag{4}$$

Table 9.4 summarizes trends in the spatial segregation of residences in the Stockholm region during the past two decades. Part A presents trends in residential location by occupation and industry. Part B presents trends in residential location by sex and level of education. The categories are the same as those introduced in

Section 9.2 above. They include five occupational categories, seven industries, and three educational categories.

For each classification, the table presents the average entropy level (based on the 26 geographical areas described earlier) and the maximum entropy level for the region as a whole. The table also reports the index of segregation for each classification, S_i. Finally, we report an explicit statistical test of the difference of entropy levels (and the increasing similarity of geographic regions) between 1970 and 1980 and between 1980 and 1990.

The table indicates that the levels of residential segregation by occupation and industry are low, and they have been declining steadily in the past three decades. The maximum entropy of occupation and industry is declining (as the region is becoming more specialized), but the average entropy has declined more rapidly, and residential segregation has declined.

The table also indicates that there is essentially no residential segregation by sex. In contrast, segregation by level of education has increased rather substantially. The maximum entropy of education has increased (as levels of schooling have become more equalized), but the average entropy has increased more rapidly – hence residential segregation by education level has increased.

Table 9.5 presents comparable information on the spatial segregation of workplaces.

Part A presents trends in workplace heterogeneity by occupation and industry. The 26 geographical areas have become more similar according to the occupational mix of employment. In contrast, the industrial composition of the subareas has become more differentiated – as industries have become more segregated through their location decision.

In contrast, part B of the table reports that there is only a small tendency for employment in these areas to be differentiated by sex. There is a stronger tendency for employment segregation by level of education, and this tendency is increasing over time.

Table 9.6 compares the concentration of residence sites and work sites within the region.

If residence places or workplaces were equally dispersed throughout the region, the corresponding entropy level would be 3.258 (i.e. 26 ln [1/26]). The concentration of residences in 1970 reduced the entropy of residence sites to 2.689, a reduction of 17.59 percent. The concentration of workplaces in 1970 reduced the entropy of workplaces to 2.161 or by 33.7 percent. Over time, the decentralization of residential population and employment has led to increases in both residential and employment entropy. This dispersion has been most rapid and pronounced for employment locations. Nevertheless in 1990, residential sites were still much more widely dispersed throughout the metropolitan area than employment sites were.

In summary, when compared to the distribution of population, the distribution of employment has decentralized more rapidly. But the relative segregation of employment sites, by the characteristics of workers, is substantially higher than the segregation of the residences chosen by workers classified according to the same criteria.

Table 9.4. Trends in heterogeneity by residential location for Stockholm, 1970–1990

A. Have residential areas become more similar by occupational and industrial mix over time?

	1970	1980	1990
Occupation:			
Average entropy	1.339	1.321	1.263
Maximum entropy	1.361	1.342	1.281
Relative reduction (percent)	1.618	1.540	1.446
t-ratio		44.338	64.589
Industry:			
Average entropy	1.821	1.840	1.815
Maximum entropy	1.850	1.862	1.834
Relative reduction (percent)	1.532	1.170	1.060
t-ratio		36.926	26.271
Occupation and industry:			
Average entropy	2.930	2.958	2.933
Maximum entropy	2.974	2.997	2.968
Relative reduction (percent)	1.486	1.288	1.172
t-ratio		33.085	16.691

B. Have residential areas become more differentiated by sex and education level?

	1970	1980	1990
Sex:			
Average entropy	0.680	0.692	0.693
Maximum entropy	0.685	0.692	0.693
Relative reduction (percent)	0.645	0.120	0.067
t-ratio		179.723	17.970
Education:			
Average entropy	0.720	0.795	0.952
Maximum entropy	0.733	0.815	0.980
Relative reduction (percent)	1.857	2.530	2.828
t-ratio		96.545	167.183
Sex and Education:			
Average entropy	1.389	1.479	1.643
Maximum entropy	1.408	1.501	1.672
Relative reduction (percent)	1.394	1.526	1.734
t-ratio		115.570	173.308

Note: The t-ratio tests the difference in mean entropy between the earlier and the later time period. $t \approx (I_T - I_\tau) / (Var I_\tau / \alpha)^{1/2}$ where α is the sampling fraction. The variance of entropy at time T is approximately $\sum_j \frac{N_{.j}}{N_{..}^2} \left\{ \sum_i \frac{N_{ij}}{N_{.j}} \ln^2 \frac{N_{ij}}{N_{.j}} - I_i \right\} + \frac{(M-1)L}{2N_{..}^2}$ where L is the number of categories.

Table 9.5. Trends in heterogeneity by workplace for Stockholm, 1970–1990

A. Have workplaces become more similar by occupational and industrial mix?

	1970	1980	1990
Occupation:			
Average entropy	1.328	1.308	1.263
Maximum entropy	1.361	1.344	1.292
Relative reduction (percent)	2.412	2.670	2.171
t-ratio		45.808	45.205
Industry:			
Average entropy	1.784	1.780	1.737
Maximum entropy	1.849	1.865	1.827
Relative reduction (percent)	3.499	4.564	4.956
t-ratio		6.681	32.752
Occupation and industry:			
Average entropy	2.877	2.891	2.845
Maximum entropy	2.972	3.008	2.957
Relative reduction (percent)	3.197	3.891	3.830
t-ratio		14.434	24.886

B. Have workplaces become more differentiated by sex and education?

	1970	1980	1990
Sex:			
Average entropy	0.683	0.690	0.669
Maximum entropy	0.685	0.693	0.693
Relative reduction (percent)	0.359	0.335	0.528
t-ratio		92.733	5.543
Education:			
Average entropy	0.725	0.816	0.969
Maximum entropy	0.731	0.825	0.982
Relative reduction (percent)	0.798	1.076	1.304
t-ratio		115.224	153.630
Sex and Education:			
Average entropy	1.393	1.497	1.655
Maximum entropy	1.406	1.511	1.674
Relative reduction (percent)	0.912	0.918	1.118
t-ratio		130.098	154.999

Table 9.6. Residential and employment entropies within the Stockholm Region, 1970–1990

	1970	1980	1990
A. By residence			
Average entropy	2.689	2.880	2.925
Maximum	3.258	3.258	3.258
Relative reduction (percent)	17.465	11.602	10.221
B. By workplace			
Average entropy	2.161	2.449	2.540
Maximum	3.258	3.258	3.258
Relative reduction (percent)	33.671	24.831	22.038

Table 9.A1 in the appendix presents the disaggregated data which form the basis for these trends. It reports the entropies of occupation, industry and education separately for residences and workplaces for 1970, 1980 and 1990. Higher entries in this table are associated with larger deviations from the regional average of the distribution of occupation, industry and education respectively. The table indicates that there is a strong persistence over time in the geographical areas which are outliers in each of these dimensions.

Table 9.7 summarizes the geographical areas which were most atypical in 1990.

Table 9.7. Most atypical geographical areas in workplace and residence place distribution for the Stockholm Region, 1990

	Workplace	**Residence**
Occupation	Norrtälje	Norrtälje
	Värmdö	Danderyd
	Södertälje	Södertälje
	Sigtuna	Nynäshamn
	Nynäshamn	Stockholm-inner
Industry	Sigtuna	Norrtälje
	Vaxholm	Södertälje
	Huddinge	Nynäshamn
	Danderyd	Vaxholm
	Värmdö	Sigtuna
Education	Norrtälje	Danderyd
	Nynäshamn	Stockholm-inner
	Värmdö	Norrtälje
	Haninge	Lidingö
	Danderyd	Nynäshamn

The analysis so far has concentrated on geographical areas and has analyzed the extent to which locational patterns vary by geographical area. In a parallel fashion, one might concentrate on demographic or individual characteristics and analyze the extent to which the spatial dispension of each category deviates from the regional average. Consider the following index of demographic heterogeneity.

$$J_i = \sum_j \frac{N_{ij}}{N_{i.}} \ln\left(\frac{N_{ij}}{N_{i.}} / \frac{N_{.j}}{N_{..}}\right) \tag{5}$$

Equation (5) indicates how the distribution of a given category i varies with respect to the geographical areas. It investigates the spatial segregation of a category of individuals rather than the spatial segreagation of geographical areas.

The weighted average of the index over individual categories is

$$\bar{J} = \sum_j \frac{N_{i.}}{N_{..}} J_i \tag{6}$$

It is easy to show that $\bar{J} = \bar{I}$, which is not surprising since the same information concerning subareas is used in both cases concerning the distribution of employment by categories and subareas. However, it is important to notice that the relative segregation will differ. If expression (6) is rearranged we get the following expression corresponding to (3).

$$\bar{J} = \sum_j \frac{N_{.j}}{N_{..}} \ln\left(1/\frac{N_{.j}}{N_{..}}\right) - \sum_i \frac{N_{i.}}{N_{..}} \left\{ \sum_j \frac{N_{ij}}{N_{i.}} \ln\left(1/\frac{N_{ij}}{N_{i.}}\right) \right\} \tag{7}$$

$$= I_j - \Sigma \frac{N_{i.}}{N_{..}} I_i$$

$$= I_j - \bar{I}_i$$

Note that the second term on the right hand side of (7) is identical to the second term on the right hand side of (3), and hence $\bar{I}_i = \bar{I}_j$ Since the first term in (7) differs from the first term in (3), however the measure of segregation will differ

$$S = \bar{J} / I_j = 1 - \bar{I}_i / I_j \neq 1 - \bar{I}_j / I_i \tag{8}$$

This is because the absolute segregation is related to the entropy of the regional distribution over categories in (3) and over subareas in (7).

Table 9.8 reports the spatial distribution of the various categories of education, occupation, and industry over workplaces and residence places during thirty year period, 1960–1990.

The differences in the pattern of worksites of the group with the highest education are much larger than for the other two groups. The differences in the pattern of residence sites for the highly educated are still larger. The residential concentration of the highest education group is much more pronounced than for the other education groups, or for any occupational or industrial concentration.

Planners are familiar with concepts of "balance" between the residence and workplaces contained in geographical subareas, and the ratio of jobs to residences in a geograhical area is often used as a measure of this balance. The entropy measures discussed here suggest a more general measure of the balance between jobs and residences:

$$B_j = \sum_j \frac{N_{.j}}{N_{..}} \ln\left(\frac{N_{.j}}{N_{..}} / \frac{D_{.j}}{D_{..}}\right) \tag{9}$$

where $D_{.j}$ is the number of workplaces in subarea j.

Analogously

$$B_i = \sum_j \frac{N_{ij}}{N_{i.}} \ln\left(\frac{N_{ij}}{N_{i.}} / \frac{D_{ij}}{D_{i.}}\right) \tag{10}$$

where D_{ij} is the number of worksites in subarea j with employment in category i.

Table 9.8. Segregation for various categories of education, occupation and industry for Stockholm Region, by 1970–1990

	Worksites			Residence sites		
	1970	1980	1990	1970	1980	1990
A. Education						
Low	0.002	0.004	0.009	0.004	0.008	0.020
Medium	0.011	0.012	0.010	0.015	0.018	0.009
High	0.033	0.038	0.030	0.100	0.108	0.092
Average	0.006	0.009	0.013	0.014	0.021	0.028
Relative reduction	0.008	0.011	0.013	0.019	0.025	0.028
B. Occupation						
Knowledge	0.009	0.008	0.009	0.021	0.021	0.019
Administrative	0.068	0.075	0.063	0.026	0.016	0.009
Services	0.003	0.006	0.005	0.002	0.001	0.002
Goods handling	0.060	0.088	0.087	0.048	0.067	0.087
Average	0.033	0.036	0.028	0.022	0.021	0.018
Relative reduction	0.024	0.027	0.022	0.016	0.015	0.014
C. Industry						
Manufacturing	0.072	0.119	0.118	0.041	0.049	0.042
Local service	0.009	0.016	0.019	0.005	0.006	0.020
Nat/reg services	0.122	0.110	0.109	0.038	0.027	0.026
Public adm	0.074	0.211	0.192	0.025	0.026	0.018
Education/research	0.022	0.045	0.039	0.029	0.015	0.014
Health/social care	0.101	0.079	0.096	0.015	0.004	0.005
Other	0.022	0.032	0.065	0.029	0.019	0.016
Average	0.065	0.085	0.091	0.028	0.022	0.019
Relative reduction	0.035	0.046	0.050	0.015	0.012	0.011

Note: Table entries are segregation indices derived in equation (8).

By analogy to equation (1), (2) and (3), it is possible to calculate the relative concentration of origin-destination combinations or commuting patterns for various categories. The following measure is a straightforward application of this reasoning.

$$C_i = \sum_m \sum_n \frac{N_{imn}}{N_{i..}} \ln\left(\frac{N_{imn}}{N_{i..}} \Big/ \frac{N_{.mn}}{N_{...}} \right) \tag{11}$$

where N_{imn} is the number of workers of category i commuting between subareas m and n.

Table 9.9. Relative concentration of commuting patterns for various categories of sex, education, occupation and industry for Stockholm Region, 1990

		Relative concentration
Sex	Men	0.016
	Women	0.018
	Average	0.017
	Relative reduction	0.004
Education	Low	0.027
	Medium	0.019
	High	0.125
	Average	0.040
	Relative reduction	0.009
Occupation	Knowledge	0.035
	Administration	0.076
	Service	0.012
	Goods handling	0.132
	Average	0.046
	Relative reduction	0.010
Industry	Manufacturing	0.150
	Local services	0.045
	Nat/reg services	0.125
	Public administration	0.249
	Education/research	0.106
	Health/social care	0.157
	Other	0.098
	Average	0.126
	Relative reduction	0.028

Note: Table entries are concentration indices derived in equations (12) and (13).

The weighted average over categories is

$$\overline{C} = \sum_i \frac{N_{i..}}{N_{...}} C_i \tag{12}$$

and a measure of relative concentration is obtained by dividing $\overline{C}$ by

$$\sum_m \sum_n \frac{N_{mn.}}{N_{...}} \ln\left(1 / \frac{N_{mn.}}{N_{...}}\right) \tag{13}$$

Calculations of this kind have been made both for one-way and for two-way classifications into categories of the employment. In the latter case we have simply defined new categories and used the same measures.

Table 9.9 presents information on the relative concentration of origin-destination (OD) combinations for various categories. Each measures the deviation of the OD pattern for a particular category from the OD pattern for all categories according to expressions (11) and (12).

Note that there are quite substantial differences. For example, the worktrip pattern of highly educated workers is more than five times as concentrated as is the pattern of those with less education. These workers live in areas with higher concentrations of similar people and commute to areas with higher concentrations of more highly educated workers.

Particularly striking is the far more concentrated worktrip pattern of public sector employees.

Table 9.10 gives the same kind of information when employment is classified both according to industry and occupation and according to education and sex.

Table 9.10. Relative concentration of commuting patterns for cross-classifications of industry/occupation and education/sex

Industry	Knowledge	Adm	Service	Goods handling
Mannufacturing	0.25161	0.15402	0.14748	0.28499
Local services	0.21347	0.09442	0.06608	0.18159
Nat/reg services	0.20356	0.19724	0.11351	0.18082
Public adm	0.35040	0.32041	0.30923	1.53281
Education/research	0.13818	0.25888	0.21933	0.54626
Health/social care	0.18776	0.26919	0.18482	0.40265
Other	0.16489	0.13518	0.15799	0.24439
Average	0.18422			
Relative reduction	0.04034			
Education	**Men**	**Women**		
Low	0.04670	0.04802		
Medium	0.04020	0.04423		
High	0.17562	0.12770		
Average	0.06238			
Relative reduction	0.01366			

Note: Table entries are concentration indices derived in equations (12) and (13).

9.4 Conclusion

The patterns of workplace location and residential location of workers are highly interrelated. These two distributions give rise to the pattern of commuting and work trip behavior. This paper has provided a description of trends in workplace and residence patterns in Stockholm over the past two decades, and has analyzed the joint distribution of workplace-residence place patterns (and hence commuting patterns) for 1990.

The paper emphasizes the linkage between the site-specific demands for human capital in production (as measured by industry and occupation) and the consequent residential choices of those workers (disaggregated by sex and education, as well as industry and occupation). The analysis shows that the relative deconcentration of worksites has been accompanied by a modest increase in the segregation of workplaces by industry and occupation. In contrast, there has been a substantial desegregation of residence places by the occupation and industry of workers. Despite this, there has been a pronounced increase in the segregation of the resident population by level of education. Particularly striking is the increase in the relative segregation of residences chosen by those with the highest levels of education. The analysis of these joint distributions for 1990 finds substantial differences in work trip patterns of workers, especially when classified by industry and education levels. The concentration of similar workers among particular origin-destination patterns is quite large, especially for certain types of workers, for example those working in the public sector.

Acknowledgments

A previous version of this paper was presented at the Symposium on Network Infrastructure and the Urban Environment, Stockholm, August 1994. We are grateful to Leonid Engelson for computational assistance.

Financial support for this research has been provided by the Transportation Center and the Center for Real Estate and Urban Economics, Universtiy of California, Berkeley and the Department for Infrastructure and Planning, the Royal Institute of Technology, Stockholm.

References

Theil, H. (1967), *Economics and Information Theory*, North-Holland Publishing Company, Amsterdam.

Theil, H. (1972), *Statistical Decomposition Analyses*, North-Holland Publishing Company, Amsterdam.

Appendix

Table 9.A1 Intra-regional variation in entropy levels for residences and worksites

Occupation		Worksites			Residence sites		
		1970	**1980**	**1990**	**1970**	**1980**	**1990**
1	Upplands-Väsby	0.090	0.055	0.034	0.029	0.013	0.011
2	Vallentuna	0.157	0.088	0.042	0.032	0.006	0.007
3	Österåker	0.080	0.051	0.037	0.015	0.006	0.004
4	Värmdö	0.223	0.154	0.065	0.113	0.046	0.026
5	Järfälla	0.042	0.042	0.024	0.012	0.003	0.004
6	Ekerö	0.051	0.030	0.022	0.020	0.003	0.000
7	Huddinge	0.058	0.027	0.019	0.003	0.004	0.003
8	Botkyrka & Salem	0.061	0.026	0.018	0.009	0.009	0.012
9	Haninge	0.065	0.040	0.032	0.011	0.016	0.017
10	Tyresö	0.037	0.027	0.027	0.006	0.000	0.003
11	Upplands-Bro	0.081	0.057	0.015	0.033	0.021	0.021
12	Täby	0.020	0.013	0.004	0.030	0.026	0.022
13	Danderyd	0.045	0.053	0.033	0.076	0.079	0.064
14	Sollentuna	0.014	0.014	0.004	0.009	0.010	0.008
15	Södertälje	0.064	0.079	0.060	0.061	0.062	0.052
16	Nacka	0.032	0.022	0.014	0.003	0.003	0.004
17	Sundbyberg	0.017	0.017	0.005	0.008	0.010	0.003
18	Solna	0.002	0.005	0.005	0.006	0.006	0.005
19	Lidingö	0.025	0.017	0.017	0.051	0.038	0.032
20	Vaxholm	0.066	0.086	0.049	0.020	0.011	0.004
21	Norrtälje	0.177	0.143	0.108	0.186	0.128	0.099
22	Sigtuna	0.054	0.058	0.057	0.030	0.036	0.021
23	Nynäshamn	0.096	0.082	0.052	0.100	0.064	0.044
24	Stockholm-west	0.009	0.006	0.006	0.008	0.003	0.002
25	Stockholm-south	0.009	0.006	0.011	0.005	0.005	0.002
26	Stockholm-inner	0.027	0.042	0.037	0.020	0.037	0.040
	Average	0.033	0.036	0.028	0.022	0.021	0.019

Industry		Worksites			Residence sites		
		1970	1980	1990	1970	1980	1990
1	Upplands-Väsby	0.127	0.114	0.108	0.048	0.009	0.009
2	Vallentuna	0.097	0.143	0.118	0.043	0.014	0.005
3	Österåker	0.209	0.159	0.113	0.025	0.016	0.013
4	Värmdö	0.325	0.178	0.176	0.152	0.044	0.023
5	Järfälla	0.173	0.117	0.091	0.008	0.007	0.014
6	Ekerö	0.128	0.207	0.102	0.025	0.011	0.004
7	Huddinge	0.074	0.170	0.250	0.009	0.005	0.006
8	Botkyrka & Salem	0.142	0.096	0.132	0.015	0.010	0.017
9	Haninge	0.115	0.178	0.121	0.021	0.019	0.016
10	Tyresö	0.082	0.027	0.063	0.003	0.002	0.004
11	Upplands-Bro	0.086	0.060	0.174	0.046	0.008	0.010
12	Täby	0.078	0.071	0.021	0.037	0.023	0.018
13	Danderyd	0.080	0.311	0.225	0.084	0.060	0.045
14	Sollentuna	0.024	0.031	0.022	0.013	0.005	0.004
15	Södertälje	0.109	0.135	0.158	0.098	0.097	0.079
16	Nacka	0.100	0.058	0.032	0.009	0.005	0.004
17	Sundbyberg	0.096	0.061	0.028	0.012	0.008	0.006
18	Solna	0.073	0.050	0.055	0.012	0.006	0.005
19	Lidingö	0.151	0.034	0.019	0.042	0.031	0.030
20	Vaxholm	0.205	0.434	0.315	0.235	0.125	0.064
21	Norrtälje	0.133	0.120	0.126	0.144	0.104	0.092
22	Sigtuna	0.097	0.208	0.388	0.075	0.056	0.049
23	Nynäshamn	0.081	0.208	0.174	0.157	0.119	0.077
24	Stockholm-west	0.034	0.044	0.050	0.001	0.001	0.003
25	Stockholm-south	0.021	0.044	0.040	0.004	0.004	0.002
26	Stockholm-inner	0.049	0.069	0.082	0.029	0.031	0.028
	Average	0.065	0.085	0.091	0.028	0.022	0.019

Education		Worksites			Residence sites		
		1970	1980	1990	1970	1980	1990
1	Upplands-Väsby	0.005	0.018	0.016	0.002	0.006	0.018
2	Vallentuna	0.011	0.016	0.020	0.003	0.000	0.005
3	Österåker	0.017	0.010	0.019	0.007	0.001	0.006
4	Värmdö	0.023	0.045	0.059	0.023	0.022	0.034
5	Järfälla	0.006	0.001	0.008	0.009	0.001	0.002
6	Ekerö	0.015	0.002	0.007	0.017	0.007	0.000
7	Huddinge	0.017	0.007	0.004	0.004	0.010	0.016
8	Botkyrka & Salem	0.001	0.005	0.008	0.003	0.018	0.034
9	Haninge	0.021	0.011	0.026	0.003	0.022	0.037
10	Tyresö	0.008	0.002	0.010	0.003	0.004	0.012
11	Upplands-Bro	0.017	0.026	0.015	0.007	0.015	0.034
12	Täby	0.005	0.000	0.000	0.046	0.036	0.036
13	Danderyd	0.001	0.011	0.026	0.120	0.159	0.161
14	Sollentuna	0.000	0.003	0.004	0.011	0.009	0.006
15	Södertälje	0.002	0.021	0.020	0.012	0.027	0.037
16	Nacka	0.009	0.002	0.001	0.006	0.006	0.006
17	Sundbyberg	0.023	0.019	0.010	0.019	0.020	0.011
18	Solna	0.006	0.004	0.008	0.001	0.000	0.002
19	Lidingö	0.001	0.006	0.012	0.080	0.069	0.062
20	Vaxholm	0.006	0.002	0.003	0.002	0.001	0.000
21	Norrtälje	0.045	0.058	0.076	0.065	0.055	0.068
22	Sigtuna	0.008	0.008	0.017	0.005	0.019	0.024
23	Nynäshamn	0.008	0.030	0.060	0.026	0.029	0.055
24	Stockholm-west	0.001	0.001	0.001	0.001	0.000	0.002
25	Stockholm-south	0.001	0.005	0.009	0.012	0.012	0.006
26	Stockholm-inner	0.004	0.008	0.012	0.004	0.042	0.068
	Average	0.006	0.009	0.013	0.014	0.021	0.028

10 Parameter Estimation for Combined Travel Choice Models

David E. Boyce
Department of Civil and Materials Engineering
University of Illinois at Chicago
842 W Taylor Street
Chicago, IL 60607, USA

Yu-Fang Zhang
New York Metropolitan Transportation Council
Suite 82 East
One World Trade Center
New York, NY 10048, USA

10.1 Introduction

Parameter estimation for individual travel choice models of the logit type has become a relatively routine exercise with the availability of modern statistical software (Ben-Akiva and Lerman, 1985; Oppenheim, 1995). For example, relatively complex mode choice models with nested submode models are now widely applied. Models which combine trip assignment with other steps of the travel forecasting procedure, however, remain an academic curiosity, despite their successful application to planning problems in Chicago and Stockholm (Boyce et al., 1992; Abrahamsson and Lundqvist, 1998). These combined models present estimation problems, moreover, which have not been addressed in the context of logit model estimation.

In estimating both individual and combined travel choice models, the analyst is presented with a dilemma. While surveys often include questions about travel times and costs of journeys taken, the responses are typically observed to cluster at intervals of 5, 10 or 15 minutes, suggesting that the responses are rounded. Responses for auto operating costs for the trip may simply be a rough estimate. Moreover, travel times and costs for alternative modes not used may be only a guess. Practitioners faced with this situation generally ignore these survey responses and use network-based travel times and costs for parameter estimation. The problem with this approach is that the auto travel times are necessarily based on an trip assignment model for the survey trip table, or some other, hopefully similar trip table. Since network travel times can vary substantially among assignments for various trip tables, an issue of the consistency between the survey data and the modal times and costs arises.

In our research on combined models of travel choice, we elected to explore an alternative approach. We conjectured that it would be possible to estimate the model parameters in a manner that is fully consistent with the network travel times and costs. We realized that this estimation problem can be formulated as a two-level optimization problem, but one that may be difficult to solve for a global optimum. Since little is known about the solution characteristics of this problem, we decided to explore its behavior computationally. In the course of this investigation, we also became interested in the issue of the functional form of the nested logit function representing origin-destination and mode choices (Lundqvist and Mattsson, 1992; Abrahamsson, 1996).

In its final form, this research examined these two questions:

1. How can the problem of estimating parameters of logit models be formulated and solved in the context of combined travel choice models?
2. How different are the parameter estimates for the alternative functional forms of the nested logit model?

In this chapter we seek to summarize our answers to these two questions. The chapter is organized in the following way. First, we consider the formulation of the parameter estimation problem in the context of a combined model stated as an optimization problem. Next, we describe our approach to solving the problem computationally, and present some typical results. Then, we consider three alternative functional forms for the nested O-D and mode choice problem, and present our estimation results for these forms. We conclude the chapter with a discussion of some policy implications of our findings. For related papers, see Boyce and Zhang (1993; 1997) and Zhang (1995).

10.2 Parameter Estimation Procedure

We begin with a simplified statement of our combined travel choice model. Then we state the bilevel estimation problem. Finally, we state a heuristic solution procedure and provide some illustrative results of its application.

10.2.1 A Simplified Combined Travel Choice Model

To motivate the estimation problem, we first state one version of the model we seek to estimate. This model combines three of the four steps of the conventional travel forecasting procedure: trip distribution, mode choice and trip assignment. For additional details, see Boyce et al. (1988) or Boyce and Daskin (1997).

The assumptions underlying the model are as follows:

1. In choosing routes, travelers minimize their generalized costs of travel, about which they have perfect information. For the auto network with its flow-dependent costs, this assumption results in user-optimal route choices: all used routes between each origin-destination (OD) pair have equal costs and no unused route has a lower cost. For the composite transit network with its fixed service schedule and fares, this assumption corresponds to a single, minimal-cost route choice for each OD pair (all-or-nothing assignment).
2. In choosing modes and destinations, given their trip origins, travelers' benefits and costs are not fully described by the generalized cost function, resulting in a randomly distributed perception error. The aggregate effect of these errors may be represented by an entropy function. Such a function has been shown to equal the direct utility of a representative traveler (Oppenheim, 1995). In addition, a constraint on the total flows arriving at each destination is also imposed to represent the attractiveness of destinations.

The simplified model may be formulated as follows. Alternative models are stated in Section 10.3.

$$\min_{h,p} Z(h,p) = \frac{1}{T} \sum_{a\in A} \int_0^{f_a} c_a(x)dx + \sum_{ij} c_{ijt}p_{ijt} + \frac{1}{\mu} \sum_{ijm} p_{ijm} \ln p_{ijm}$$

s.t.

$$\sum_{r \in R_{ij}} h_r = Tp_{ija}/\alpha, \quad \forall ij$$

$$\sum_{jm} p_{ijm} = o_i, \quad \forall i$$

$$\sum_{im} p_{ijm} = d_j, \quad \forall j$$

$$h_r \geq 0, \quad \forall r$$

where $f_a \equiv \sum_{ij} \sum_{r \in R_{ij}} h_r \delta_{ra}, \quad a \in A$

The notation is defined as follows:
T - total origin-destination flow during the peak period (person-trips/hour);
f_a - vehicle flow on link $a, a \in A$, the set of links in the highway network (vehicles/hour);
$c_a(f_a)$ - generalized travel cost for auto trips on link a, a weighted function of in-vehicle travel time, out-of-vehicle time and operating cost (generalized cost/vehicle-trip);
c_{ijt} - generalized travel cost for transit trips from zone i to zone j (generalized cost/person-trip);
p_{ijm} - proportion of total flow T traveling from zone i to zone j by mode m, $m = a$ (auto) or t (transit);
μ - deterrence parameter for destination-mode choice;
h_r - vehicle flow on route $r, r \in R_{ij}$, the set of routes from zone i to zone j (vehicles/hour);
α - auto occupancy (person-trips/vehicle-trip);
o_i - proportion of all trips T departing from zone i;
d_j - proportion of all trips T terminating at zone j;
δ_{ra} - 1, if link a belongs to route r; 0 otherwise.

By forming the Lagrangian and taking partial derivatives with respect to h_r, the following optimality conditions for auto route choice are obtained:

$$\sum_a c_a(f_a)\delta_{ra} - u_{ij} \geq 0, \quad \forall r \in R_{ij}, \quad \forall ij$$

$$h_r(\sum_a c_a(f_a)\delta_{ra} - u_{ij}) = 0, \quad \forall r \in R_{ij}, \quad \forall ij$$

$$\sum_{r \in R_{ij}} h_r - Tp_{ija}/\alpha = 0, \quad \forall ij$$

These conditions correspond to the route choice assumptions stated above, where u_{ij} is the minimum travel cost from zone i to zone j. Now, let $c_{ija} \equiv u_{ij}$. Then, the following destination-mode choice function may be derived as an optimality condition for p_{ijm}:

$$p_{ijm} = a_i o_i b_j d_j \exp(-\mu c_{ijm}), \quad \forall ijm$$

where

$$a_i = \sum_{jm} b_j d_j \exp(-\mu c_{ijm}), \forall i$$

$$b_j = \sum_{im} a_i o_i \exp(-\mu c_{ijm}), \forall j$$

These balancing factors assure that the origin and destination constraints are satisfied.

While it is possible to motivate the entropy function as a dispersion constraint (Erlander, 1977), the value of its associated Lagrange multiplier $(1/\mu)$ cannot be determined numerically because the entropy of the actual travel choice pattern is not observable. Therefore, we must estimate μ using statistical methods, given a sample of observed travel choices. In addition, the cost functions $c_a(f_a)$ and c_{ijt} include unknown coefficient values defined as follows:

$$c_a(f_a) = \gamma_1 t_a(f_a) + \gamma_2 w_a + \gamma_3 k_a(f_a)$$

$$c_{ijt} = \gamma_0 + \gamma_1 t_{ijt} + \gamma_2 w_{ijt} + \gamma_3 k_{ijt}$$

where
$t_a(f_a)$ - in-vehicle auto travel time on link a at flow f_a;
w_a - out-of-vehicle time on link a, if link a is an access/egress link;
$k_a(f_a)$ - auto operating cost on link a at flow f_a;
t_{ijt} - in-vehicle travel time by transit from i to j;
w_{ijt} - out-of-vehicle travel time by transit from i to j;
k_{ijt} - transit fare from i to j;
γ_0 - transit bias coefficient;
$\gamma_1, \gamma_2, \gamma_3$ - generalized cost coefficients representing the relative importance of the corresponding times and costs to the population of travelers.

Note that in the travel choice function, c_{ijm} appears as a product of μ:

$$\mu c_{ijm} = \mu(\gamma_0 \delta_m + \gamma_1 t_{ijm} + \gamma_2 w_{ijm} + \gamma_3 k_{ijm})$$

where $\delta_m = 1$, if $m = t$, and 0 otherwise. Hence, it is not possible to estimate μ and γ separately. Henceforth, we absorb μ into γ by setting $\mu = 1$.

10.2.2 The Estimation Problem

Our estimation problem, therefore, is to find the values of $\gamma = (\gamma_0, \gamma_1, \gamma_2, \gamma_3)$ for which the observed travel choices are most likely, given the hypothesized model. A standard procedure for solving this problem is the maximum likelihood method (Sen and Smith, 1995). Assuming that the travel choices are distributed according to the multinomial distribution, the problem of estimating γ by maximum likelihood may be stated as follows:

$$\max_{\gamma} \mathcal{L} = \frac{N!}{\prod_{ijm} N_{ijm}!} \prod_{ijm} p_{ijm}(\gamma)^{N_{ijm}}$$

where N_{ijm} is the number of trips observed to be made from zone i to zone j by mode m, $N = \sum_{ijm} N_{ijm}$, and $p_{ijm}(\gamma)$ is the destination-mode choice model stated earlier. According to the standard estimation procedure, we maximize the natural logarithm of this likelihood function.

$$\max_{\gamma} \ln \mathcal{L} = \sum_{ijm} N_{ijm} \ln p_{ijm}(\gamma) + \text{constants}$$

Assuming that the travel costs (c_{ijm}) are known and fixed, the optimality conditions for this problem are the following:

$$\gamma_0 : \quad \sum_{ijm} \delta_m N_{ijm} = N \sum_{ijm} \delta_m p_{ijm}(\gamma)$$

$$\gamma_1 : \quad \sum_{ijm} t_{ijm} N_{ijm} = N \sum_{ijm} t_{ijm} p_{ijm}(\gamma)$$

$$\gamma_2 : \quad \sum_{ijm} w_{ijm} N_{ijm} = N \sum_{ijm} w_{ijm} p_{ijm}(\gamma)$$

$$\gamma_3 : \quad \sum_{ijm} k_{ijm} N_{ijm} = N \sum_{ijm} k_{ijm} p_{ijm}(\gamma)$$

A variant of these conditions is the case where γ_n is mode-specific, or γ_{nm}. In this case we have the following general condition:

$$\gamma_{nm} : \sum_{ij} c_{ijm}^{(n)} N_{ijm} = N \sum_{ij} c_{ijm}^{(n)} p_{ijm}(\gamma), \quad m = (a, t)$$

where $c_{ijm}^{(n)}$ refers to $\delta_m, t_{ijm}, w_{ijm}$ and k_{ijm}. We also note that the balancing factors a_i and b_j are not considered in this estimation problem. Rather, we solve them as Lagrange multipliers using the standard iterative technique. The results do correspond to their maximum likelihood estimates, however.

Now we are ready to state the overall problem faced in implementing the model of Section 10.2.1. The travel costs for the auto mode, t_{ija} and k_{ija}, are unknown, since they are flow-dependent and not readily observable. In a more general sense, the travel costs for the transit mode are also not known, since transit headways ultimately depend on demand; however, we take these costs to be fixed at their scheduled values.

The problem we wish to solve, then, may be stated as follows:

$$U : \max_{\gamma} \ln \mathcal{L}(\gamma) = \sum_{ijm} N_{ijm} \ln p_{ijm}(\gamma)$$

where $p_{ijm}(\gamma)$ solves

$$L : \min_{h,p} Z(h, p) = C(h, p, \gamma) - S(p)$$

s.t. conservation of flow and non-negativity constraints

As indicated by the notation (U, L), this is a bilevel programming problem. The upper problem U is to find γ that maximizes a function of $p(\gamma)$, given the network travel costs. The lower problem L is to solve the combined travel choice problem, including the travel costs, given γ. $C(h, p, \gamma)$ represents the travel costs and $S(p)$ is the entropy function.

Such bilevel programming problems are known, in general, to be computationally hard. The maximization problem is not concave because of the implicit constraints of the lower problem; therefore, a unique maximum is not guaranteed. Rather, it is generally characterized by multiple local optima. Moreover, the upper objective function is known to be quite flat, so that changes in the value of γ may result in negligible changes in $\ln \mathcal{L}$. This might not be a problem were it not for the fact that the lower problem is not solved exactly with our solution algorithm, but only to some stated convergence level. Fortunately, a lower bound is available to monitor its convergence. The likelihood function may be observed to decrease slightly as the lower problem converges towards the optimum. Hence, the termination point for the lower problem may critically affect the resulting likelihood value, and at best the comparison of results for alternative solutions is problematic.

10.2.3 A Heuristic Solution Method

The problem which we faced was how to search for good values of the coefficient vector γ. Realizing that the objective function of the upper problem is an implicit function of the lower problem, we rejected gradient search procedures, perhaps inadvisedly. In an earlier effort (Lee, 1987), the search procedure of Hooke and Jeeves (1961) was applied. This method is appropriate if derivatives of the objective function with respect to the unknown coefficients are unavailable, which in principle is the case here. The method systematically explores the upper objective function, using the value of the log likelihood function to guide the search. For the reason just noted above, the function itself does not provide decisive guidance.

Learning from this experience, we sought to find an improved method. If the travel costs are held temporarily fixed, then the optimality conditions stated above are valid. In preliminary experiments, we observed that we could use these conditions to guide the search, as follows:

1. Given an initial solution of the coefficient vector γ, solve the lower problem to obtain (p^e_{ijm}) by applying an algorithm proposed by Evans (1976), and evaluate $\ln \mathcal{L}(\gamma)$;

2. Adjust the values of γ as follows:

   $\gamma'_n = \gamma_n(X^e_n / X^o_n)$

   where

   $$X^e_n = \sum_{ijm} p^e_{ijm} c^{(n)}_{ijm} \text{ (estimated mean)}$$

   $$X^o_n = \sum_{ijm} p^o_{ijm} c^{(n)}_{ijm} \text{ (observed mean)}$$

   In the case of the flow-dependent auto travel times and operating costs, t_{ija} and k_{ija}, the predicted values are used together with the observed trip proportions to calculate the observed costs X^o_n. p^o_{ijm} is defined as N_{ijm}/N.

3. Re-solve the lower problem for (p^e_{ijm}) and $\ln \mathcal{L}(p^e_{ijm})$. Check if $|\,(X^e_n/X^o_n) - 1\,| < \epsilon$ for all n, where $\epsilon = 0.005$, for example. If yes, stop. If no, return to step 2. (Note that when $X^e_n = X^o_n$, or $X^e_n/X^o_n = 1$, the optimality conditions for γ_n are satisfied.)

The solution to this problem is substantially more difficult than the usual likelihood estimation problem with fixed travel costs. In each iteration, the values of t_{ija} and k_{ija} change. If these values were fixed, the procedure is expected to converge, although such convergence was not actually tested. (Abrahamsson (1996) applied MINOS, an optimization software system, to solve a similar problem with the travel costs assumed to be fixed.)

In initial experiments with the above scheme, relatively good performance was observed for γ_1, γ_2 and γ_3, but not for γ_0, for the case of non-mode specific coefficients. As the values came close to convergence, however, the adjustment in step 2 tended to be too small. If the values were far away from convergence, the adjustment was too large. Therefore, a modified adjustment scheme was investigated, as follows:

1. If $\mid (X_n^e/X_n^o) - 1 \mid \leq 0.01$, then $\gamma_n' = \gamma_n (X_n^e/X_n^o)^2$;
2. If $0.01 < \mid (X_n^e/X_n^o) - 1 \mid \leq 0.1$, then $\gamma_n' = \gamma_n (X_n^e/X_n^o)$;
3. If $\mid (X_n^e/X_n^o) - 1 \mid > 0.1$, then $\gamma_n' = \gamma_n (X_n^e/X_n^o)^{0.5}$.

This scheme results in larger adjustments in the vicinity of the desired ratio ($< \pm 0.01$), and smaller adjustments for values far from the desired ratio ($> \pm 0.1$). This procedure definitely improved the convergence of the coefficient values.

Even with this improvement, some coefficients did not converge as well as we desired, especially for the mode-specific coefficients. Therefore, a smoothing procedure was applied using regression analysis. A multiple regression equation was fitted to each γ_n and the corresponding values of the ratios for all coefficients (X_l^e/X_l^o) obtained from a sequence of iterations of the estimation procedure:

$$\gamma_n = a_n + \sum_l b_{nl} \frac{X_l^e}{X_l^o}$$

where a_n and b_{nl} are regression coefficients for γ_n. One equation was fitted for each cost coefficient. By setting the values of the ratios (X_n^e/X_n^o) to 1.0, a smoothed value of each coefficient was calculated from the regression equation. These values were then re-inserted into the iterative estimation procedure in order to further refine them. An example of the results of 20 iterations of the estimation procedure is shown in Fig. 10.1 for the case where the coefficients are mode-specific. In the final results, most of the ratios lay in the range $0.995 < (X_n^e/X_n^o) < 1.005$. Many were closer to 1.0, but a few were just outside the desired limits.

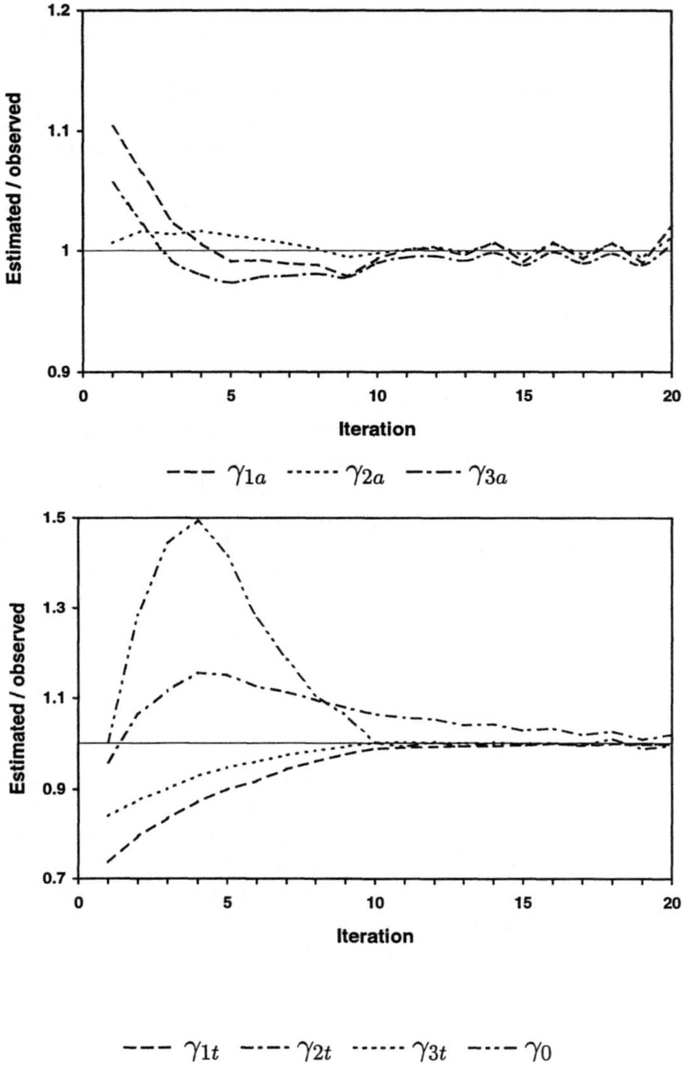

Fig. 10.1. Convergence of the estimation procedure

What can be concluded about the above procedure? We offer the following comments:

1. The procedure does converge approximately to one vector which maximizes the likelihood function consistent with the choice proportions and auto travel times and costs given by the combined model.
2. There may be additional vectors which satisfy the conditions. This issue was not addressed in our research.
3. The estimation procedure is computationally intensive. For each set of parameter estimates, the combined model was solved over 100 times, at 20 iterations per solution, or 1000s of Evans algorithm iterations per set of estimates. By using a more exact algorithm for the route choice (trip assignment) model, it may be possible to reduce the number of iterations, as well as improve on the convergence properties (Lundgren and Patriksson, 1996).
4. The observed travel choices used in these experiments were not a survey of travel choices. Rather, a predicted peak period trip matrix of auto and transit trips prepared by the Chicago Area Transportation Study for 1990 was used. For this matrix the population of trip makers T equals the sample size N. For this reason it is not possible to calculate the standard errors of the estimated coefficients, as is normally done.

10.3 Alternative Functional Forms

The combined model of origin-destination-mode and route choice formulated in Section 10.2.1 is one of three functional forms studied in our research. In this section the three forms are stated, and the estimation results are presented.

10.3.1 Nested Route Choices

The choice model of Section 10.2.1 assumes that choices of origin-destination and mode are made jointly. An alternative hypothesis is that the choices are conditional. In stating these functions, the destination is arbitrarily selected as one choice, and the origin is made conditional on it. We could have reversed origins and destinations and obtained the same results, since all of the models are origin and destination constrained.

The alternative functional forms may be stated as follows:

1. Mode choice conditional on origin-destination choice, referred to as OD/M:

   $$p_{ijm} = d_j p_{i|j} p_{m|ij}$$

2. Origin choice conditional on mode-destination choice, referred to as DM/O:

   $$p_{ijm} = d_j p_{m|j} p_{i|jm}$$

3. Joint origin-destination-mode choice model (Section 10.2.1), referred to as OD-M:

   $$p_{ijm} = d_j p_{im|j}$$

The functional forms of these models, and the optimality conditions for estimating their coefficients may be stated as follows.

OD/M:

$$p_{ijm} = a_i o_i b_j d_j \exp(-\eta \tilde{c}_{ij}) \frac{\exp(-\mu c_{ijm})}{\sum_m \exp(-\mu c_{ijm})} \tag{1}$$

where

$$a_i = \frac{1}{\sum_j b_j d_j \exp(-\eta \tilde{c}_{ij})}$$

$$b_j = \frac{1}{\sum_i a_i o_i \exp(-\eta \tilde{c}_{ij})}$$

$$\tilde{c}_{ij} = -\frac{1}{\mu} \ln \sum_m \exp(-\mu c_{ijm})$$

An alternative choice function corresponding to functional form (1) is:

$$p_{ijm} = d_j \frac{a_i o_i \exp(-\eta \tilde{c}_{ij})}{\sum_i a_i o_i \exp(-\eta \tilde{c}_{ij})} \frac{\exp(-\mu c_{ijm})}{\sum_m \exp(-\mu c_{ijm})}$$

For this combined model we assume that $0 \leq \eta/\mu \leq 1$, which implies that the dispersion of origin choices from the cost minimizing solution is greater than the corresponding dispersion of mode choices. The entropy term in the objective function is:

$$S(p) = -\frac{1}{\mu} \sum_{ij} p_{ij} \ln p_{ij} - \frac{1}{\eta} \sum_{ij} p_{ij} \sum_m p_{m|ij} \ln p_{m|ij}$$

The optimality conditions for the likelihood function are:

$$\gamma_{nm}: \quad \sum_{ij} c_{ijm}^{(n)} N_{ijm} = \sum_{ij} c_{ijm}^{(n)} [(1-\lambda) N_{ij} + \lambda N_j p_{i|j}] p_{m|ij}$$

$$\eta: \quad \sum_{ij} \tilde{c}_{ij} N_{ij} = \sum_{ij} \tilde{c}_{ij} N_j p_{i|j}$$

$$\mu: \quad \sum_{ijm} c_{ijm} N_{ijm} = \sum_{ijm} c_{ijm} [(1-\lambda) N_{ij} + \lambda N_j p_{i|j}] p_{m|ij}$$

where $\lambda = \eta/\mu$. Since μ is set equal to 1, the last condition is satisfied jointly with the condition on γ_{mn}.

DM/O:

$$p_{ijm} = a_i o_i b_j d_j \exp(-\eta c_{ijm}) \frac{\exp(-\mu \tilde{c}_{jm})}{\sum_m \exp(-\mu \tilde{c}_{jm})} \tag{2}$$

where

$$a_i = \frac{1}{\sum_{jm} b_j p_{jm} \exp(-\eta c_{ijm})}$$

$$b_j = \frac{1}{\sum_i a_i o_i \exp(-\eta c_{ijm})}$$

$$\tilde{c}_{jm} = -\frac{1}{\eta} \ln \sum_i a_i o_i \exp(-\eta c_{ijm})$$

$$p_{jm} = d_j p_{m|j}$$

An alternative choice function corresponding to functional form (2) is:

$$p_{ijm} = d_j \frac{\exp(-\mu \tilde{c}_{jm})}{\sum_m \exp(-\mu \tilde{c}_{jm})} \frac{a_i o_i \exp(-\eta c_{ijm})}{\sum_i a_i o_i \exp(-\eta c_{ijm})}$$

For this combined model we assume that $0 \leq \mu/\eta \leq 1$. The entropy term in the objective function is:

$$S(p) = -\frac{1}{\mu} \sum_{jm} p_{jm} \ln p_{jm} - \frac{1}{\eta} \sum_{jm} p_{jm} \sum_i p_{i|jm} \ln p_{i|jm}$$

The optimality conditions for the likelihood function are:

$$\begin{aligned}
\gamma_{nm}: &\quad \sum_{ij} c_{ijm}^{(n)} N_{ijm} = \sum_{ij} c_{ijm}^{(n)} [(1-\lambda')N_{jm} + \lambda' N_j p_{m|j}] p_{i|jm} \\
\mu: &\quad \sum_{jm} \tilde{c}_{jm} N_{jm} = \sum_{jm} \tilde{c}_{jm} N_j p_{m|j} \\
\eta: &\quad \sum_{ijm} c_{ijm} N_{ijm} = \sum_{ijm} c_{ijm} [(1-\lambda')N_{jm} + \lambda' N_j p_{m|j}] p_{i|jm}
\end{aligned}$$

where $\lambda' = \mu/\eta$. In this case η is set equal to 1, so the last condition is satisfied jointly with the condition on γ_{mn}. To facilitate comparisons, we re-state the conditions for model D-M.

OD-M:

$$p_{ijm} = a_i o_i b_j d_j \frac{\exp(-\mu c_{ijm})}{\sum_m \exp(-\mu c_{ijm})} \tag{3}$$

where

$$a_i = \frac{1}{\sum_{jm} b_j d_j \exp(-\mu c_{ijm})}$$

$$b_j = \frac{1}{\sum_{im} a_i o_i \exp(-\mu c_{ijm})}$$

An alternative choice function corresponding to functional form (3) is:

$$p_{ijm} = d_j \frac{a_i o_i \exp(-\mu c_{ijm})}{\sum_{im} a_i o_i \exp(-\mu c_{ijm})}$$

For this combined model, the entropy term in the objective function is:

$$S(p) = -\frac{1}{\mu} \sum_{ijm} p_{ijm} \ln p_{ijm}$$

The optimality conditions for the likelihood function are:

$$\gamma_{nm}: \quad \sum_{ij} c_{ijm}^{(n)} N_{ijm} = \sum_{ij} c_{ijm}^{(n)} N_j p_{im|j}$$

$$\mu: \quad \sum_{ijm} c_{ijm} N_{ijm} = \sum_{ijm} c_{ijm} N_j p_{im|j}$$

Since μ is set equal to 1, the last condition is satisfied jointly with the condition on γ_{mn}.

10.3.2 Estimation Results

The three functional forms proposed above were estimated for two cases: equal coefficient values across the auto and transit modes, and mode-specific coefficient values. The first case was found to perform poorly in predicting mean modal travel times and costs. Therefore, we focus our description of findings on the mode-specific case. As indicated above, the coefficients of each model were estimated with the same convergence criterion. However, the actual convergence achieved varies slightly from case to case, as shown by the overall goodness-of-fit measure ρ^2.

The results of the estimation of four models are summarized in Table 10.1. The top half of the table gives the estimated coefficient values for four models: OD-M(0), common values across modes, OD-M(1), mode-specific values, and OD/M and DM/O , as defined in the Section 10.3.1. Values are reported for four cost coefficients and two deterrence parameters. As indicated previously, the deterrence parameter for the conditional choice is set to 1; the remaining parameter is then equal to λ in the case of model OD/M , or λ' in the case of model DM/O .

The lower half of the table gives various evaluation measures. Most are self-explanatory. Some of these regional means correspond directly to cost coefficients: regional transit share (γ_0); in-vehicle travel time (γ_1); out-of-vehicle travel time (γ_2); and auto operating cost/transit fare (γ_3). Other measures are less directly related to cost coefficients: transit share to the CBD, the most important destination served by transit; travel distance, which partly determines auto operating cost; and space-mean-speed, which is the ratio of travel distance to travel time.

As noted previously, the "observed" values of auto travel time and auto operating cost are based on the interzonal travel times predicted by the model, since no observations on these variables exist. The values shown in the table are computed from the predicted travel times for model OD-M(1), using the observed interzonal trip proportions.

The final evaluation measure is the goodness-of-fit statistic ρ^2, which is defined as follows:

$$\rho^2 = \frac{L_M - L_N}{L_P - L_N}$$

Table 10.1. Estimated coefficient values and evaluation measures

Choice Functions		OD-M(0)	OD-M(1)	OD/M	DM/O
Cost Coefficients		Estimated Values			
Transit bias		0.7756	4.9015	4.4818	4.2716
In-vehicle time		0.0680	-	-	-
auto		-	0.0900	0.0940	0.0974
transit		-	0.0308	0.0329	0.0309
Out-of-vehicle time		0.0119	-	-	-
auto		-	0.2530	0.2278	0.2698
transit		-	0.0052	0.0083	0.0011
Operating cost/fare		0.0203	-	-	-
auto		-	0.0412	0.0127	0.0105
transit		-	0.0081	0.0085	0.0078
Dispersion Parameters					
mode (μ)		1.0	1.0	1.0	1.0755
O-D (η)		-	-	0.9942	1.0
Evaluation Measures	Observed	Predicted			
Regional transit share - %	15.97	15.96	15.99	16.06	16.07
Transit share to CBD - %	63.20	65.62	61.69	61.15	60.65
Mean in-vehicle time (min)					
auto	17.20	21.50	17.18	17.00	17.00
transit	33.54	24.82	33.54	33.38	33.53
Mean out-vehicle time (min)					
auto	4.84	4.93	4.86	4.86	4.84
transit	10.87	10.40	10.84	10.87	10.80
Mean travel distance (km)					
auto	13.07	14.52	13.34	13.28	13.27
transit	20.94	15.82	22.58	22.48	22.51
Space-mean-speed (km/h)					
auto	35.58	32.96	36.32	36.45	36.46
transit	28.29	26.95	30.53	30.48	30.47
Operating cost/fare (cents)					
auto	58.14	68.51	58.09	57.81	57.87
transit	94.33	79.36	94.24	94.27	94.20
Goodness-of-fit (ρ^2)	-	85.74	89.37	89.31	89.33

Note: The observed values of mean auto in-vehicle travel time and mean operating cost/fare shown in this table are calculated from the predicted O-D auto in-vehicle travel times for choice function OD-M(1).

where

$$L_M = \sum_{ijm} N_{ijm} \ln p_{ijm} \text{ (log likelihood value for the model)}$$

$$L_N = \sum_{ijm} N_{ijm} \ln o_i d_j \bar{p}_m \text{ (log likelihood value for the null hypothesis)}$$

where $\bar{p}_m =$ the observed regional share for mode m

$$L_P = \sum_{ijm} N_{ijm} \ln \frac{N_{ijm}}{N} \text{ (log likelihood value for the model that predicts perfectly)}$$

In our computation of ρ^2, we set $\bar{p}_m$ equal to the observed modal share, not to 0.5, as is sometimes done for the case of two modes. Hence, our values of ρ^2 are more conservative than if we had used $p_m = 0.5$.

In examining the results, we first consider model OD-M(0), shown on the left side of the table. In this model, the cost coefficients are estimated across modes. While this model fits the observed means across modes very well, it fits the modal means rather poorly. In particular, see the predicted vs. observed means for in-vehicle travel time. For this reason, this model was rejected, despite its parsimony. Nevertheless, the value of ρ^2 for this model is quite high, which serves as a warning concerning the utility of this statistic.

The three models with mode-specific coefficients are shown next. Despite their different functional forms, these three models have very similar coefficient values and values of ρ^2. Note that the regional transit shares for models OD/M and DM/O are not predicted as well as for model OD-M. This difference probably explains the slightly lower values of ρ^2 for the two nested models.

The ratio $\lambda = \eta/\mu$ for model OD/M is effectively 1.0. In this case, this model simplifies to model OD-M, as can be inferred from its functional form and from its optimality conditions for the estimates. In contrast, the ratio $\lambda' = \mu/\eta$ for model DM/O equals 1.0755. Since for model DM/O , we assumed that $0 \leq \lambda' \leq 1$, the estimates contradict this assumption. Therefore, we must reject this model. Research based on Stockholm data, however, has found that model OD/M is rejected in favor of model DM/O (Abrahamsson, 1996; Lundqvist and Mattsson, 1992). Therefore, this "reverse" model form should continue to be considered in future studies.

The estimates for each of the models indicate problems of multicollinearity between the bias coefficient and the transit time and fare coefficients. In effect, transit share is being largely predicted by a rather large bias coefficient. It is unclear whether this is a reflection of the quality of the data or the model specification.

10.4 Policy Implications

The model that is addressed in this chapter actually represents a family of models of urban location and travel choice. It is closely related to the model described by Anderstig and Mattsson (1998). In this section we discuss some broader implications of advances in this class of models for network infrastructure and environmental planning.

Decisions about future investments in network infrastructure increasingly require improved predictions about the economic, social and environmental consequences of those investments. Although it has been the ambition of regional transportation planners since the 1950s to make such long term predictions, only in recent years has their possibility become a reality. Two reasons may be identified for this advance.

First, the understanding of the formulation and solution of network-based models has improved very substantially during the past 25 years. Most of these improvements have emerged from academic research, sometimes undertaken in collaboration with transportation planning agencies, especially with regard to the provision of network and socio-economic data.

Second, huge advances in computing technology, especially since the early 1980s, have allowed network modelers to solve models in more detail and for more comprehensive systems than previously contemplated. Likewise, advances in software systems for transportation planning have brought some of these advances into use in planning agencies.

Despite these advances, transportation planning practice is severely constrained at present by two factors. First, most planning agencies rely on consultants to provide their transportation models. These consultants largely base their recommendations on what has worked for them in the past, and not on what is known today. They represent a serious impediment to progress in the field. Second, software developers mainly respond to the demands of planning agencies, as defined by the agencies themselves and their consultants, and not as defined by what constitutes best, or even good, practice. Were it not for the environmental lobby, and the requirements for model update programs in recent legislation, the situation would be even worse, as it was during the 1980s.

Now, we are facing a situation where the research state of the art is very far ahead of the state of practice, and the gap is widening. Conventional graduate degree programs and short courses alone will not narrow this gap, although every well-trained person entering the field potentially helps. More concerted efforts are necessary.

The state of best practice needs to be defined, and computer-based handbooks and software prepared to support its implementation. This statement needs to be broad-based and comprehensive, implying that a substantial undertaking is needed. Modern tools such as compact disks and the Internet need to be used in its distribution.

Case studies illustrating best practice need to be performed for dissemination to training programs and planning agencies. These should include software and data for solving problems by the reader. A third idea is to offer an annual award for the best planning model implementation and application so that motivated professionals can have a forum for showing what they can do.

Finally, despite ongoing advances in the state of the art, more research will yield even better and more useful methods. Ten major problems could easily be identified for substantial, ongoing research projects. These projects should be targeted to solving specific problems that are evident in planning agencies, as well as innovations being proposed, but not yet explored, by academic researchers. In the United States, there is presently no national funding program for such research.

10.5 Conclusions

The findings reported in this chapter represent one small advance in an ongoing effort to modernize regional transportation planning methods. Gradually, the idea of solving transportation planning models in an integrated or simultaneous way, rather than the obsolete four-step travel forecasting procedure, is coming to the forefront. Even so, modern software for solving such models by planning agencies remains unavailable.

This computational approach followed in this research was exploratory. As a result of these studies, discussions with mathematicians have identified alternative, potentially more powerful approaches. Additional research is required to implement and evaluate these methods. More advanced mathematical training of graduate students in transportation engineering is also warranted.

Future research will also benefit from improved network and travel choice data. For example, at present no network data set exists on route choices of urban travelers. While excellent route choice models have been formulated and solved, no data set is available for estimating their parameters, and evaluating their adequacy, as compared with the present standard approach which assumes that drivers have perfect information about their route choices. Data on departure time choice is also urgently needed to estimate models needed to test the benefits and costs of peak-period pricing. Additional data sets pertaining to residential and employment choices of multi-worker households are also critically needed. Most urban location models still assume that each household has only one worker, at best an embarrassment.

In summary, much remains to be done. The need is clearly evident. Better ways will be found. The question is whether they will be used sooner or later. If better decisions are the goal, the sooner the better.

Acknowledgements

The preparation of this chapter was undertaken while the first author was a visiting professor at the Royal Institute of Technology and Linkoping University in Sweden. He gratefully acknowledges the support of both institutions and fruitful discussions with colleagues there. The assistance of Ms. Xin Tian in preparing the camera-ready copy and solving numerous technical issues is appreciated.

Financial support of this research by the National Science Foundation through grants to the National Institute of Statistical Sciences and the National Center for Supercomputing Applications, University of Illinois at Urbana-Champaign, is also appreciated. The support of the Illinois Department of Transportation and the Federal Highway Administration through various projects is also gratefully acknowledged.

References

Abrahamsson, T. (1996), *Network Equilibrium Approaches to Urban Transportation Markets*, Ph.D. thesis, Department of Infrastructure and Planning, TRITA-IP FR 96-13, Royal Institute of Technology, Stockholm.

Abrahamsson, T. and Lundqvist, L. (1998), 'Formulation and estimation of combined network equilibrium models with applications to Stockholm', *Transportation Science*, in press.

Anderstig, C. and Mattsson, L.-G. (1998), 'Modelling land-use and transport interaction: Policy analyses using the IMREL model', Chapter 17, this volume.

Ben-Akiva, M. and Lerman, S. (1985), *Discrete Choice Analysis*, MIT Press, Cambridge, MA.

Boyce, D.E. and Daskin M.S. (1997), 'Urban transportation', in ReVelle, C. and McGarity, A. (eds), *Design and Operation of Civil and Environmental Engineering Systems*, John Wiley & Sons, New York, NY.

Boyce, D.E., LeBlanc L.J. and Chon, K.S. (1988), 'Network equilibrium models of urban location and travel choices: A retrospective survey', *Journal of Regional Science*, vol. 28, pp. 159-183.

Boyce, D.E. and Zhang, Y. (1993), 'Estimation and testing of a combined network equilibrium model of travel choice for the Chicago region', in Kohno, H. and Nijkamp, P. (eds), *Potentials and Bottlenecks in Spatial Development*, Springer-Verlag, Heidelberg, pp. 45–69.

Boyce, D.E. and Zhang, Y. (1997), 'Calibrating a combined model of trip distribution, modal split and assignment', *Transportation Research Record*, in press.

Boyce, D.E., Tatineni, M. and Zhang, Y. (1992), *Scenario Analyses for the Chicago Region with a Sketch Planning Model of Origin-Destination, Mode and Route Choice*, Final Report to the Illinois Department of Transportation, Springfield, IL.

Erlander, S. (1977), 'Accessibility, entropy and the distribution and assignment of traffic', *Transportation Research*, vol. 11, pp. 149-153.

Evans, S.P. (1976) 'Derivation and analysis of some models for combining trip distribution and assignment,' *Transportation Research*, vol. 10, pp. 37–57.

Hooke, R. and Jeeves, T.A. (1961), 'Direct search solution of numerical and statistical problems', *Journal of the Association of Computing Machinery*, vol. 8, pp. 212–229.

Lee, C.-K. (1987), *Implementation and Evaluation of Network Equilibrium Models of Urban Residential Location and Travel Choices*, Ph.D. thesis in civil engineering, University of Illinois, Urbana-Champaign, IL.

Lundgren, J. and Patriksson, M. (1996), 'An algorithm for the combined distribution and assignment model', presented at the EURO96 Working Group on Transportation, University of Newcastle.

Lundqvist, L. and Mattsson, L.-G. (1992) 'Modelling travel demand in an extended metropolitan region', paper presented at the 32nd European Congress of the Regional Science Association International, Brussels.

Oppenheim, N. (1995), *Urban Travel Demand Modeling*, John Wiley & Sons, New York, NY.

Sen, A. and Smith, T.E. (1995), *Gravity Models of Spatial Interaction Behavior*, Springer-Verlag, Heidelberg.

Zhang, Y. (1995), *Parameter Estimation for Combined Models of Urban Travel Choices Consistent with Equilibrium Travel Costs*, Ph.D. thesis in civil engineering, University of Illinois, Chicago, IL.

11 Discrete Spatial Price Equilibrium

Sven Erlander
Department of Mathematics
Linköping University
S-581 83 Linköping
Sweden

Jan T. Lundgren
Department of Mathematics
Linköping University
S-581 83 Linköping
Sweden

11.1 Introduction

In this paper we consider how the flow of a single commodity is distributed between regional markets in a transportation network. We assume that there is an unlimited number of price-taking 'shippers' who can enter the market and ship a single unit of commodity between some supply and demand points whenever there is a profit to be made. A common assumption made for this problem is that the shippers in the system behave rationally. In this case, a shipment of the commodity will usually take place only if the procurement costs at the supply market plus the transportation costs are less than, or equal, to the price obtained for the commodity at the demand market. If perfect competition prevails, no shipments will be made if procurement costs plus transportation costs are greater then the price obtained at the demand market. These are the equilibrium conditions defining a *spatial price equilibrium* model. The classical way to derive the conditions, according to Samuelson (1952) and Takayama and Judge (1971), is to formulate a mathematical program, and to obtain the equilibrium conditions as the optimal conditions of the mathematical program.

The classical model is a continuous model, i.e., the flow of the commodity can take any feasible real value. In this paper we will treat the shipments of the commodity as discrete entities. Thus, the flows will be integer valued and we consider a *discrete version* of the spatial price equilibrium model. It is common in transportation and regional science to define network flows using continuous real variables. From a practical point of view, the difference between the discrete and continuous cases is very small due to the usually very large number of shipments. However, from a theoretical point of view, it is of value to maintain the integer character of the flows when possible.

The integer approach allows us to make an alternative derivation of the equilibrium conditions and to interpret the model in economic terms. The justification for the classical formulation is that the optimal conditions for the optimization problem are the equilibrium conditions - the equilibrium conditions are derived from the mathematical program by solving for an optimal solution. However, the mathematical program itself has no economic interpretation. In our approach, as we shall see, there is such an economic interpretation.

A spatial price equilibrium is a situation such that no shipper can make a profit by changing his shipment plan (defined by the choice of origin, destination and route for a unit of the commodity). This idea assumes that shippers behave rationally and each shipper has complete knowledge not only of the different price schedules but also of the actions taken by the other shippers. The assumption of complete knowledge is however rather unrealistic. Instead, we introduce a probabilistic approach to derive the equilibrium conditions. We define the collection of shipment plans for all shippers as a shipment pattern. Our main result is that the *most probable shipment patterns are discrete spatial price equilibria* (Theorem 6). Thus, our approach preserves the rationality of the shippers in the more realistic framework of a probabilistic model that does not presuppose complete knowledge by the shippers.

By relaxing the integral conditions of our discrete model, we obtain the classical continuous optimization formulation (Theorem 10). Hence, the classical formulation can be viewed as an approximation to our formulation. In this sense, there is an economic interpretation of the classical optimization formulation.

Using the probabilistic approach, we also derive (in Theorem 9), an expression for the probability of a trade pattern (or, in terms of the words used in the continuous version, a flow pattern). By maximizing this expression we obtain the *most probable trade pattern*. This model can also be approximated by a continuous model. In this way we obtain an entirely new continuous model - *the continuous dispersed spatial price equilibrium model* (Theorem 11).

The outline of the paper is as follows: In Section 11.2 we introduce the basic concepts, including the network structure and the cumulative cost function, and we present the definition and derivation of the discrete spatial price equilibrium conditions. In the following two sections, we derive the most probable shipment pattern and the most probable trade pattern (Section 11.3), and in addition, develop the corresponding continuous approximation models (Section 11.4). The content in Sections 11.2.1-11.2.3 is mainly based on Smith (1983), and the "active shipper interpretation" in Section 11.2.1 is according to Smith (1993).

11.2 Discrete Spatial Price Equilibrium

We assume that the shipment pattern is a function of a large number of decisions made about the commodity, and we treat the shippers as the active, decision-taking, participants in the system. The decision for each unit of the commodity, the shipment plan, will involve the choice of a supply market at which the unit will be procured, a route in the transportation network on which the unit will be shipped, and a demand market to which the unit will be delivered. A basic decision will also be whether it is beneficial to ship the unit or not. Hence, the total number of shipments in the system is not given. The shipment plans for all shippers form the shipment pattern. The decisions are made depending on prices given from the passive participants in the system - the "suppliers", "purchasers" and "carriers". At each supply market a supplier will announce the minimum unit price at which each given (integer) quantity of the commodity will be supplied. Similarly, at each demand market a purchaser will announce the maximum price at which each given quantity of the commodity will be purchased. Finally, for each transportation link a carrier will announce the minimum unit price at which each given quantity of the commodity will be shipped. Given all the price schedules (price functions), the profits for each shipment plan, and for the whole shipment pattern, can be calculated. We will have an equilibrium if and only if, for each active shipper, the profit is non-negative and no other plan yields a higher profit. In addition, at equilibrium, no inactive shipper will be motivated to enter the market since no positive profit can be earned. In this way we define the *discrete spatial price equilibrium conditions*. Note that, due to the integer requirements, the profits will not be exactly zero for all active shippers.

Given the shipment pattern we can define a *cumulative cost function*, which is the negative of the cumulative profit. This function corresponds to the objective function (the sum of integrals) used in the classical derivation of the continuous spatial price equilibrium model. Minimizing the cumulative cost function gives a discrete spatial price equilibrium (Theorem 2). This is a new result, but in no way a trivial result. Rather, it permits an interesting, new interpretation of the classical formulation.

We start this section with some basic concepts and definitions, and then we define the discrete spatial price equilibrium conditions. We introduce the cumulative cost function and show that the equilibrium conditions can be derived by minimizing this function. This minimization also allows us to make a behavioral interpretation of the conditions. Finally, we give conditions for the existence of equilibria.

11.2.1 Basic Concepts and Definitions

Consider a graph consisting of a set of nodes and a set of directed links $l \in L$ connecting the nodes. The set of nodes will contain a set of supply nodes $i = 1, \ldots, I$ representing the supply points (markets) for the commodity of interest, a set of demand nodes $j = 1, \ldots, J$, representing the demand points (markets), as well as intermediate nodes. Each supply point i is characterized by a 'supplier' whose role is to announce a *supply price schedule*, $S_i : \mathrm{Z}_+ \to \mathrm{R}_+$, with values $S_i(A_i)$ denoting the minimum unit price at which A_i units of the commodity will be supplied at i. Similarly, each demand point j is characterized by a 'purchaser' whose role is to announce a *demand price schedule*, $D_j : \mathrm{Z}_+ \to \mathrm{R}_+$, with values $D_j(B_j)$ denoting the maximum price at which B_j units of the commodity will be purchased at j. Finally, each transportation link l is characterized by a 'carrier' whose role is to announce a *shipment price schedule*, $c_l : \mathrm{Z}_+ \to \mathrm{R}_+$, with values $c_l(v_l)$ denoting the minimum unit price at which v_l units of the commodity will be shipped on link l.

Remark. We use the letters S_i and D_j to denote price schedules, not supply and demand.

The supply and demand price schedules $S_i(A_i)$ and $D_j(B_j)$ correspond to the inverse supply and demand functions in the classical continuous case. In the continuous case the functions are assumed to be strictly monotone. However, in the present discrete case there is no need to assume strict monotonicity. Instead we assume that $S_i(A_i)$ is a nondecreasing function of A_i, and that $D_j(B_j)$ is a nonincreasing function of B_j (in fact we need only require $S_i(A_i)$ and $D_j(B_j)$ to be monotone upper-semicontinuous functions). The shipment-price schedules $c_l(v_l)$ are assumed to be nondecreasing functions for each $l \in L$.

The shippers are treated as the active participants in the system, and suppliers, purchasers and carriers are treated as passive participants. Each shipper t is required to announce his *shipment plan*, $r_t = (i, j, l_1, \ldots, l_m)$, which consists of a unit-commodity order at some supply point i, a unit-commodity offer at some demand point j, and a route (loop-free sequence of links) choice $(l_1, \ldots, l_m)$, for shipment between i and j. Note that i is uniquely defined by l_1 and j is defined by l_m, and hence that the shipment plan r for shipper t is representable by the *route* $r_t = (l_1, \ldots, l_m)$. Let the set of all possible routes (shipment plans) be denoted by R. The list of announced shipment plans, $\mathbf{u} = (r_1, \ldots, r_T)$, by some finite set of shippers, $t = l, \ldots, T$, is called a *shipment pattern*, and an *equilibrium* is then taken to be a situation in which no active shipper t is motivated to change his current announcement, and in which no inactive shippers are motivated to enter the market. In our definitions we have assumed that each shipper decides to ship exactly one unit of commodity. Of course, we can allow the shippers to decide to ship several units each, the important point is that there is one decision for each unit of commodity.

11.2.2 Definition of Discrete Spatial Price Equilibrium

To formalize the equilibrium concept, observe first that if the announced actions of all shippers are summarized by $A_i(\mathbf{u})$ orders at each supply point $i, i = 1, \ldots, I$, $B_j(\mathbf{u})$ offers at each demand point $j, j = 1, \ldots, J$, and $v_l(\mathbf{u})$ units of shipment flow on each link $l, l \in L$, in the network, then the *profits* for each plan, $r_t = (i, j, l_1, \ldots, l_m) \in \mathbf{u}$, $t = 1, \ldots, T$, are given by

$$\pi(r_t \mid \mathbf{u}) = D_j(B_j(\mathbf{u})) - S_i(A_i(\mathbf{u})) - \sum_{l \in r_t} c_l(v_l(\mathbf{u})).$$

Also, let $\mathbf{u}_t$ denote the list $(r_1, \ldots, r_{t-1}, r_{t+1}, \ldots, r_T)$ obtained from $\mathbf{u}$ by deleting the t-th plan.

The following definition is an immediate extension to our case of the definition of equilibrium in integers introduced by Rosenthal (1973).

Definition. The shipment pattern $\bar{\mathbf{u}} = (\bar{r}_1, \ldots, \bar{r}_T)$ is a *discrete spatial price equilibrium* at T if and only if

$$\pi(\bar{r}_t \mid \bar{\mathbf{u}}) = \max_{r \in R} \pi[r \mid (r, \bar{\mathbf{u}}_t)] \geq 0 \text{ for all } t = 1, \ldots, T, \text{ and}$$

$$\pi[r_{T+1} \mid (r_{T+1}, \bar{\mathbf{u}})] \leq 0 \text{ for all } r_{T+1} \in R.$$

The first set of inequalities says that each active shipper t, with plan $r_t \in \mathbf{u} = (r_t, \mathbf{u}_t)$, will be in equilibrium if and only if his profits are non-negative and no other plan yields higher profits. Similarly, the second set of inequalities says that each inactive shipper will be in equilibrium (i.e., will not be motivated to enter the market) if and only if there are no additional positive profits to be earned, i.e., each additional shipment plan, $r_{T+1} \in R$, has a non-negative profit.

The definition is the discrete version of the classical spatial price equilibrium conditions (Samuelson, 1952; Takayama and Judge, 1971). The discrete version differs from the continuous formulation in that the profits for the active shippers are not required to be exactly equal to zero as in the continuous model.

Alternatively, we can define spatial price equilibrium in terms of the *net costs* for the shippers. The net costs for each shipment plan $r_t = (i, j, l_1, \ldots, l_m) \in \mathbf{u}$, $t = 1, \ldots, T$, which are the negative of the profits, are given by

$$c(r_t \mid \mathbf{u}) = S_i(A_i(\mathbf{u})) + \sum_{l \in r_t} c_l(v_l(\mathbf{u})) - D_j(B_j(\mathbf{u})).$$

Definition. The shipment pattern $\bar{\mathbf{u}} = (\bar{r}_1, \ldots, \bar{r}_T)$ is a *discrete spatial price equilibrium* at T if and only if,

$$c(\bar{r}_t \mid \bar{\mathbf{u}}) = \min_{r \in R} c[r \mid (r, \bar{\mathbf{u}}_t)] \leq 0 \text{ for all } t = 1, \ldots, T, \text{ and}$$

$$c[r_{T+1} \mid r_{T+1}, \bar{\mathbf{u}}] \geq 0 \text{ for all } r_{T+1} \in R.$$

The definition of equilibrium in terms of costs instead of profits has historic reasons. The mathematical program used to derive the continuous equilibrium conditions is usually formulated in terms of cost. Also, as will be seen in the following section, a cumulative cost function is used in the derivation of the equilibrium conditions.

11.2.3 Derivation of Discrete Spatial Price Equilibrium

We start this section by defining a cumulative cost function, which will be the main tool for deriving and interpreting the equilibrium conditions. We will make use of the separability properties of both route costs and the 'inverse' supply and demand schedules and establish the fundamental *exchangeability property* in Theorem 1. Our main result will be Theorem 2 showing that a discrete spatial price equilibrium is obtained if we minimize the cumulative cost function.

Let $\mathbf{u}(t) = (r_1, \ldots, r_t)$ denote the *partial shipment pattern*, consisting of the announced shipment plans for shippers $1, \ldots, t$. We can then define the *net cost* $c(r_t \mid r_1, \ldots, r_{t-1})$ for shipper t by

$$c(r_t \mid r_1, \ldots, r_{t-1}) = S_i(A_i(\mathbf{u}(t)) + \sum_{l \in r_t} c_l(v_l(\mathbf{u}(t))) - D_j(B_j(\mathbf{u}(t))).$$

Clearly, this would be the negative of the profit for shipper t, if there were no more than t shipments in the network. Due to the assumed form of the functions defining net cost, it is easy to verify that net cost is a nondecreasing function of t.

Furthermore, let $c(\mathbf{u}(t))$ denote the *cumulative cost function* for the partial shipment pattern $\mathbf{u}(t)$:

$$c(\mathbf{u}(t)) = c(r_1, \ldots, r_t) = \sum_{k=1}^{t} c(r_k \mid r_1, \ldots, r_{k-1}).$$

In particular, for $t = T$, we obtain the cumulative cost function $c(\mathbf{u})$ for the total shipment pattern $\mathbf{u} = \mathbf{u}(T)$,

$$c(\mathbf{u}) = c(\mathbf{u}(T)) = c(r_1, \ldots, r_T) = \sum_{t=1}^{T} c(r_t \mid r_1, \ldots, r_{t-1}).$$

One interesting property of the cumulative cost function is the following: if one shipment is added, the cost incurred by the new shipment, i.e., the *incremental cost*, can be expressed

$$c(r_1, \ldots, r_t) - c(r_1, \ldots, r_{t-1}) = c(r_t \mid r_1, \ldots, r_{t-1}),$$

which is exactly the incremental net cost discussed above.

The order of the shipments in the shipment pattern **u** is not relevant - the individual shipments r_t are *exchangeable* - as will be seen from Theorem 1 and its corollary. This follows from the fact that the cumulative cost function $c(\mathbf{u})$ is *separable* in all functions A_i, B_j and c_l.

Theorem l. *Exchangeability property.* The cumulative cost function can be written

$$c(\mathbf{u}) = \sum_{i=1}^{I} \sum_{h=1}^{A_i(\mathbf{u})} S_i(h) + \sum_{l\in L} \sum_{h=1}^{v_l(\mathbf{u})} c_l(h) - \sum_{j=1}^{J} \sum_{h=1}^{B_j(\mathbf{u})} D_j(h).$$

Proof. See Erlander and Lundgren (1992).

Remark. The right hand side of the expression in Theorem 1 corresponds to the objective function used by Rosenthal (1973) for the integer network problem. The fact that the cumulative cost function can be written in this form follows from the separability of the cost elements and was first shown by Smith (1983) for the fixed demand trip assignment problem in a transportation network. Results by Bernstein and Smith (1994) indicate that the separability condition can be lifted only at the price of very restrictive assumptions.

The following corollary is an immediate consequence of Theorem 1.

Corollary. Let $\mathbf{u}' = (r'_1, \ldots, r'_T)$ be any permutation of $\mathbf{u} = (r_1, \ldots, r_T)$. Then the cumulative cost functions

$$c(\mathbf{u}) = c(r_1, \ldots, r_T) = \sum_{t=1}^{T} c(r_t \mid r_1, \ldots, r_{t-1}), \text{ and}$$

$$c(\mathbf{u}') = c(r'_1, \ldots, r'_T) = \sum_{t=1}^{T} c(r'_t \mid r'_1, \ldots, r'_{t-1}), \text{ are equivalent, i.e., } c(\mathbf{u}) = c(\mathbf{u}').$$

From the Corollary it is clear that the cumulative cost function $c(\mathbf{u})$ can always be written in such a way that the last term represents the cost incurred by the last added shipment irrespective of which this may be. This fact will be of great importance when we go further and derive the equilibrium conditions by minimizing the cumulative cost function.

Let U_T denote the set of all possible shipment patterns for a given number of shipments T, and let $c_T(\mathbf{u}(T))$ denote the value of the cumulative cost function when there are T shipments to be made. Let, furthermore, $\bar{\mathbf{u}}(T) = (\bar{r}_1(T), \ldots, \bar{r}_T(T))$ denote an optimal solution to the problem

$$\min_{\mathbf{u}\in U_T} c_T(\mathbf{u}(T)).$$

A spatial price equilibrium is obtained if we minimize the cumulative cost function $c(\mathbf{u})$ not only over all possible shipment patterns, $\mathbf{u} \in U_T$, for a given T, but also over all possible values on T, as can be seen from the following theorem.

Theorem 2. *Discrete Spatial Price Equilibrium.* Let

$$c_{\bar{T}}(\bar{\mathbf{u}}(\bar{T})) = \min_{T} c_T(\bar{\mathbf{u}}(T)) = \min_{T} \min_{\mathbf{u}\in U_T} c_T(\mathbf{u}(T)).$$

Then $\bar{\mathbf{u}}(\bar{T})$ is a spatial price equilibrium at $\bar{T}$.

Proof. See Erlander and Lundgren (1992).

Note that in the optimization problem in Theorem 2, both the trip pattern $\mathbf{u}$ and the total number of shipments T are variables. The continuous version of this problem (as will be shown later in Theorem 10) is exactly the problem most often used to derive the equilibrium conditions.

11.2.4 Behavioral Interpretation

The cumulative cost function can be written in such a way that the last term represents the cost incurred by the last added shipment irrespective of which this may be (Corollary to Theorem 1). Hence, minimizing $c(\mathbf{u})$ implies minimizing the cost for the last added shipment whichever this may be. Thus, minimizing $c(\mathbf{u})$ implies that, at the optimal solution, no single shipment can be sent in a less costly way and no inactive shipper can make a profit by entering the market - we have a Nash equilibrium.

The justification for the usual equilibrium conditions is that the shippers are assumed to behave rationally. Each shipper is assumed to minimize his cost (maximize his profit). However, the cost of the last shipper entering the market is the last term of the cumulative cost function. Hence, if the last entering shipper is trying to minimize his cost, he is at the same time trying to minimize the cumulative cost function. However, the cumulative cost function can be rewritten in such a way that the last term of the sum is the cost of anyone of the active shippers. Hence, *any* shipper trying to minimize his cost, is at the same time trying to minimize the cumulative cost function. In other words, the cumulative cost function represents the *simultaneous behavior* of all shippers in the sense that minimizing the cumulative cost function implies minimizing the cost of each shipper. *This is the new interpretation.*

11.2.5 Existence of Equilibria

We now address the question under which conditions there exists a discrete spatial price equilibrium, i.e., under which conditions there exists a minimum to the optimization problem stated in Theorem 2.

Let the set of all routes with supply point i and demand point j be denoted by R_{ij}.

Theorem 3. Let $S_i : \mathrm{Z}_+ \to \mathrm{R}_+, i = 1, \ldots, I$, be nondecreasing functions, let $D_j : \mathrm{Z}_+ \to \mathrm{R}_+, j = 1, \ldots, J$, be nonincreasing functions, and let $c_l : \mathrm{Z}_+ \to \mathrm{R}_+, l \in L$, be nondecreasing functions. Assume that
(i) $S_i(1) + \sum_{l\in r} c_l(1) - D_j(1) < 0$ for some $(i, j) \in \{1, \ldots, I\} \times \{1, \ldots, J\}$ and some $r \in R_{ij}$, and that
(ii) for each (i, j) and some $r \in R_{ij}, i = 1, \ldots, I$ and $j = 1, \ldots, J$, there exists some

$$\hat{h}_{ijr} \in \mathrm{Z}_+ \text{ such that } S_i(h) + \sum_{l\in r} c_l(h) > D_j(h) \text{ for all } h \geq \hat{h}_{ijr}.$$

Then there exists a discrete spatial price equilibrium.

Remark. We consider only the case where the number of shipments is nonzero. A situation where no shipments are made, may also be a spatial price equilibrium.

Proof. See Erlander and Lundgren (1992).

Theorem 3 says that if (i) there exists a first route such that its incremental cost (net cost) is negative, and (ii) by increasing the number of routes sufficiently the incremental cost can be made strictly positive, this implies that the cumulative cost function will start by decreasing its value and eventually, for increasing number of routes, increase its value. Hence, there must be a minimum in between.

11.3 A Probabilistic Approach

A spatial price equilibrium is a situation such that no shipper can make a profit by changing his shipment plan. This idea assumes that shippers behave rationally and that each shipper has complete knowledge not only about the different price schedules, but also about the actions taken by the other shippers. The assumption of complete knowledge is however rather unrealistic. Instead, we introduce a probabilistic approach to derive the equilibrium conditions. The approach is based on the efficiency principle, see Smith (1978), Erlander (1985) and Erlander and Smith (1990), and is an extension of the theory developed for the fixed demand trip assignment problem in a transportation network (Smith, 1983; 1988), and for the combined trip distribution and assignment problem (Erlander, 1990). For an introduction to the efficiency principle, see also Erlander and Stewart (1990).

At the root of our approach is the underlying assumption that the incline towards a rational behavior usually ascribed to economic agents, in our case the shippers, can be described by a probability distribution. The probability distribution is here over the shipment pattern, i.e., over the combined outcome of all the simultaneous decisions about all the units of the commodity, and not over the choice of the individual shipper. We utilize the fact that the cumulative cost function represents the simultaneous behavior of all shippers. We assume that the behavior of the shippers is *efficient* in the sense that less costly shipment patterns are at least as likely as more costly shipment patterns. This is the *efficiency principle*. The assumption of efficient behavior completely determines the form of the probability distribution and it can be shown (Theorem 7) that it has an exponential form, i.e., is an exponential function of the cost measure, which in our case is the cumulative cost function.

It should be observed that we need not assume that the decisions about individual units of the commodity be independent. There may be any kind of interaction between individual decisions as long as the overall behavior can be described probabilistically as efficient, which explains the power of the use of efficiency in this sense.

Assuming efficiency implies that shipment patterns with lower value of the cumulative cost function are more probable. Hence, the most probable shipment patterns can be obtained by finding the shipment pattern that maximizes the probability. The efficiency principle takes care in a probabilistic way of the fact that a shipment pattern is a spatial price equilibrium if it minimizes the cumulative cost function. One of our main results is therefore: Let the probability distribution be efficient. Then *the most probable shipment patterns are spatial price equilibria* (Theorem 6). Thus, our approach preserves the rationality of the shippers in the more realistic framework of a probabilistic model that does not presuppose complete knowledge by the shippers. Our probabilistic model has the realistic interpretation that equilibria are likely to be observed in practice, since the most probable shipment patterns are equilibria, but it does not exclude other shipment patterns.

Since the shipments consist of a homogeneous commodity, it is of interest to consider the number of shipments on each possible route. This trade pattern (or, in the words used in the continuous version, flow pattern) is a function of the shipment pattern, and many shipment patterns correspond to the same trade pattern. Assuming efficient behavior and efficient probability distribution, we obtain an expression for the probability of the trade pattern (Theorem 9). This probability can then be maximized in order to find the *most probable trade patterns*. This is, however, a very difficult optimization problem in integer variables. The model can be interpreted as a *discrete dispersed spatial price equilibrium* model. To learn about the structure of the solutions we formulate a continuous approximation to the problem. In this way we obtain an entirely new model, the *continuous dispersed spatial price equilibrium* (Theorem 11).

11.3.1 Most Probable Shipment Pattern

Let the shipment pattern $\mathbf{u} = (r_1, \ldots, r_T)$ be a random vector taking values in the discrete sample space $U = \bigcup_T U_T$ with probabilities given by the probability distribution $\mathbf{p}$. The sample space U consists of all feasible selections of routes including different supply points and demand points for the T units of the commodity that will be shipped (previously defined as U_T), and all selections of the value of T. We assume that the number of elements in the sample space U is finite, and that each element has positive probability. The justification for this is that in the discrete case treated here each shipment occupies some positive volume in physical space. Hence, the number of shipments, i.e., the value of T, is bounded from above due to the finite total physical volume in which the network is represented. It will never be necessary in practical applications to enumerate all sample elements. They are used only in the theoretical derivation.

Let $\mathbf{u}_k \in U, k = 1, \ldots, K$, denote the different shipment patterns and let the probability distribution $\mathbf{p}$ be defined by $\mathbf{p} = (p_1, \ldots, p_K)$, where $p_k = \mathrm{P}\{\mathbf{u} = \mathbf{u}_k\} > 0, k = 1, \ldots, K$, and $\sum_{k=1}^{K} p_k = 1$.

Our objective is to characterize the probability distribution $\mathbf{p} = (p_1, \ldots, p_K)$. Note that $\mathbf{u}$ denotes the whole shipment pattern, $\mathbf{u} = (r_1, \ldots, r_T)$, so that the probability distribution can also be written $\mathrm{P}\{\mathbf{u}\} = \mathrm{P}\{r_1, \ldots, r_T\}$. It should be observed that there is nothing in the assumptions made so far that excludes any kind of interaction between the different shipments that are contained in a shipment pattern $\mathbf{u}$.

Let us now consider independent samples from the probability distribution $\mathbf{p}$. Let $\mathbf{u}^1, \ldots, \mathbf{u}^N$ be an independent sample of size N from the probability distribution $\mathbf{p}$. The sample can for instance be the shipment pattern for N different days. The probability of this sample can, because of the independence, be written

$$\mathrm{P}\{\mathbf{u}^1, \ldots, \mathbf{u}^N\} = \mathrm{P}\{\mathbf{u}^1\} \ldots \mathrm{P}\{\mathbf{u}^N\} = \prod_{k=1}^{K} p_k^{z_k(\mathbf{u}^1, \ldots, \mathbf{u}^N)},$$

where $z_k(\mathbf{u}^1, \ldots, \mathbf{u}^N)$ denotes the number of occurrences of the pattern $\mathbf{u}_k$ in the sequence $\mathbf{u}^1, \ldots, \mathbf{u}^N$.

Efficiency is generally defined using a set of activity measures and a set of cost measures for each random variable. However, in this special case where the shipment pattern is our random variable we will not have any activity measure and only one cost measure. The definition of *activity equivalence* reduces to be the requirement that the samples should have the same size N.

The cost measure will be the cumulative cost function, introduced in Section 11.2.3, which according to Theorem 1 can be written

$$c(\mathbf{u}) = \sum_{i=1}^{I} \sum_{h=1}^{A_i(\mathbf{u})} S_i(h) + \sum_{l \in L} \sum_{h=1}^{v_l(\mathbf{u})} c_l(h) - \sum_{j=1}^{J} \sum_{h=1}^{B_j(\mathbf{u})} D_j(h).$$

We are now ready to define efficiency in our special case.

Definition. The probability distribution $\mathbf{p}$ is *efficient* if and only if for every N and every pair of samples $\mathbf{u}^1, \ldots, \mathbf{u}^N$ and $\mathbf{w}^1, \ldots, \mathbf{w}^N$ we have

$$[c(\mathbf{u}^1) + \ldots + c(\mathbf{u}^N) \leq c(\mathbf{w}^1) + \ldots + c(\mathbf{w}^N)] \Longrightarrow$$

$$[\mathrm{P}\{\mathbf{u}^1, \ldots, \mathbf{u}^N\} \geq \mathrm{P}\{\mathbf{w}^1, \ldots, \mathbf{w}^N\}].$$

Efficiency prevails when, among samples of shipment patterns of the same size, those shipment patterns with lower total cost over the sample - as measured by the cumulative cost function based on the supply price schedules, the demand price schedules and the shipment price schedules - are at least as probable as those with higher total cost. Assuming efficiency thus means assuming that for each shipment the route selection, including supply point and demand point, is biased towards less costly routes, that is towards routes which could be candidates for an equilibrium. We have defined activity equivalence in the simplest way, imposing sample size as the only constraint. Other definitions are possible. One example is given by Erlander and Smith (1990), who define activity equivalence in terms of total supply activity and total demand activity at supply points and demand points, respectively. This formulation is more restrictive in the sense that the total activity in the system, i. e., the value of T, is given exogenously.

The following two theorems are immediate from the definition of efficiency.

Theorem 4. Let the probability distribution $\mathbf{p}$ be efficient. Then all shipment patterns $\mathbf{u}$ with identical values on $c(\mathbf{u})$ are equally probable.

Theorem 5. Let the probability distribution $\mathbf{p}$ be efficient. Then

$$[c(\bar{\mathbf{u}}) = \min_{\mathbf{u} \in U} c(\mathbf{u})] \Longleftrightarrow [\bar{\mathbf{u}} \text{ is a most probable shipment pattern}].$$

We are now ready to state one of the main results of this section.

Theorem 6. Let the probability distribution $\mathbf{p}$ be efficient. Then the most probable shipment patterns are discrete spatial price equilibria.

Remark. Nothing is said about the reverse implication.

Proof. The theorem is an immediate consequence of Theorem 2 and Theorem 5, by identifying the feasible set U as consisting of all possible T and all route selections U_T.

What makes Theorem 5 and Theorem 6 so interesting is that the spatial price equilibrium conditions are derived using very basic and natural assumptions about efficient behavior. If $\mathbf{p}$ is efficient, maximizing the probability is equivalent to minimizing the cumulative cost function $c(\mathbf{u})$, which implies the equilibrium conditions. Usually we start by making an assumption about the conditions themselves and then we construct an optimization problem whose optimal conditions can be shown to satisfy the spatial price equilibrium conditions. Using the efficiency approach, we derive both the conditions and the optimization formulation. Note that the integer assumption is necessary for coming to this result.

Theorem 6 says that the most probable shipment pattern satisfies the spatial price equilibrium conditions, but it does not tell how probable this shipment pattern is. Many shipment patterns that do not satisfy the spatial price equilibrium conditions can have a probability very close to the most probable one. Also, due to the assumption $p_k > 0, k = l, \ldots, K$, all possible shipment patterns will have a strictly positive probability, though for most of them the probability will be extremely small.

The relation between the assumption of efficiency and the form of the probability distribution is given by the following representation theorem.

Theorem 7. *Representation theorem.* The probability distribution $\mathbf{p}$ is efficient if and only if there exist constants $\mu \in \mathrm{R}$ and $\gamma \in \mathrm{R}_+$ such that for all $k = 1, \ldots, K$, and $\mathbf{u}_k \in U, p_k = \exp(\mu - \gamma c(\mathbf{u}_k))$. The parameters μ and γ are unique unless $c(\mathbf{u}_1) = \ldots = c(\mathbf{u}_K)$.

Proof. The theorem can be proved using the proof of the general representation theorem in Erlander (1985).

Theorem 7 gives the form of the probability distribution in terms of the parameters μ and γ. The parameters cannot be chosen entirely at will. The probability distribution has to satisfy the constraint $\sum_{k=1}^{K} p_k = 1$ and a constraint on the cost level, for example, defined as the expected cumulative cost. It is assumed that the specified cost level C is chosen such that a feasible probability distribution exists, i.e., $C \geq C_{min}$. The solution has the form given in Theorem 7 and the parameters μ and γ have the interpretation of Lagrange multipliers corresponding to the normalizing constraint and the constraint on the cost level, respectively.

How sensitive the probability is to the value of the cumulative cost function is given by the value of γ. For small values of γ, $c(\mathbf{u})$ will not discriminate much between the various shipment patterns $\mathbf{u}$, and a larger proportion of the shipment patterns may not satisfy the spatial price equilibrium conditions. In the limit, when $\gamma \to 0$, the probability will be independent of the value of $c(\mathbf{u})$ and all shipment patterns will be equally probable. For large values of γ, $c(\mathbf{u})$ will discriminate strongly between the various shipment patterns. The parameter γ can be interpreted as the Lagrange multiplier corresponding to the constraint on the expected cumulative cost C. A large value of γ corresponds to a small value of the expected cumulative cost, other things being equal. Since minimizing $c(\mathbf{u})$ implies spatial price equilibrium, only shipment patterns that are spatial price equilibria will have a positive probability if C is chosen sufficiently close to the lower bound C_{min}.

The discrete mathematical programming problem of minimizing the cumulative cost will be replaced by a continuous mathematical programming problem in continuous variables in Section 11.4.

11.3.2 Most Probable Trade Pattern

We have seen that the most probable shipment patterns satisfy the spatial price equilibrium conditions. The definition of a shipment pattern requires specification of the supply point, demand point and route in the transportation network for each unit of shipment. However, since the shipments consist of a homogeneous commodity it is of interest to consider the number of shipments on each possible route as well.

Let $\mathbf{v} = [v_{ijr}]$ be a specific *trade pattern*, where v_{ijr} denotes the discrete number of shipments from i to j on route r. The ijr-frequency function, $v_{ijr} : U \to \mathrm{Z}_+$, defined for all $\mathbf{u} = (r_1, \ldots, r_T) \in U$, represents the frequency $v_{ijr}(\mathbf{u})$ of shipment plan $r \in R_{ij}$ in each shipment pattern $\mathbf{u}$.

The frequency profile $\mathbf{v}(\mathbf{u}) = [v_{ijr}(\mathbf{u}) : i = 1, \ldots, I, j = 1, \ldots, J,\ r \in R_{ij}]$ can be viewed as the trade pattern generated by shipment pattern $\mathbf{u} \in U$, and many shipment patterns correspond to the same trade pattern. Let $U_{\mathbf{v}}$ be the set of all shipment patterns $\mathbf{u}$ that correspond to trade pattern $\mathbf{v}$, i.e., $U_{\mathbf{v}} = \{\mathbf{u} \mid [v_{ijr}(\mathbf{u})] = [v_{ijr}]\}$.

The number of elements in $U_{\mathbf{v}}$ is $(T!)/(\prod_i \prod_j \prod_r v_{ijr}!)$, where $T = \sum_i \sum_j \sum_r v_{ijr}$ is the total number of shipments in trade pattern $\mathbf{v}$.

According to Theorem 7 the probability p_k is a function of shipment patterns. We will now show that it is in fact a function of the trade patterns generated by shipment patterns.

First, note that the functions $A_i : U \to \mathrm{Z}_+$, $B_i : U \to \mathrm{Z}_+$ and $v_l : U \to \mathrm{Z}_+$, defined as

$$A_i(\mathbf{u}) = \sum_{j,r} v_{ijr},\quad B_j(\mathbf{u}) = \sum_{i,r} v_{ijr},\quad \text{and } v_l(\mathbf{u}) = \sum_{i,j,r} v_{ijr}\delta_{ij,lr},$$

where

$$\delta_{ij,lr} = \begin{cases} 1 & \text{if route } r \text{ from } i \text{ to } j \text{ uses link } l, \\ 0 & \text{otherwise,} \end{cases}$$

are all constant for all $\mathbf{u} \in U_{\mathbf{v}}$, and can be replaced by functions of $\mathbf{v}$, namely

$$A_i(\mathbf{v}) = \sum_{j,r} v_{ijr},\quad B_j(\mathbf{v}) = \sum_{i,r} v_{ijr},\quad \text{and } v_l(\mathbf{v}) = \sum_{i,j,r} v_{ijr}\delta_{ij,lr}.$$

This means that the cumulative cost function

$$c(\mathbf{u}) = \sum_{i=1}^{I} \sum_{h=1}^{A_i(\mathbf{u})} S_i(h) + \sum_{l \in L} \sum_{h=1}^{v_l(\mathbf{u})} c_l(h) - \sum_{j=1}^{J} \sum_{h=1}^{B_j(\mathbf{u})} D_j(h),$$

can, for $\mathbf{u} \in U_{\mathbf{v}}$, be replaced by the function

$$c(\mathbf{v}) = \sum_{i=1}^{I} \sum_{h=1}^{A_i(\mathbf{v})} S_i(h) + \sum_{l \in L} \sum_{h=1}^{v_l(\mathbf{v})} c_l(h) - \sum_{j=1}^{J} \sum_{h=1}^{B_j(\mathbf{v})} D_j(h).$$

Theorem 8. The probabilities $p_k = \exp(\mu - \gamma c(\mathbf{u}_k))$, $k = 1, \ldots, K$, are constant for all $\mathbf{u}_k \in U_{\mathbf{v}}$. In other words, p_k is a function of $\mathbf{v}$.

Proof. The theorem follows from Theorem 7 since $c(\mathbf{u}) = c(\mathbf{v})$ for $\mathbf{u} \in U_{\mathbf{v}}$.

Let $\mathrm{P}\{\mathbf{v}\}$ denote the probability of the trade pattern $\mathbf{v}$.

Theorem 9. Let the shipment pattern distribution $\mathrm{P}\{\mathbf{u}\}$ be efficient. Then the probability of the specific trade pattern $\mathbf{v} = [v_{ijr}]$ is given, for $\mu \in \mathrm{R}$ and $\gamma \in \mathrm{R}_+$, and for any $\mathbf{u}_k \in U_{\mathbf{v}}$, by

$$\mathrm{P}\{\mathbf{v}\} = \frac{T!}{\prod_{i,j,r} v_{ijr}!} \exp(\mu - \gamma c(\mathbf{u}_k)).$$

Proof. The probability p_k is constant for all $\mathbf{u}_k \in U_{\mathbf{v}}$ according to Theorem 8. To obtain $\mathrm{P}\{\mathbf{v}\}$ we simply have to multiply p_k by the number of elements in $U_{\mathbf{v}}$.

A large value of the cumulative cost function $c(\mathbf{u}_k)$ implies that the shipment pattern $\mathbf{u}_k \in U_{\mathbf{v}}$ has a small probability p_k. However, if the number of elements in $U_{\mathbf{v}}$ is large, the probability of the specific trade pattern $\mathbf{v}$, $\mathrm{P}\{\mathbf{v}\}$, may be high in spite of the fact that the probability p_k of each element $u_k \in U_{\mathbf{v}}$ may be low. This means that most probable trade patterns $\mathbf{v}$ do not automatically correspond to most probable shipment patterns $\mathbf{u}_k$.

The most probable trade patterns can be obtained by maximizing the probability $\mathrm{P}\{\mathbf{v}\}$. This is, however, a very difficult optimization problem. In order to learn about the structure of the solutions to this discrete problem, we can formulate and solve a continuous approximation of the problem. This will be done in Section 11.4.

11.4 The Continuous Spatial Price Equilibrium

Until now we have treated the shipment of the commodity in the network as a shipment of separate units, i.e., the flow has been defined to take only integer values. Now we relax the integer condition and assume that there will be a continuous flow of the commodity in the network. In this way we will obtain the continuous spatial price equilibrium as the continuous approximation to the discrete formulation. We will see that the results are quite analogous to the integer case.

In previous sections, we used the concept of shipment pattern to define our variables, and the number of shipments on specific routes in the network were given as a frequency function $v_{ijr}(\mathbf{u})$ of the shipment pattern $\mathbf{u}$. In the continuous model, the concept of shipment pattern has no meaning, and the frequencies are represented by continuous 'flow' variables (which correspond roughly to aggregate bundles of infinitesimally small 'shipment plans'). Hence, each nonnegative vector $\mathbf{v} = [v_{ijr}]$ is taken to define a possible *flow pattern*, where each $v_{ijr} \in \mathrm{R}_+$ represents a (continuous) flow on route r from i to j. In these terms, the quantities

$$A_i(\mathbf{v}) = \sum_{j,r} v_{ijr} \text{ and } B_j(\mathbf{v}) = \sum_{i,r} v_{ijr},$$

denote the associated (continuous) flows from supply point i and to demand point j, respectively. Also, the *volume* v_l on link l generated by flow pattern $\mathbf{v} = [v_{ijr}]$ is given by

$$v_l(\mathbf{v}) = \sum_{i,j,r} v_{ijr}\delta_{ij,lr}.$$

Clearly, the total flow $T = T(\mathbf{v})$ is given by $T(\mathbf{v}) = \sum_{i=1}^{I}\sum_{j=1}^{J}\sum_{r\in R_{ij}} v_{ijr}$.

To develop the continuous version of cumulative costs it is convenient to strengthen our assumptions regarding supply and demand functions. In particular we now assume the existence of continuous, increasing supply functions $s_i : \mathrm{R}_+ \to \mathrm{R}_+$, $i = 1,\ldots,I$, where quantities as well as prices are taken to be continuous variables. In this case the associated *supply price functions* $S_i : \mathrm{R}_+ \to \mathrm{R}_+$, $i = 1,\ldots,I$, are given for all quantities $q \in \mathrm{R}_+$ by $S_i(q) = s_i^{-1}(q)$ and are by construction both increasing and continuous. Similarly, if demand functions $d_j : \mathrm{R}_+ \to \mathrm{R}_+$ are also taken to be continuous and decreasing, the associated *demand price functions* $D_j : \mathrm{R}_+ \to \mathrm{R}_+$, $j = 1,\ldots,J$, are given for all $q \in \mathrm{R}_+$ by $D_j(q) = d_j^{-1}(q)$ and are also continuous and decreasing. Finally, if the *shipment price schedules* $c_l : \mathrm{R}_+ \to \mathrm{R}_+$, $l \in L$, are here taken to be continuous nondecreasing functions, then the appropriate *cumulative cost function* $c : \mathrm{R}_+^{|R|} \to \mathrm{R}$, now has the integral form

$$c(\mathbf{v}) = \sum_{i=1}^{I}\int_0^{A_i(\mathbf{v})} S_i(h)dh + \sum_{l\in L}\int_0^{v_l(\mathbf{v})} c_l(h)dh - \sum_{j=1}^{J}\int_0^{B_j(\mathbf{v})} D_j(h)dh,$$

first introduced by Beckmann et al. (1956) for a different problem.

Next, let the *route cost* on each $r \in R_{ij}$ be denoted by $c_{ijr} = \sum_{l\in L} c_l(v_l)\delta_{ij,lr}$.

Definition. A flow pattern $\mathbf{v} = [v_{ijr}]$ corresponds to a *spatial price equilibrium* if and only if for all i, j, r

$$\begin{aligned} v_{ijr} > 0 \;&\Rightarrow\; S_i(A_i) + c_{ijr} - D_j(B_j) = 0, \\ v_{ijr} = 0 \;&\Rightarrow\; S_i(A_i) + c_{ijr} - D_j(B_j) \geq 0. \end{aligned}$$

These are of course the standard spatial price equilibrium conditions, first formulated by Samuelson (1952) and Takayama and Judge (1971).

Theorem 10. Let $\bar{\mathbf{v}} = [\bar{v}_{ijr}]$ be a solution to the convex minimization problem

$$\min c(\mathbf{v}) = \sum_{i=1}^{I}\int_0^{A_i(\mathbf{v})} S_i(h)dh + \sum_{l\in L}\int_0^{v_l(\mathbf{v})} c_l(h)dh - \sum_{j=1}^{J}\int_0^{B_j(\mathbf{v})} D_j(h)dh,$$

s.t. $v_{ijr} \geq 0$, $\forall i, j, r$,

where

$$A_i(\mathbf{v}) = \sum_{j,r} v_{ijr}, \quad B_j(\mathbf{v}) = \sum_{i,r} v_{ijr}, \quad \text{and } v_l(\mathbf{v}) = \sum_{i,j,r} v_{ijr}\delta_{ij,lr}.$$

Then $\bar{\mathbf{v}} = [\bar{v}_{ijr}]$ is a spatial price equilibrium. In addition, if all shipment-price functions are increasing, this equilibrium is unique.

Remark. Observe that the equalities above are just definitions and do not imply that $A_i(\mathbf{v})$ and $B_j(\mathbf{v})$ are given, but rather define these quantities as endogenous variables. This means that, except for the nonnegativity constraints, the problem is unbounded.

Proof. See Erlander and Lundgren (1992).

The optimization problem in Theorem 10 is the continuous version of the discrete optimization problem formulated in Theorem 2, and we have arrived at the continuous version by relaxing the integer constraints and replacing summation by integration. The continuous version of the optimization problem was given by Florian and Los (1982).

In Section 11.2.4, we saw that the discrete optimization problem has a nice interpretation with respect to the behavior of the shippers. The objective function (which is equal to the cumulative cost function) represents the simultaneous behavior of all shippers in the sense that minimizing the objective function implies minimizing the cost of each shipper, and the equilibrium conditions are satisfied. With this in mind, it should be no surprise that a continuous flow $\bar{\mathbf{v}}$ minimizing the corresponding continuous function satisfies the equilibrium conditions for the continuous case.

We have seen that under the assumption of efficiency the most probable shipment patterns are spatial price equilibria (Theorem 6). Now, what can be said in the continuous case? Do the optimal solutions $\bar{\mathbf{v}}$ to the optimization problem in Theorem 10 correspond to the most probable shipment patterns? Clearly, this conclusion cannot hold in general, since there is no guarantee that the optimal solutions are integer. On the other hand, under suitable conditions we may take the continuous solutions $\bar{\mathbf{v}}$ to represent the discrete solutions $\bar{\mathbf{u}}$, in the sense that $\bar{\mathbf{v}}$ will give an approximative picture of $\bar{\mathbf{u}}$, with precision good enough for most practical purposes. *In this sense we may say that continuous spatial price equilibria correspond approximately to most probable shipment patterns.* Note, however, that the same flow value $\bar{\mathbf{v}}$ may arise from many different shipment patterns $\bar{\mathbf{u}}$, a fact that will be used in the following.

In Section 11.3.2, we discussed the concept of trade patterns, and, taking into account that many shipment patterns $\mathbf{u}_k$ correspond to the same trade pattern, we derived an expression for the probability $\mathrm{P}\{\mathbf{v}\}$ of a specific, discrete valued, trade pattern $\mathbf{v}$,

$$\mathrm{P}\{\mathbf{v}\} = \frac{T!}{\prod_{i,j,r} v_{ijr}!} \exp(\mu - \gamma c(\mathbf{u}_k)).$$

The most probable trade patterns can now be obtained by maximizing the probability $\mathrm{P}\{\mathbf{v}\}$. This is, however, a very difficult optimization problem. Relaxing the integer constraint on the elements of $\mathbf{v}$, we obtain the continuous analogue. Thus we replace the probability $\mathrm{P}\{\mathbf{v}\}$ with the expression

$$\frac{T(\mathbf{v})!}{\prod_{i,j,r} v_{ijr}!} \exp(\mu - \gamma c(\mathbf{v})).$$

Maximizing this expression can be done by taking logarithms, neglecting the constant μ, approximating the factorials by Stirling's formula and using the fact that maximization of $f(\mathbf{x})$ is equal to minimization of $-f(\mathbf{x})$. In this way we obtain the convex continuous minimization problem

$$\min_{\mathbf{v}\geq 0} Z(\mathbf{v}) = -T(\mathbf{v})(\log T(\mathbf{v}) - 1) + \sum_{i,j,r} v_{ijr}(\log v_{ijr} - 1) + \gamma c(\mathbf{v}).$$

The solution obtained from this continuous model will be called a *continuous dispersed spatial price equilibrium*, and corresponds in the discrete case to the most probable trade patterns under the assumption of efficient shipment behavior. The solution has the form of a logit model over the simultaneous choice of origin, destination and route, as will be seen from the following theorem.

Theorem 11. *Continuous dispersed spatial price equilibrium.* Let the shipment pattern distribution $\mathbf{p}$ be efficient with $\gamma > 0$. Then the optimal solutions $\bar{\mathbf{v}}$ to the problem

$$\min_{\mathbf{v}\geq 0} Z(\mathbf{v}) = -T(\mathbf{v})(\log T(\mathbf{v}) - 1) + \sum_{i,j,r} v_{ijr}(\log v_{ijr} - 1) + \gamma c(\mathbf{v})$$

define continuous dispersed spatial price equilibria.
Furthermore, the optimal solutions are of the form

$$\bar{v}_{ijr} = T(\bar{\mathbf{v}}) \exp(-\gamma[S_i(A_i(\bar{\mathbf{v}})) + c_{ijr}(\bar{\mathbf{v}}) - D_j(B_j(\bar{\mathbf{v}}))]) > 0,$$

where

$$A_i(\bar{\mathbf{v}}) = \sum_{j,r} \bar{v}_{ijr},\ i = 1, \ldots, I,\ B_j(\bar{\mathbf{v}}) = \sum_{i,r} \bar{v}_{ijr},\ j = 1, \ldots, J,\ T(\bar{\mathbf{v}}) = \sum_{i,j,r} \bar{v}_{ijr},$$

$$v_l(\bar{\mathbf{v}}) = \sum_{i,j,r} \bar{v}_{ijr}\delta_{ij,lr}, \text{ and } c_{ijr}(\bar{\mathbf{v}}) = \sum_{l\in L} c_l(\bar{v}_l)\delta_{ij,lr}.$$

If condition (ii) of Theorem 3 holds, then there always exists an equilibrium.
If the link functions $c_l(v_l)$ are strictly monotonically increasing, the optimal solution is unique.
Finally, the probability of $\bar{\mathbf{v}}$ can be expressed

$$\mathrm{P}\{\bar{\mathbf{v}}\} \approx \exp(\mu + \gamma[\sum_{i=1}^{I} \bar{A}_i S_i(\bar{A}_i) + \sum_{i,j,r} \bar{c}_{ijr}\bar{v}_{ijr} - \sum_{j=1}^{J} \bar{B}_j D_j(\bar{B}_j) - c(\bar{\mathbf{v}})]).$$

Remark. The optimal solutions are of logit form

$$v_{ijr} = T(\mathbf{v}) \frac{\exp(-\gamma[S_i(A_i(\mathbf{v})) + c_{ijr}(\mathbf{v}) - D_j(B_j(\mathbf{v}))])}{\sum_{i,j,r} \exp(-\gamma[S_i(A_i(\mathbf{v})) + c_{ijr}(\mathbf{v}) - D_j(B_j(\mathbf{v}))]},$$

since

$$\sum_{i,j,r} \exp(-\gamma[S_i(A_i(\bar{\mathbf{v}})) + c_{ijr}(\bar{\mathbf{v}}) - D_j(B_j(\bar{\mathbf{v}}))]) = 1.$$

Proof. See Erlander and Lundgren (1992).

It should be observed that μ and γ are the parameters of the probability distribution $\mathbf{p}$. They are the same as the parameters in the expression of p_k given in Theorem 7. Also, the coefficient γ in the objective function in the optimization problem in Theorem 11 can be interpreted as the Lagrange multiplier to a cost constraint specifying, in this case, the value of $c(\mathbf{v})$. Thus, the optimization problem in Theorem 11 can be reformulated as

$$\min_{\mathbf{v}\geq 0} Z(\mathbf{v}) = -T(\mathbf{v})(\log T(\mathbf{v}) - 1) + \sum_{i,j,r} v_{ijr}(\log v_{ijr} - 1),$$

subject to

$$c(\mathbf{v}) = \sum_{i=1}^{I} \int_0^{A_i(\mathbf{v})} S_i(h)dh + \sum_{l\in L} \int_0^{v_l(\mathbf{v})} c_l(h)dh - \sum_{j=1}^{J} \int_0^{B_j(\mathbf{v})} D_j(h)dh \leq C,$$

where $C \geq C_{\min}$ is the specified value of $c(\mathbf{v})$. The previous formulation can be interpreted as a Lagrangian relaxation subproblem to this latter problem.

The formula for the optimal solution in Theorem 11 is a nice form of writing the optimal solution. However, the system of equations has to be solved to give the explicit value of $\bar{\mathbf{v}}$, which is only implicitly defined by the formula. The expression in the exponential, $S_i(A_i(\bar{\mathbf{v}})) + c_{ijr}(\bar{\mathbf{v}}) - D_j(B_j(\bar{\mathbf{v}}))$, is nothing but the incremental cost for one additional shipment along route r in supply-demand-pair (i, j). Since $\bar{v}_{ijr} > 0$ for all i, j, r, we also know that $\bar{v}_{ijr} < T(\bar{\mathbf{v}})$ for all i, j, r, and that *the incremental cost for all routes in the optimal solution will be strictly positive.*

The optimization problem formulated in Theorem 11 defines a dispersed spatial price equilibrium. In the mathematical sense, this is the result of introducing the entropy function in the objective function, in this way allowing flows also on routes which have a positive incremental cost.

In the formulation used in Theorem 10 we derived the continuous version of a spatial price equilibrium using an optimization formulation where $c(\mathbf{v})$ defined the objective function. In Theorem 11, for large values of the parameter γ, the contribution from $c(\mathbf{v})$ will dominate in the objective function and in the limit, when $\gamma \to \infty$, we will obtain the continuous spatial price equilibrium model. Looking at the expression for the optimal solution, this means that the exponent $S_i(A_i(\bar{\mathbf{v}}))+c_{ijr}(\bar{\mathbf{v}})-D_j(B_j(\bar{\mathbf{v}}))$, i.e., the incremental cost, will discriminate strongly among the routes. When γ grows, only routes with an incremental cost close to zero will have a significant flow. Since we know that, as $\gamma \to \infty$, only shipment patterns that are spatial price equilibria will have a positive probability, it is easy to understand that only trade patterns obtained from such shipment patterns will have a positive probability.

The term *dispersed network equilibrium* was introduced by Smith (1988), and *dispersed assignment models* have been suggested by Wilson (1970), Dial (1971) and Fisk (1980) for the trip assignment problem. These models make the assignment according to a logit formula where the proportion of the flows on each route in a given origin-destination pair follows a negative exponential function of the route costs, i.e., not only the least-cost routes will have a positive flow. Translated to our case, this means that there will be a flow of the commodity also on routes with a positive incremental cost. In this way, the uncertainty facing the decision maker is taken into account. Dispersed network equilibrium models have also been formulated by Erlander (1990) considering the combined distribution and assignment problem. Sheffi (1985) treats logit models for route flows under the name of 'Stochastic User Equilibrium'.

Harker (1988) used the term dispersed spatial price equilibrium for a model where the distribution of shipments between regions follows a logit formula, but where the assignment of shipments over the transportation network is made according to the usual equilibrium conditions. Our model is different and also derived in a completely different way. Roy (1990) also discusses several models considering dispersed spatial price equilibria

Acknowledgments
The authors wish to thank Tony E. Smith and David Bernstein for valuable comments on an early version of the paper, and Tony E. Smith for several contributions to the final version. This research has been supported by the Swedish Transport Research Board (Dnr 91-05-63 and 92-128-63) and The Swedish Institute. It was performed in part while the second author was guest researcher at Centre for Research on Transportation (CRT) in Montreal, Canada.

References

Beckmann, M.J. and McGuire, C.B. and Winsten, C.B. (1956), *Studies in the Economics of Transportation*, Yale University Press, New Haven, CT.

Bernstein, D. and Smith, T.E. (1994), 'Programmability of discrete network equilibria', SOR-94-19, Princeton University, NJ.

Dial R.B. (1971), 'A probabilistic multipath traffic assignment model which obviates path enumeration', *Transportation Research*, vol. 5, pp. 83–111.

Erlander, S. (1985), 'On the principle of monotone likelihood and loglinear models', *Mathematical Programming*, vol. 21, pp. 137–151.

Erlander, S. (1990), 'Efficient population behavior and the simultaneous choices of origins, destinations and routes', *Transportation Research*, vol. 24B, pp. 363–373.

Erlander, S. and Lundgren, J.T. (1992), 'Spatial price equilibrium and efficiency', Department of Mathematics, LiTH-MAT-R-1992-39, Linköping University, Linköping.

Erlander, S. and Smith, T.E. (1990), 'General representation theorems for efficient population behavior', *Applied Mathematics and Computation*, vol. 36, pp. 173–217.

Erlander, S. and Stewart, N.F. (1990), *The Gravity Model in Transportation Analysis - Theory and Extensions*, VSP, Utrecht, The Netherlands.

Fisk, C. (1980), 'Some developments in equilibrium traffic assignment', *Transportation Research*, vol. 14B, pp. 243–255.

Florian, M. and Los, M. (1982), 'A new look at spatial price equilibrium models', *Regional Science and Urban Economics*, vol. 12, pp. 579–597.

Harker, P.T. (1988), 'Dispersed spatial price equilibrium', *Environment and Planning A*, vol. 20, pp. 353–368.

Rosenthal, R.W. (1973), 'The network equilibrium problem in integers', *Networks*, vol. 3, pp. 53–59.

Roy, J.R. (1990), 'A dispersed equilibrium commodity trade model',*The Annals of Regional Science*, vol. 24, pp. 13–28.

Samuelson, P.A. (1952), 'Spatial price equilibrium and linear programming', *The American Economic Review*, vol. 42, pp. 283–303.

Sheffi, Y. (1985), *Urban Transportation Networks: Equilibrium Analysis with Mathematical Programming Methods*, Prentice-Hall, Englewood Cliffs.

Smith, T.E. (1978), 'A cost-efficiency principle of spatial interaction behavior', *Regional Science and Urban Economics*, vol. 8, pp. 313–337.

Smith, T.E. (1983), 'A cost-efficiency approach to the analysis of congested spatial-interaction behavior', *Environment and Planning A*, vol. 15, pp. 435–464.

Smith, T.E. (1988), 'A cost-efficiency theory of dispersed network equilibria', *Environment and Planning A*, vol. 20, pp. 231–266.

Smith, T.E. (1993), Private communication.

Takayama, T. and Judge, G.G. (1971), *Spatial and Temporal Price Allocation Models*, North-Holland, Amsterdam.

Wilson, A.G. (1970), *Entropy in Urban and Regional Modelling*, Pion, London.

12 Integration of Freight Network and Computable General Equilibrium Models

Terry L. Friesz
Departments of Systems Engineering
and Operations Research & Engineering
George Mason University
Fairfax, VA 22030, USA

Zhong-Gui Suo
PB Farradyne Inc
Rockville, MD 20852
USA

Lars Westin
Department of Economics
University of Umeå
S-907 87 Umeå
Sweden

12.1 Introduction

Over the past 30 years very significant progress has occurred in the understanding and modeling of passenger trip making behavior over networks. Yet, corresponding advances in understanding and modeling of freight transportation decision making over inter-regional, inter-modal networks have not occurred. In fact the most recent large scale U.S. freight network model is able to predict equilibrium network link volumes agreeing with Federal Railway Administration (FRA) density codes (reported data describing annual tonnages on every physical link of the rail system) with a frequency of only about 60% (Friesz et al., 1981; 1983a; 1983b; 1985). This is poor performance since density codes denote upper and lower bounds for link volumes; the difference between those upper and lower bounds is frequently of the same order of magnitude as the predicted volumes themselves. Poor as this accuracy is, it is substantially greater (about three times greater) than that reported for earlier models (Bronzini, 1980) and was achieved by straight-forward extensions of the urban passenger network modeling paradigm. Still greater accuracy may be obtained from a model designed specifically for freight applications from the outset.

A highly accurate inter-regional, inter-modal freight network forecasting tool which employs a detailed representation of the actual freight transportation network (rather than a highly aggregate abstraction) is critical to federal policy and decision making related to regulation/deregulation, for it allows volumes, costs, modal splits and the like to be estimated. It also allows the region-specific impacts of transportation policies to be determined. Furthermore this type of tool can prove valuable to private freight companies interested in their competitive posture vis-a-vis other companies. Finally, the methods discussed in this paper allow the transportation sector to be represented in a detailed and theoretically precise manner within a general equilibrium model of the entire U.S. economy.

The accuracy disparity between predictive urban passenger network models and predictive inter-regional, inter-modal freight network models may be traced to the following considerations cited by Friesz et al. (1983a):

1. Freight-related databases necessary to calibrate and validate predictive network models are not as extensive and probably not as accurate as those maintained for passenger travel;
2. Freight transportation decisions are inherently more complex and difficult to model than passenger travel decisions;

Table 12.1. Typology of predictive freight network models

Model	Criteria‡‡																
	1	2	3	4	5	6	7	8	9	10	11	12	13	14	15	16	17
Harvard-Brookings	Y	Y	Y	Y	N	Y	Y	N	*	*	Y	N	N	N	N	N	N
CACI	Y	Y	Y	N	Y	N	Y	N	*	*	Y	N	N	N	N	N	N
Peterson	N	N	*	*	Y	N	N	Y	N	N	Y	N	N	N	N	N	N
Lansdowne	N	Y	Y	N	N	N	Y	Y	Y	N	Y	N	N	N	N	N	N
Princeton	N	Y	Y	N	N	N	Y	Y	Y	N	Y	N	N	N	N	N	N
Boston	Y	N	*	*	N	N	N	Y	N	N	Y	N	N	N	N	N	N
NETLAB (FNEM)	Y	Y	Y	Y	Y	Y	Y	Y	Y	N	Y	N	Y	N	N	N	N
Montreal (STAN)	N	Y	N	Y	Y	N	N	Y	N	N	Y	N	N	Y	N	Y	N

where Y = Yes, N = No, and * = Not Applicable.

‡‡ Criteria:

1. Multiple modes
2. Multiple commodities
3. Sequential Loading of commodities
4. Simultaneous loading of commodities
5. Explicit congestion
6. Elastic transportation demand
7. Explicit shippers
8. Explicit carriers
9. Sequential shipper and carrier submodels
10. Simultaneous shipper and carrier submodels
11. Sequential general equilibrium and network model
12. Simultaneous general equilibrium and network model
13. Nonmonotonic functions
14. Explicit backhauling
15. Blocking strategy
16. Fleet constraints
17. Imperfect competition

3. The predictive freight network models which have been proposed have not adequately broken from the passenger network paradigm, whose assumptions are simply erroneous for many freight applications;
4. Sufficiently efficient and inexpensive algorithms for solving theoretically precise freight network problems have not been commonly available or well understood by practitioners; and
5. Large scale predictive freight network models have not been integrated with computable general equilibrium models to forecast consistent national/regional economic activities and prices on the one hand and detailed freight flows on the other.

This paper proposes a spatial computable general equilibrium (SCGE) model in direct response to item 5 above.

12.2 Typology of Predictive Freight Models

In order to place the SCGE model presented subsequently into perspective, it is useful to consider the previous predictive freight models reported in the literature. These are reviewed in detail by Friesz et al. (1983a). Here we only present a typology and summarize the key features of these models in tabular form (Table 12.1). For brevity we will suppress any discussion of these models, and the interested reader is referred to Friesz et al. (1983a) for details. In Table 12.1 we differentiate and evaluate existing models according to a set of 13 criteria developed by and discussed at length by Friesz et al. (1983a). The list of references to this paper also includes key papers which describe the theory and application of the models presented in Table 12.1.

It will be noted in reviewing Table 12.1 that no current model achieves the simultaneous solution of a general equilibrium model and a freight network model. It is precisely this weakness which is addressed in the next section.

12.3 Integrating CGE and Freight Network Models

One of the key ingredients for integration of freight network and general equilibrium models is the treatment of transportation demands as endogenous variables as opposed to exogenously determined commodity supplies and demands found in virtually all the freight models applied to date. Such a treatment is based on existing work extending the spatial price equilibrium problem to incorporate arbitrary networks with nonlinear asymmetric transportation cost, commodity supply and commodity demand functions. This so-called generalized spatial price equilibrium problem does not employ explicit transportation demand functions; instead it derives transportation demand from the production and consumption characteristics of spatially separated markets (nodes). As such, a generalized spatial price equilibrium model can be used as a replacement for the Wardropian shippers' model; because such a replacement model may employ elastic commodity supply and demand functions for each node, trip generation becomes endogenous and the problem of consistency is avoided. If trip generation is introduced into freight flow analysis via a spatial price equilibrium submodel, one has in principle created a specific computable general equilibrium model which represents the transportation sector in a detailed manner. This is because such a model not only determines freight flows and transportation demands, but it also determines the equilibrium prices (and quantities) of all commodities, which feature is the defining attribute of a general equilibrium model. Yet such a model is of limited practical significance, for it requires commodity supply and demand functions for each spatially distinct market and each commodity as model inputs. Although it is generally possible to estimate a demand function for a given commodity market of known spatial extent, the estimation of commodity supply functions can be very problematic. For one thing, the presence of imperfect competition in some commodity markets makes the very meaning of a supply function unclear. More importantly, the use of supply functions as opposed to an input-output matrix masks the rich detail of regional and sectoral interdependencies which is so valuable in policy studies.

In fact, computable general equilibrium (CGE) models have tended to employ a formulation of a perfectly competitive equilibrium based on a constant input-output matrix derived from Leontieff technology. However, before turning to a formulation based on a Leontieff technology, let us first state a more general perfectly competitive CGE model. To this end, the following notation (Mathiesen, 1985) is employed:

M	$\equiv$	the activity index set
N	$\equiv$	the commodity index set
y_j	$\equiv$	the level of activity j
$y \in \Re^{\lvert M \rvert}$	$\equiv$	the activity vector
π_i	$\equiv$	the price of commodity i
$\pi \in \Re^{\lvert N \rvert}$	$\equiv$	the price vector
b_i	$\equiv$	the initial endowment of commodity i in the economy
$b \in \Re^{\lvert N \rvert}$	$\equiv$	the endowment vector
$d_i(\pi)$	$\equiv$	the demand function for commodity i
$d(\pi) \in \Re^{\lvert N \rvert}$	$\equiv$	the commodity demand vector

$A(\pi) \equiv (a_{ij}) \equiv$ the activity analysis matrix for the economy: a_{ij} is the output/input of commodity i per unit of activity j, where $a_{ij} > 0 \Longrightarrow$ activity j produces commodity i, $a_{ij} < 0 \Longrightarrow$ activity j consumes commodity i

Furthermore, we will say that a vector (π^*, y^*) solves the computable general equilibrium model CGE (b,d,A) if and only if it satisfies the following:

(a) No activity earns a positive profit:

$$-A(\pi^*)^T \pi^* \geq 0 \tag{1}$$

(b) No commodity is in excess demand:

$$b + A(\pi^*)y^* - d(\pi^*) \geq 0 \tag{2}$$

(c) No prices or activity levels are negative:

$$\pi^* \geq 0, \quad y^* \geq 0 \tag{3}$$

(d) An activity earning negative profit is not operated and an operated activity earns zero profit:

$$[-A(\pi^*)^T \pi^*]^T y^* = 0 \tag{4}$$

(e) A commodity in excess supply is free, and a positive price implies market clearing via Walras' Law:

$$[b + A(\pi^*)y^* - d(\pi^*)]^T \pi^* = 0 \tag{5}$$

Evidently (1)-(5) have the form of a nonlinear complementarity problem. In fact the definitions

$$z \equiv \begin{pmatrix} \pi & y \end{pmatrix}^{\mathrm{T}} \tag{6}$$

$$F(z) \equiv \begin{bmatrix} b + A(\pi)y - d(\pi) \\ -A(\pi)^T \pi \end{bmatrix} \tag{7}$$

lead to the concise nonlinear complementarity formulation:

$$\begin{aligned} z^T F(z) &= 0 \\ F(z) &\geq 0 \\ z &\geq 0 \end{aligned} \tag{8}$$

As one normally assumes the demand functions $d(\pi)$ are homogeneous of degree zero, the prices π are relative. One can either elect a numeraire or, as we do here, restrict prices to belong to the $(|N| - 1)$ - dimensional unit simplex, denoted as

$$S^{|N|-1} = \{\pi \in \Re^{|N|} : \sum_{i=1}^{|N|} \pi_i = 1\} \tag{9}$$

The case of a Leontieff technology corresponds to the case of a constant activity analysis matrix: $A(\pi) = A$. As Harker and Pang (1990) and Goldsman and Harker (1990) have pointed out, the Leontieff assumption allows (1)-(5) to be cast in the form of a variational inequality problem VIP (b,d,A): find $\pi^* \in X$, such that

$$[b - d(\pi^*)]^T (\pi - \pi^*) \geq 0 \quad , \quad \forall \pi \in X \tag{10}$$

where

$$X \equiv \{\pi \in \Re_+^{|N|} : -A^T\pi \geq 0\} \bigcap S^{|N|-1} \tag{11}$$

In the event $A(\pi)$ is not a constant matrix, successive linearizations may in principle be performed to create a sequence of VIPs like (10).

We now turn to the question of how to integrate a detailed freight network model with (1)-(5). For simplicity of exposition, let us take the case of *inelastic transportation demand and a single transportation mode.* Consequently, we may make the following definitions, which supplement and generalize those given previously :

R	$\equiv$	set of all regions comprising the economy
T^i_{rs}	$\equiv$	the demand for transportation of commodity i between origin region r and destination region s
$T \in \Re^{\|R\|\times\|R\|\times\|N\|}$	$\equiv$	the transportation demand vector
y^s_j	$\equiv$	level of activity j in region s
$y \in \Re^{\|M\|\times\|R\|}$	$\equiv$	the activity vector
π^r_i	$\equiv$	price of commodity i in region r
$\pi \in \Re^{\|N\|\times\|R\|}$	$\equiv$	the price vector
b^r_i	$\equiv$	the initial endowment of commodity i in region r
$b \in \Re^{\|N\|\times\|R\|}$	$\equiv$	the endowment vector
$d^r_i(\pi)$	$=$	the demand function for commodity i in region r
$d \in \Re^{\|N\|\times\|R\|}$	$\equiv$	the commodity demand vector
$A(\pi) \equiv (a^{sr}_{ij})$	$\equiv$	activity analysis matrix
a^{sr}_{ij}	$\equiv$	input-output coefficient for commodity i in region s relative to activity j in region r, where $a^{sr}_{ij} > 0 \Longrightarrow$ activity j in region r produces commodity i for consumption in region s $a^{sr}_{ij} < 0 \Longrightarrow$ activity j in region r consumes commodity i supplied by region s
$A^{rs}(\pi)$	$\equiv$	the submatrix of $A(\pi)$ pertaining to origin-destination pair (r, s)
$\Gamma[A(\pi)]$	$\equiv$	diag $[A^{rs}(\pi)]$, a diagonal matrix formed from the submatrices A^{rs}

However, in light of the preceding development

$$T^i_{rs} = \sum_{j=1}^{|M|} a^{sr}_{ij} y^r_j \tag{12}$$

As $T \in \Re^{|R|\times|R|\times|N|}$, we may rewrite (12) in the more compact form

$$T = \Gamma[A(\pi)]y \tag{13}$$

Furthermore, if we define z and $F(z)$ as in (6) and (7), but employing the above regional generalizations, a complementarity problem of the form (8) again results. If one has already solved this complementary problem, transportation origin-destination flows could be calculated *ex post* using (13).

To continue our development of an integrated model, we need to select a routing principle for the detailed transportation network flows. For illustrative purposes, let us select a simple Wardropian user equilibrium to describe transportation flows. To articulate this transportation model, we introduce the following additional definitions:

L	$\equiv$	the set of all network links
P_{rs}	$\equiv$	the set of network paths connecting origin region r to destination region s
P	$\equiv$	$\bigcup_{rs} P_{rs}$, the set of all network paths
Δ	$\equiv$	(δ_{ap}), the link path incidence matrix
δ_{ap}	$\equiv$	an element of Δ, where $\delta_{ap} = 1$ if link a belongs to path p and $\delta_{ap} = 0$, otherwise
Λ	$\equiv$	$(\lambda_{p,rs})$, the path OD pair incidence matrix
$\lambda_{p,rs}$	$\equiv$	an element of Λ, where $\lambda_{p,rs} = 1$ if path p connects origin node r to destination node s and $\lambda_{p,rs} = 0$, otherwise
h_p^i	$\equiv$	the flow of commodity i on path $p \in P$
$h \in \Re^{\lvert P\rvert \times \lvert N\rvert}$	$\equiv$	the commodity path flow vector
f_a^i	$\equiv$	$\sum_{p \in P} \delta_{ap} h_p^i$, the flow of commodity i on link $a \in L$
$f \in \Re^{\lvert L\rvert \times \lvert N\rvert}$	$\equiv$	the commodity link flow vector
$c_a^i(f)$	$\equiv$	the unit cost of transporting commodity i over link $a \in L$
$\tilde{c}\ (f)$	$\equiv$	$(\cdots, c_a^i(f), \cdots) \in \Re^{\lvert L\rvert \times \lvert N\rvert}$
$c_p^i(h)$	$\equiv$	$\sum_{a \in A} \delta_{ap} c_a^i(f)$ = the unit cost of transporting commodity i over path p
c	$\equiv$	$(\cdots, c_p^i, \cdots) \in \Re^{\lvert P\rvert \times \lvert N\rvert}$
u_{rs}^i	$\equiv$	$\min_{p \in P_{rs}} c_p^i$, the minimum cost of transporting commodity i between origin region r and destination region s
$u \in \Re^{\lvert R\rvert \times \lvert R\rvert \times \lvert N\rvert}$	$\equiv$	the vector of minimum transportation costs

A flow-cost pattern (f^*, u^*) is a Wardropian user equilibrium if and only if :

(a) Excess path cost is never negative:

$$c(h) - \Lambda u \;\geq\; 0 \tag{14}$$

(b) Transport flows are conserved:

$$\Lambda^{\mathrm{T}} h - T \;=\; 0 \tag{15}$$

(c) Utilized paths have zero excess cost and paths with positive excess cost are not used:

$$h(c(h) - \Lambda u) \;=\; 0 \tag{16}$$

(d) Path flows and origin-destination costs are nonnegative:

$$h \geq 0, \quad u \geq 0 \tag{17}$$

It is well known that under the assumption of cost positivity (14)-(17) may be cast in the form of nonlinear complementary problem. In particular under cost positivity, the following is equivalent to (14)-(17):

$$\begin{aligned} h(c(h)-\Lambda u) &= 0 \\ u(\Lambda^{\mathrm{T}}h-T) &= 0 \\ c(h)-\Lambda u &\geq 0 \\ \Lambda^{\mathrm{T}}h-T &\geq 0 \\ (h \;\; u) &\geq 0 \end{aligned} \tag{18}$$

Recalling from (13) that $T = \Gamma[A(\pi)]y$, it is evident that the definitions

$$w \equiv \begin{pmatrix} h & u \end{pmatrix}^{\mathrm{T}}$$

$$G(w,z) \equiv \begin{pmatrix} c(h)-\Lambda u \\ \Lambda^{\mathrm{T}}h-\Gamma[A)\pi)]y \end{pmatrix} \tag{19}$$

lead to a concise nonlinear complementarity formulation of Wardropian user equilibrium:

$$\begin{aligned} w^T G(w,z) &= 0 \\ G(w) &\geq 0 \\ w &\geq 0 \end{aligned} \tag{20}$$

If we knew the commodity price vector π^* and activity vector y^*, then (20) could be readily solved to yield the network flow pattern h^* and associated costs u^*.

As a consequence of the preceding development, we will say that the vector $(z^*; w^*) = (\pi^*, y^*; h^*, u^*)$ solves the spatial computable general equilibrium model SCGE (b, d, A, c) if and only if both system (1)-(5), using regional variables and parameters, and system (13)-(17) are satisfied together. We may combine the two complementarity problems (8) and (20) to create a single problem describing SCGE (b, d, A, c):

$$x = \begin{pmatrix} \pi & y & h & u \end{pmatrix}^{\mathrm{T}} \tag{21}$$

$$H(x) = \begin{bmatrix} b + A(\pi)y - d(\pi) \\ -A(\pi)^T\pi \\ c(h)-\Lambda u \\ \Lambda^{\mathrm{T}}h - \Gamma[A(\pi)]y \end{bmatrix} \tag{22}$$

$$\begin{aligned} x^T H(x) &= 0 \\ H(x) &\geq 0 \\ x &\geq 0 \end{aligned} \tag{23}$$

Rather evidently, variational inequality formulations of the SCGE (b, d, A, c) problem are also possible, allowing one to use results successfully employed for traffic and spatial price equilibrium problems to establish existence and uniqueness and to efficiently compute solutions (Harker and Pang, 1990). Some experience with such problems allows one to conclude that one variational inequality representation of SCGE(b, d, A, c) for the case of a Leontieff technology is: find $(\pi^*, f^*) \in \Omega$ such that

$$[b - d(\pi^*)]^T(\pi - \pi^*) + \tilde{c}(f^*)(f - f^*) \geq 0 \quad \forall(\pi, f) \in \Omega, \tag{24}$$

where

$$\begin{aligned} \Omega \;\equiv\; & \{\pi \in S^{|N|-1} \cap \Re_+^{|N|\times|R|}, f \in \Re^{|L|\times|N|}, s \in \Re^{|N|} : -A^T\pi \geq 0, \Delta h = f, h \geq 0, \\ & \Lambda h - \Gamma(A)A^{-1}[d(\pi) - b - s] = 0\} \end{aligned} \tag{25}$$

Note that because $\pi \in S^{|N|-1} \cap \Re_+^{|N|\times|R|}$, we know $\pi > 0$. From (2) we also know $b + Ay - d + s = 0$ where $s \in \Re^{|N|}$. It follows that $y = A^{-1}(d - b - s)$, which expression was used in (25) to rewrite the flow conservation condition : $\Lambda h - T = \Lambda h - \Gamma(A)y = \Lambda h - \Gamma(A)A^{-1}(d - b - s) = 0$.

The importance of (23) and (24) is that they allow the simultaneous and consistent calculation of economic activities and detailed transportation network flows. This model, by necessity, brings with it the computational disadvantages of both the standard computable general equilibrium models and traffic network equilibrium models. Thus, methods must be devised to avoid complete enumeration of network paths and to utilize effect decomposition based on the structure of the activity matrix A.

One particular drawback of formulation (24) is the fact that the flow conservation constraints are potentially nonlinear in prices. If linear demand functions are employed, this difficulty does not arise and the set Ω is convex. It is likely that the computational disadvantages found by Goldsman and Harker (1990) for nonspatial CGE models with Leontieff technologies will carry over to (24) when the demand functions are non-linear.

Finally, it is important to recognize that SCGE models more general and realistic than (23) or (24) can be created using the same types of arguments presented above. In particular, elastic transportation demand, simultaneous treatment of shippers and carriers, backhauling and endogenous network aggregation can all be added to the above structure.

12.4 Imperfect Competition

One of the key issues in SCGE modeling is representation of the tendency of carriers to collude with one another and to bargain with shippers in setting rates. This is a very difficult topic for it leads one into the theory of cooperative and bargaining games. There are many disparate concepts for the solution of such games (nucleus, Shapely value, propensity to disrupt, etc.). Generally there are not tractable methods for calculation of the solution of large scale cooperative/bargaining games, owing primarily to the fact that such games are usually not reducible to convex mathematical programs or single level variational inequalities, as are Cournot-Nash noncooperative games.

One approach which has been used by Harker (1983) and Harker and Friesz (1985; 1986a; 1986b) to treat bargaining and collusion in freight systems is to employ the concept of a rate reaction function. Such a function is empirically determined and estimates the rate a carrier will charge given other inputs describing the economic environment, especially the volume of traffic between the relevant origin-destination pair. Such functions, because of their empirical basis, are implicitly capable of capturing the collusion and bargaining that go on in real world freight transportation. These functions may be combined with either the Stackelberg leader-follower or the Cournot-Nash view of shipper-carrier interactions to create simultaneous models with essentially the same mathematical and computational characteristics as the models presented in Section 12.3. Unfortunately no tests with real data of the accuracy of the rate reaction function approach have been carried out.

Daughety and Inaba (1981) found in their study of midwestern agricultural goods movements that there was very little difference between rail rates in a perfectly competitive market and rates when railroads acted as one firm (perfect collusion, the polar extreme of perfect competition). This raises the question of whether models treating other types of imperfect competition lying between these two extremes will predict noticeably different behavior. To answer this question such models will have to be created and solved using real data.

Argonne National Laboratory (1985) employed the FNEM software to create a simulation scheme for setting freight rates under conditions of imperfect competition, specifically duopoly and monopolistic competition. This scheme assumes that the rates charged are a percentage of the carrier's cost. This percentage will generally be set to yield freight rates very slightly less than those of the competition. The problem can be quite complicated because of shared infrastructure, congestion externalities and other interdependencies among shippers and carriers. By calculating a cost minimizing flow pattern, multiplying a given route's cost by the aforementioned percentage to obtain an approximate rate, and recalculating the flow pattern based on this rate, one can estimate rates, flows and cost on any route. The Argonne approach was to repeat this sequence of calculations several times, producing an iterative mapping whose fixed point is a consistent rate-flow-cost triple. This approach is quite similar to ideas advanced by Lederer (1989) for determining rate structures in economic networks. However, a variational inequality representation of this rate setting model and conditions ensuring convergence of the iterative scheme have not yet been developed.

Another important aspect of imperfect competition relevant to spatial economic systems is spatial arbitrage. Friesz and Bernstein (1992) have pointed out that the tendency of shippers to promote and take advantage of differentials between local and delivered prices requires the traditional spatial price equilibrium model be modified. In particular, constraints preventing such spatial arbitrage need to be imposed. In their full generality, these constraints can present daunting theoretical and computational challenges.

12.5 Future Research

In this section we briefly describe further research necessary to develop both theoretically sound and computationally tractable SCGE models.

From the point of view of more realistic formulations, one has to consider inclusions of imperfect competition and more general carriers' cost function in formulating a SCGE model. Specifically, such inclusions are:

1. The cost markup strategy for setting rates described above;
2. Constraints to control the extent of spatial arbitrage; and
3. The nature and effect of constraints which endogenously aggregate shippers' delays and costs from carriers' delays and costs.

Analytical and computational implications of such inclusions should be addressed. One of the analytical implications is the difficulty in establishing the uniqueness property of SCGE models when cost functions are nonmonotonic. One may have to explore possible adaptations of the pseudo-monotonicity property of costs functions employed. In particular, the concept of strict pseudo-monotonicity seems to hold potentials to allow the use of cost functions of lesser unrealistic restrictions while ensuring uniqueness of solutions.

Development of efficient solution algorithms is obviously another research topic. Friesz et al. (1994) point out that, although the above formulated SCGE models can be solved using general purpose nonlinear program solvers, the key to efficiently obtaining equilibrium solutions is to develop algorithms which make the best use of the special structures of SCGE models. We briefly discuss the main philosophy in creating such algorithms for the SCGE variational inequality formulations developed in Section 12.3.

There are two classes of iterative algorithms for a variational inequality problem, $VIP(H,\Omega)$. The first class of methods, see for example, Harker and Pang (1990), consists of creating a sequence $\{x^k\} \subseteq \Omega$ such that each $\{x^{k+1}\}$ solves problem $VIP(H^k,\Omega)$:

$$H^k(x^{k+1})^{\mathrm{T}}(y - x^{k+1}) \quad \forall\, y \in \Omega, \tag{26}$$

where $H^k(x)$ is some approximation of $H(x)$. If the approximation takes a linear form, then we have

$$H^k(x) = H(x^k) + A(x^k)(x - x^k) \tag{27}$$

where $A(x^k)$ is and $(n \times n)$ matrix and may take one of the following forms:

$$\begin{aligned}
A(x^k) &= \nabla H(x^k) && \text{(Newton's method)}\\
&\approx \nabla H(x^k) && \text{(Quasi-Newton's method)}\\
&= D(x^k) && \text{(Linearized jacobi method)}\\
&= \begin{cases} L(x^k) + D(x^k)/\omega^* \\ U(x^k) + D(x^k)/\omega^* \end{cases} && \text{(Successive Overrelaxation method)}\\
&= \tfrac{1}{2}[\nabla H(x^k) + \nabla H(x^k)^{\mathrm{T}}] && \text{(Symmetrized Newton's method)}\\
&= G && \text{(Projection method)}
\end{aligned} \tag{28}$$

with

$$\begin{aligned}
D(x^k) &\equiv \text{the diagonal part of } \nabla H(x^k)\\
L(x^k) &\equiv \text{the lower triangular part of } \nabla H(x^k)\\
U(x^k) &\equiv \text{the upper triangular part of } \nabla H(x^k)\\
\omega^* &\equiv \text{a parameter } \in (0,2)\\
G &\equiv \text{a constant, symmetric and positive definite matrix}
\end{aligned}$$

The second class of methods generally transform a variational inequality problem into an equivalent mathematical program with an objective function (generally non-differentiable) called *gap function* first introduced by Auslender (1976). Fukushima (1992) shows that a differentiable gap function can also be constructed for the equivalent mathematical program. Such a differentiable gap function will surely offer greater computational advantages.

The major utility of these algorithms is the fact that they generate a sequence of quadratic programs which can be approached in a variety of ways, including much studied decomposition methods which exploit the special structure of network optimization problems and, hence, SCGE models.

Lastly, regarding the appropriateness of using a Wardropian user equilibrium to describe the freight transportation market, we comment that in fact any other transportation equilibrium model which may be expressed as a nonlinear complementarity problem could be employed instead by following the philosophy for integrating models presented here. In that virtually any equilibrium model may be so expressed, this restriction of a nonlinear complementarity structure for an equilibrium model of the transportation network is not significant. More significant perhaps is the question of whether the paradigm of equilibrium/comparative statics is itself well founded; that question is however beyond of scope of this paper. The interested reader is referred to Friesz et al. (1996; 1998) for a discussion of our emerging perspectives on *disequilibrium* modeling. We anticipate in the near future constructing a "computable general disequilibrium model" which includes considerable transportation network detail and which might be thought of as the disequilibrium counterpart of the model of the present paper.

References

Argonne National Laboratory (1985), *Rate Simulation Software User's Guide*, Environmental Research Division, Report ANL/ER-TM-84-1, vol. VI, Argonne, IL.

Auslender, A. (1976), *Optimization: Numerical Methods*, Masson, Paris.

Bronzini, M. (1980), 'Evolution of a multimodal freight transportation model', *Proceedings Transportation Research Forum*, vol. 21, pp. 475–485.

Daughety, A.F. and Inaba, F.S. (1981), 'An analysis of regulatory change in the transportation industry', *Review of Economics and Statistics*, vol. 53, pp. 246–255.

Friesz, T.L. Gottfield, J.A., Books, R.E., Zielen, A.J., Tobin, R.L. and Meleski, S.A. (1981), *The Northeast Regional Environmental Impact Study: Theory, Validation and Application of a Freight Network Equilibrium Model*, Monograph ANL/ES-120, Argonne National Laboratory, Argonne, IL.

Friesz, T.L., Tobin, R.L. and Harker, P.T. (1983a), 'Predictive intercity freight network models: The state of the art', *Transportation Research*, vol. 17A, pp. 409–417.

Friesz, T.L., Gottfield, J.A. and Tobin, R.L. (1983b), 'The transportation impacts of increased coal haulage: Two case studies', *Transportation Research*, vol. 17A, pp. 505–525.

Friesz, T.L., Gottfield, J.A. and Morlok, E.K. (1985), 'A sequential shipper-carrier network model for predicting freight flows', *Transportation Science*, vol. 20, pp. 80–91.

Friesz, T.L. and Bernstein, D.H. (1992), 'Imperfect competition and arbitrage in separated Markets', in Griffiths, W.E., Lũtkepohl, H. and Bock, M.E. (eds), *Readings in Econometric Theory and Practice in Honor of George Judge*, North Holland, Amsterdam.

Friesz, T.L., Westin, L. and Suo, Z.-G. (1994), 'A nonlinear complementarity formulation of SCGE problems', in *Proceedings of the Workshop on Transportation and Computable General Equilibrium Modeling*, Venice.

Friesz, T.L., Bernstein, D.H. and Stough, R. (1996), 'Dynamic systems, variational inequalities and control theoretic models for predicting time-varying urban network flows', *Transportation Science*, vol. 30, pp. 14–31.

Friesz, T.L., Bernstein, D.H. and Suo, Z.-G. (1998), 'A dynamic disequilibrium interregional commodity flow model', forthcoming in *Transportation Research B*.

Fukushima, M. (1992), 'Equivalent optimization problems and descent methods for asymmetric variational inequalities', *Mathematical Programming*, vol. 53, pp. 99–110.

Goldsman, L. and Harker, P.T. (1990), 'A note on solving general equilibrium problems with variational inequality techniques', *Operations Research Letters*, vol. 9, pp. 335–339.

Harker, P.T. (1983), *Prediction of Intercity Freight Flows: Theory and Application of a Generalized Spatial Price Equilibrium Model*. Ph.D. Dissertation, University of Pennsylvania, Philadelphia, PA.

Harker, P.T. and Friesz, T.L. (1985), 'The use of equilibrium network models in logistics management: With application to the U.S. coal industry', *Transportation Research*, vol. 19B, pp. 457–470.

Harker, P.T. and Friesz, T.L. (1986a, 1986b), 'Prediction of intercity freight flows, I and II', *Transportation Research*, vol. 20B, pp. 139–153.

Harker, P.T. and Pang, J.S. (1990), 'Finite-dimensional variational and nonlinear complementarity problems: A survey of theory, algorithms and applications', *Mathematical Programming*, vol. 48B, pp. 161–220.

Lederer, P.J. (1989), 'The effect of production and transport cost on competitive spatial pricing and firm location', Working paper, University of Rochester, NY.

Mathiesen, L. (1985), 'Computation of economic equilibria by a sequence of linear complementarity problems', *Mathematical Programming Study*, vol. 23, pp. 144–162.

13 A Spatial Computable General Equilibrium Approach for Measuring Multiregional Impacts of Large Scale Transportation Projects

Toshihiko Miyagi
Faculty of Regional Studies
Gifu University
1-1 Yanagido, Gifu 501-11, Japan

13.1 Introduction

In economically lagging regions, infrastructure, transportation infrastructure in particular, is seen as an essential prerequisite for economic development. Even in industrialized regions, where the existing infrastructure's performance is deteriorating due to increasing inadequacy to cope with enlarging travel demand or superannuated transportation systems, there is still interest in the effects of new or improved transportation infrastructure. At the same time, governments are concerned with the cost of providing and maintaining infrastructure when there are pressures to reduce public expenditure. This requires governments or planners to give a rigorous explanation about the needs of the infrastructure that, in turn, implies a need for more accurate assessment of the incidence of the wider economic benefits from the infrastructure.

Recent transportation projects (such as the Channel Tunnel; the European High Speed Rail Network; the Central Bullet Train, the Second Tokai Expressway projects and the Honsyu-Shikoku Bridge Connection project in Japan, etc.) are so large in scale that they will have an impact well outside a single region. Even in the case of smaller scale projects, a major characteristic of transportation infrastructure lies in its strong long-term impact on the development of both interconnected and adjacent regions. Many of the traditional approaches have looked at the impact of individual infrastructure projects on directly affected regions. However, none of these approaches fully captures the sort of changes in which the spatial question affects the overall performance of the economy. This paper does not intend to review the literature on the traditional approaches. The reader is referred to Nijkamp et al. (1984; 1987), Rietvelt (1989) and van den Bergh et al. (1996) for useful discussions and surveys.

The first step towards multiregional analysis on the impact of transportation infrastructure is the use of interregional input-output models. As many researchers have pointed out, however, the input-output approach has some weaknesses, specifically, the assumptions of fixed linear coefficients over time (implying only short-term forecasts), of a homogeneous input (neglecting product diversity), and of pure quantity adjustments (instead of price adjustments) (Nijkamp et al., 1987). Since it usually takes several years for investment in transportation infrastructure to become operational, supply of the infrastructure cannot be readily adjusted in the short-term and, for this reason, the endowment of transportation infrastructure could also be considered as a sort of primary resource for an economic system. The only available approach that can consider quantity and price adjustments within a consistent and multiregional accounting framework is *spatial computable general equilibrium* (SCGE) or Multiregional CGE.

The purpose of this paper is to propose a SCGE model with special attention to the interactions among transportation, housing service and economic activities. Such a model would enable us not only to predict the changes in economic activities, demands for both freight and passenger transportation, and demands for housing services, but also to measure the benefits arising from such changes. The SCGE model described in the current paper is based on the SCGE model developed by Miyagi and Honbu (1993), which is essentially the same structure as the world trade model (Shoven and Whalley, 1974; 1984; Whalley, 1985). Like the other SCGE models (Bröcker, 1992; Roson and Vianelli, 1993; Roson, 1995), the present paper also adopts Armington's assumption to realize multiregional cross-hauling trade. The major difference of the proposed framework from the traditional ones lies in its treatment of the transportation market. In this paper, we propose a hierarchical model structure where the demand side of transportation is taken into account in the upper scheme (the SCGE model), and the supply side of transportation is considered in the lower scheme (the transportation submodel).

One of the technical merits of the SCGE model is that, unlike econometric models, it does not require the collection of a large amount of data and that it can be applied to regional divisions of arbitrary size: A single collective of regions for which an input-output account is available is divided into several sub-regions. Parameters included in production functions of industrial sectors and utility functions of households located in each region can be identified by calibration technique to replicate the benchmark equilibrium data obtained from the intraregional input-output table for the single collective region in question. Miyagi et al. (1997) applied the SCGE model to the estimation of the Japanese national account consisting of nine major regions, with the national account being the benchmark equilibrium data, and obtained highly compatible estimates with the surveyed interregional accounts.

The subsequent sections consist of the following: In the next section the issues that the SCGE model is supposed to be used for are specified in more detail. Section 13.3 treats the formulation of the model: the demands for both freight and passenger transportation are modeled within the SCGE framework and housing service industry is incorporated so as to reflect the interrelationship between housing demand and passenger transportation. International trade and the role of government are neglected in the current models. However, both can be clearly handled within the same framework considered here. Section 13.4 is devoted to discussion about linking the SCGE model and network equilibrium models to combine the output of the SCGE model with freight and passenger flows in transportation networks.

13.2 SCGE and Land-Use/Transportation Model

In order to evaluate the impacts of transportation improvements, the development of a land use/transportation equilibrium model is required. In this section we discuss the incorporation of land use and transportation interactions into the SCGE model in a consistent way and analyze the range of applications of the SCGE model developed in this paper.

Many of the integrated land use/transportation models so far developed have been founded on various existing theories such as urban economic theory, entropy/information theory, random utility theory and so on (Webster et al., 1988). This implies that submodels describing different economic agents may have different theoretical backgrounds. A unified theory may be needed to explain mechanisms of interaction among urban activities, land use and transportation in an economic system.

The application of general equilibrium analysis to spatial interaction may be justified for the following reasons: First, in all the traditional land use/transportation interaction models, allocation of population depends upon the distribution of employment, and so forecasts of employment location, whether exogenously prepared or endogenously estimated, are an essential component (Webster et al., 1988). Furthermore, the location of employment not only determines the distribution of the destination ends of work trip patterns, but it can also determine the distribution of trip ends for other types of travel. SCGE can determine the employment location of each industrial sector from the rational economic behavior of firms, being able to take account of interactions among various industries and technological innovation that may affect employment. Second, prices or rents are very important components in measuring the level of economic activities and the effects of policy changes. Land price and wage rate may affect the distribution of employment, thereby affecting the distribution of population. SCGE can determine a price system within the demand/supply equilibrium framework and thereby can measure the effects of the changes in economic systems on the utility levels of households by the compensated variations or the equivalent variations.

However, a direct application of the general equilibrium analysis to land use/transportation modeling would cause some difficulties. First, we have to deal with land or housing as one of the goods transacted in the market. But, unlike other commodities, since each piece of land is immobile, it is associated with a unique location in geographical space. This implies that the preferences of each household exhibit strong nonconvexity (Schweizer et al., 1976; Fujita, 1989). For establishing competitive equilibria, we usually have to assume that consumption spaces as well as preference preorderings of individuals are convex. This implies that we have to assume that every household has a fixed residential location from the beginning and can consume only those commodities available at or being transported to the location (Karmann, 1981). If we allow for locational choice of economic agents, a continuum of households (a continuum population or density distribution approach) will be necessary to avoid nonconvexity of demand (Karmann, 1981; Asami et al., 1987). In order to avoid the inherent complexity with household's residential location, I assume that both population and employment of each region are exogenously given and fixed.

The second problem is concerned with the description of space. Since its emergence thirty years ago, the equilibrium land use theory has been refined in the seminal works by Alonso (1964), Muth (1969), Mills (1972), and Solow (1973), among others. Most urban economic models have been concentrated mainly on continuous space in which location of business is restricted to only an exogenously given central market place where all trades take place (Mills, 1967; Henderson, 1977; Fujita, 1989). However, we have to handle multiregional aspects in land use/transportation interaction so that location must be discrete and business location must be multicentric.

The third problem is how to treat transportation. Differences in market prices of the same commodity between any pair of destinations would simply reflect differences in transport costs, whereas if local producers do not take market prices as given, price differences may also reflect some degree of price discrimination. Spatial price discrimination can be interpreted in either of two (effectively equivalent) ways. It occurs either as freight absorption by the producing sector or, equivalently, as the producer charging different production prices of consumers at different locations (Greenhut et al., 1987). The first approach may be called the so-called *iceberg model*. The second one may share the same meaning as *supply and demand pool concept* where all flows of a particular commodity are treated as being routed via a supply pool in the region of production and a pool in the region of absorption (Moses, 1955; Chenery, 1953). This pooling approach explicitly recognizes that the products of the same sector from different regions are likely to be more readily substitutable for each other than the products of different sectors, and corresponds to the Armington assumption. In this paper we use the demand pool concept, where the trader at each region undertakes exchange activities of the commodities sent out from all regions. *A trader* can also be called *a shipper*, following Harker (1987) and Friesz et al. (1984), which corresponds to a wholesaler or a retailer in real economic activities and plays a role in the demand-side of the freight market. Demands for passenger transportation are obtained from the household's expenditure minimization behavior. On the other hand, the supply side of the transportation market consists of *carriers* who are responsible for both freight and passenger movements. The supply-demand interactions on the transport market are dealt with in the transportation submodel to predict both freight and passenger flows in the transportation networks. We assume that while the demand for transportation between locations is not affected by congestion in the transportation network, suppliers of transportation services are affected by the level of congestion in the transportation network, which causes changes in the share of transportation modes or in shipping routes between locations.

13.3 The Basic Structure of the Model

13.3.1 SCGE model

Each region has a production structure described by industry production functions with constant returns to scale, which specify input requirements of both factors and other goods. Factor input requirements are represented by value-added functions defining substitution possibilities over capital and labor in each industry, and for housing industry land is added as one of the factors. Intermediate input requirements are represented by technical coefficients specifying substitution possibilities across production sources. All endowments are assumed to be given and fixed for each region for the moment. However, the assumption of fixed labor forces will be relaxed in the latter part of the analysis. Each region also has a system of market demand functions. We distinguish between mobile and immobile commodities.

In our model, we consider I+1 commodities of which I commodities are assumed to be transportable through space represented by a discrete set of possible locations (we call these locations *regions*) indexed by {1,2...r...S} and the (I+1)th commodity is treated as an immobile commodity, housing. We may interchangeably use the superscript h instead of I+1. In each region, the firm j produces the output j with intermediate inputs i=1,2, ...I which are bought from traders and with factors, generally indexed by $k = 1,2,\cdots,K$, and a single representative household purchases (I+1) final goods. Commodities are handled as heterogeneous across regions according to the Armington assumption. For that purpose the nested CES function (Sato, 1967) is one of the relevant functions.

Firm's Behavior. Let X_r^i, F_r^i be the output of good i in region r, and the value added in producing good i in region r, owned by the households in region r, respectively. And let x_r^{ij} denote the use of composite good i used up in producing goods j in region r. Then, for region s, the production functions on the upper level of two-level nested CES production functions can be written as

$$X_s^j = \min\left[\frac{F_s^j}{c_s^j}, \frac{x_s^{1j}}{a_s^{1j}}, \ldots, \frac{x_s^{ij}}{a_s^{ij}}, \ldots, \frac{x_s^{Ij}}{a_s^{Ij}}, \frac{x_s^{I+1,j}}{a_s^{I+1,j}}\right], \; j = 1,2,..,I+1 \tag{1}$$

where c_s^j is the value-added requirement per unit of output of good j and $\{a_s^{ij} : i = 1,2,\ldots,I+1\}$ are the fixed composite-good requirements in region s per unit production of good j. That is, we assume a perfect complementarity among intermediate goods and the value added in producing them. If we assume that F_s^j are also represented by CES functions, with substitutive factors, then the corresponding unit cost function of output j in region s, cf_s^j, can be described as

$$cf_s^j(\mathbf{q}_s, \mathbf{w}_s, x_s^j = 1 : \{\alpha^{ij}\}, \{\gamma^{kj}\}) = \sum_{i=1}^{I+1} \alpha^{ij} q_s^i + \left[\sum_{k=1}^{K} \gamma^{kj} (w_s^k)^{1-\sigma^j}\right]^{\frac{1}{1-\sigma^j}} \tag{2}$$

where:

α^{ij} technical coefficient,

q_s^i CIF-price of commodity i in region s,

w_s^k factor price in region s,

σ^j elasticity of substitution of commodity j,

γ^{kj} share parameter.

Let p_s^j be price of output j in region s. Then, in equilibrium

$$p_s^j = cf_s^j, \quad s = 1,...,S; \quad j = 1,...,I+1 \tag{3}$$

must be achieved. It follows from Shephard's lemma that the derivatives of the unit cost function with respect to the prices provide the input and the factor demands for achieving the unit production, thus, we have

$$a_s^{ij} = \frac{\partial cf_s^j}{\partial q_s^i} = \alpha^{ij} \tag{4}$$

$$c_s^{kj} = \frac{\partial cf_s^j}{\partial w_s^k} = \gamma^{kj}(w_s^k)^{-\sigma^j}\left[\sum_{k=1}^{K}\gamma^{kj}(w_s^k)^{1-\sigma^j}\right]^{\frac{\sigma^j}{1-\sigma^j}} \tag{5}$$

where c_s^{kj} is the kth value-added coefficient. With the CES being homogeneous of degree one in mind, from Euler's law (2) can be written as

$$p_s^j = \sum_{i=1}^{I+1}\alpha^{ij}q_s^i + \sum_{k=1}^{K}c_s^{kj}w_s^k, \quad j = 1,...,I+1; \quad s = 1,...,S. \tag{6}$$

Note that (6) represents cost-covering prices for commodities, which are determined independently from the quantities traded.

Trader's Behavior. We assume that there are the same number of traders as the number of traded commodities in each region, each of which transports i commodity from region r to s with *ad valorem* distortions, η_{rs}^i, from production prices:

$$v_{rs}^i = p_r^i(1+\eta_{rs}^i) \tag{7}$$

where $\eta_{rs}^i = 0\ (r \neq s)$ is assumed for immobile commodities. $\{\eta_{rs}^i\}$ should be specified as a function of interregional transport times so as to reflect the effects of transportation investments. The demand pool concept with the Armington assumption allows one to specify product differentiation by region of origin into the structure of firm's demand for intermediate goods. In other words, a trader in a region sells a single composite good to all firms in the region that require that good, where the composite good implies the commodities of the same kind coming from different regions. The price of a composite good is endogenous, since it is a function of the endogenous prices of various commodities as is expressed in (6). The degree of substitution across regions can be described by the CES function. The cost minimization behavior of the traders under the CES technology induces the following unit cost function with an elasticity of substitution for the transportation sector, σ_t:

$$ct_s^i(\mathbf{v}_s^i) = \left[\sum_{r=1}^{R} \theta_r^i (v_{rs}^i)^{1-\sigma_t}\right]^{\frac{1}{1-\sigma_t}} \tag{8}$$

In equilibrium,

$$q_s^i = ct_s^i(\mathbf{v}_s^i), \quad i = 1,\ldots,I+1; \quad s = 1,\ldots,S. \tag{9}$$

As is shown in the theory of firm's behavior, the derivatives of ct_s^i with respect to p_r^i give trade coefficients, T_{rs}^i:

$$T_{rs}^i = \frac{\theta_r^i [v_{rs}^i]^{1-\sigma_t}}{\sum_{r=1}^{R} \theta_r^i [v_{rs}^i]^{1-\sigma_t}} \cdot \frac{q_s^i}{p_r^i} \tag{10}$$

From this, equilibrium prices of composite good i are expressed as:

$$q_s^i = \sum_{r=1}^{R} p_r^i T_{rs}^i$$

Household Behavior. We assume that the representative household which is located in region r would maximize the utility consisting of pair of utilities, the ordinary utility for consumption of goods u_r^0 and the locational utility u_r^ℓ, i.e.,

$$u_r = u_r^0(\mathbf{d}_r) + u_r^\ell(d_r^\ell, \mathbf{t}_r) \tag{11}$$

under the budget constraint:

$$Y_r = \sum_{i=1}^{I} q_r^i d_r^i + q_r^h d_r^h + \sum_{s=1}^{S} t_{rs} c_{rs} \tag{12}$$

where:

Y_r the fixed amount of money available to consumers living at r,

d_r^i the amount of good i demanded by the household located at region r,

d_r^h housing services demanded in region r,

t_{rs} trips originated from r and destinated to s,

c_{rs} transportation cost between r and s.

Providing that utilities of households are completely described by the nested CES, it then follows from the property of the CES that the CES expenditure function can be described by

$$e_r(\mathbf{q}_r, \mathbf{c}_r, u_r) = u_r \bar{c}(\mathbf{q}_r, \mathbf{c}_r) \tag{13}$$

with the CES unit cost function $\bar{c}(\mathbf{q}_r, \mathbf{c}_r)$ (Varian, 1992). We will define the nested CES expenditure function as follows:

$$e_r = u_r \left[(z_r)^{1-\sigma} + (\ell_r)^{1-\sigma} \right]^{\frac{1}{1-\sigma}} \tag{14}$$

where σ is an elasticity of substitution between the expenditure associated with consumption of general goods and the expenditure arising from being located in region r. Both z_r and ℓ_r are defined by

$$z_r = \left[\sum_{i=1}^{I} \delta^i (q_s^i)^{1-\sigma_n^i} \right]^{\frac{1}{1-\sigma_n^i}} \tag{15a}$$

$$\ell_r = \left[\delta_r^h (q_r^h)^{1-\sigma^L} + \sum_{s=1}^{S} \kappa_s (c_{rs})^{1-\sigma^L} \right]^{\frac{1}{1-\sigma^L}} \tag{15b}$$

where:

$\delta^i, \delta_r^h, \kappa_s$ positive parameters,

q_r^h price of housing service in region r,

σ_n^i elasticities of substitution among commodities,

σ^L elasticity of substitution between selection of residence type and transportation.

We can find the indirect utility function by inverting the equation (14):

$$V_r(\mathbf{q}_r, \mathbf{c}_r, Y_r) = Y_r \left[(z_r)^{1-\sigma} + (\ell_r)^{1-\sigma} \right]^{-\frac{1}{1-\sigma}} \tag{16}$$

where V_r is the indirect utility function. Since households earn their income by selling exogenously fixed amounts of factors to firms, the equilibrium level of income is obtained from the equality

$$Y_r = \sum_{k=1}^{K} \sum_{s=1}^{S} F_{rs}^k w_s^k \tag{17}$$

where $\mathbf{F}_r$ is the factor endowment owned by households in region r and it is assumed that $F_{rs}^k > 0$ for k=1 (labor) and $F_{rs}^k = 0 (r \neq s)$ for k=2 (capital) and k=3 (land). Let $\{d_r^i\}$, $\{d_r^h\}$ and $\{t_{rs}\}$ denote demands for commodity i in region r, demand for housing services (floor-space demand of residence) in region r and trips from r going toward s, respectively. The demand functions for $\{d_r^i\}$, $\{d_r^h\}$ and $\{t_{rs}\}$ can be found by Roy's law:

$$d_r^i = -\frac{\partial V_r}{\partial q_r^i} / \frac{\partial V_r}{\partial Y_r} = \frac{\delta^i (q_r^i)^{1-\sigma_n^i}}{\sum_i \delta^i (q_r^i)^{1-\sigma_n^i}} \cdot \frac{(z_r)^{1-\sigma}}{(z_r)^{1-\sigma} + (\ell_r)^{1-\sigma}} \cdot \frac{Y_r}{q_r^i}$$

$$d_r^h = -\frac{\partial V_r}{\partial q_r^h} / \frac{\partial V_r}{\partial Y_r} = \frac{\delta_r^h (q_r^h)^{1-\sigma^L}}{\Gamma_r} \cdot \frac{(\ell_r)^{1-\sigma}}{(z_r)^{1-\sigma} + (\ell_r)^{1-\sigma}} \cdot \frac{Y_r}{q_r^h}$$

$$t_{rs} = -\frac{\partial V_r}{\partial c_{rs}} / \frac{\partial V_r}{\partial Y_r} = \frac{\kappa_s (c_{rs})^{1-\sigma^L}}{\Gamma_r} \cdot \frac{(\ell_r)^{1-\sigma}}{(z_r)^{1-\sigma} + (\ell_r)^{1-\sigma}} \cdot \frac{Y_r}{c_{rs}}$$

where

$$\Gamma_r = \delta_r^h (q_r^h)^{1-\sigma^L} + \sum_s \kappa_s (c_{rs})^{1-\sigma^L}$$

The demand equations above can be further simplified as:

$$d_r^i = g_{r1} \frac{\delta^i (q_r^i)^{1-\sigma_n^i}}{\sum_i \delta^i (q_r^i)^{1-\sigma_n^i}} \cdot \frac{Y_r}{q_r^i} \tag{18a}$$

$$d_r^h = g_{r2} \frac{\delta_r^h (q_r^h)^{1-\sigma^L}}{\Gamma_r} \cdot \frac{Y_r}{q_r^h} \tag{18b}$$

$$t_{rs} = g_{r2} \frac{\kappa_s (c_{rs})^{1-\sigma^L}}{\Gamma_r} \cdot \frac{Y_r}{c_{rs}} \tag{18c}$$

where:

$$g_{r1} = \frac{(z_r)^{1-\sigma}}{(z_r)^{1-\sigma} + (\ell_r)^{1-\sigma}}$$

$$g_{r2} = \frac{(\ell_r)^{1-\sigma}}{(z_r)^{1-\sigma} + (\ell_r)^{1-\sigma}}$$

g_{r1} and g_{r2} are the rate of expenditure to general goods consumption and the rate of expenditure to housing and transportation services. It is possible to find those variables from the survey of family income and expenditure. Alternatively, if some other data source on passenger travel demand is available, parameters included in (18b) and (18c) can be calibrated. More detail will be addressed in the next subsection.

Until now, we have not paid attention to the time spent in various activities. The value of time, the money values that people place on saving various forms of travel time, is one of the major components of travel demand modeling. A simple form of time budget constraint for people living in region r may be written as:

$$T_r^{\ell} + T_r^{w} + T_r^{t} = \overline{T}$$

where T_r^{ℓ}, T_r^{w} and T_r^{t} are times spent at leisure activity, work and travel, respectively. $\overline{T}$ is total time available and is assumed to be the same for all people. Replacing the left hand side of the usual income constraint (12) by the sum of nonwage income Y^N and wage income $w_r T_r^w$ (where w_r is the wage rate), this equation can be written as follows, after trivial manipulation:

$$w_r(\overline{T} - T_r^{\ell}) + Y_r^N = \sum_i q_r^i d_r^i + q_r^h d_r^h + \sum_s t_{rs}(c_{rs} + \beta_t \tau_{rs}) \tag{19}$$

In the above equation, β_t and τ_{rs} denote the value of travel time and the travel time spent on one trip between r and s, respectively. Therefore, demand models (18a)-(18c) still hold even if we replace Y_r and c_{rs} by $w_r(\overline{T} - T_r^{\ell}) + Y_r^N$ and the generalized cost $(c_{rs} + \beta_t \tau_{rs})$, respectively.

Walras Equilibrium. Note that from Euler's law for a differentiable function that is homogeneous of degree one, the following equality holds:

$$z_r = \sum_i d_r^i q_r^i \tag{20a}$$

$$\ell_r = d_r^h q_r^h + \sum_s t_{rs} c_{rs} \tag{20b}$$

$$e_r = z_r + \ell_r \tag{20c}$$

We assume that passenger transport sectors require no production input and that the expenditure on transportation by the household is offset by revenue of the passenger transport sector. This implies that (20c) must be

$$e_r = \sum_i d_r^i q_r^i + d_r^h q_r^h \tag{20c'}$$

In equilibrium, factor markets have to clear:

$$\sum_{i=1}^{I+1} \overline{E}_r^i = \sum_{i=1}^{I+1} c_r^{1i} X_r^i, \quad r = 1,\dots,S \tag{21a}$$

$$\sum_{i=1}^{I+1} \overline{K}_r^i = \sum_{i=1}^{I+1} c_r^{2i} X_r^i, \quad r = 1,\dots,S \tag{21b}$$

$$\overline{L}_r = c_r^{3h} X_r^h, \quad r = 1,\dots,S \tag{21c}$$

$$\begin{pmatrix} \textit{Factor} \\ \textit{ownership} \\ \textit{by households} \end{pmatrix} \quad \begin{pmatrix} \textit{Factor} \\ \textit{used} \\ \textit{by industries} \end{pmatrix}$$

with outputs obtained from the standard linear system of I-O equations:

$$X_r^i = \sum_s T_{rs}^i \left(\sum_j \alpha^{ij} X_s^j + d_s^i \right), \quad r = 1,\ldots,S; i = 1,\ldots,I+1. \tag{22}$$

In (21), we replace $F_r^k (k = 1,2,3)$ by $\overline{E}_r$ (the number of employed persons), $\overline{K}_r$ (capital stock) and $\overline{L}_r$ (land), respectively. Note that (22) equilibrate supplies to demands in monetary terms, i.e., zero-profit conditions hold for all industries in all regions:

$$p_r^i X_r^i = \sum_s \widetilde{T}_{rs}^i \left(\sum_j \alpha^{ij} q_s^i X_s^j + q_s^i d_s^i \right), \quad r = 1,\ldots,S; i = 1,\ldots,I+1 \tag{23}$$

where $\widetilde{T}_{rs}^i$ is the alternative form of the trade coefficient obtained from (10), being defined as:

$$\widetilde{T}_{rs}^i = T_{rs}^i \frac{p_r^i}{q_s^i} = \frac{\theta_r^i (v_{rs}^i)^{1-\sigma_i}}{\sum_r \theta_r^i (v_{rs}^i)^{1-\sigma_i}}$$

13.3.2 Calibration

In specifying parameter values in a general equilibrium model, calibration of each model to a single equilibrium data set is used, in combination with a literature search or some other selection procedure for key parameters (usually elasticities) (Whalley, 1985). More detail on the calibration procedure of the conventional SCGE or CGE models can be seen in Whalley (1985), Shoven and Whalley (1992), Brocker (1992), Miyagi and Honbu (1993). Miyagi and Honbu (1993), using the input-output accounting of the Gifu prefecture, show that the calibration method for the SCGE models developed so far provides a unique set of parameter values. However, it does not take the interaction between travel and housing demand models into account.

The model developed here needs other data sets to identify parameters included in the transportation demand model. Calibration of parameters included in the travel demand model is a little intricate, thus additional explanation may be needed. Suppose that the locational elasticity σ^L is given. Interregional trips $\{t_{rs}\}$ given by (18c) may be linked with the employment variables through the simple relation

$$\sum_r t_{rs} = \lambda_s E_s \tag{24}$$

where λ_s denotes the number of trips generated by one unit of total employment in region s. Likewise, the relationships between generated trips at each region and population or the number of households at the region can be described by the following equation:

$$\sum_s t_{rs} = \mu_r M_r = \mu_r \omega Q_r \tag{25}$$

where μ_r denotes the number of trips generated by one unit of total population in region r and M_r is the population of region r. If the number of households at each region $\{Q_r\}$ is available, then using the average size of a household, ω , the second relation may be used. Replacing separately the terms associated with region r and region s in (18c) by

$$\frac{Y_r g_{r2}}{\Gamma_r} = \alpha_r M_r \quad \text{and} \quad \kappa_s = \beta_s E_s \tag{26}$$

respectively, we have the following doubly constrained gravity model with the balancing factors α_r, β_s:

$$t_{rs} = \alpha_r \beta_s M_r E_s c_{rs}^{-\sigma^L} \tag{27}$$

Using the balancing factor method (Wilson et al., 1981) we can obtain the balancing factors as:

$$\alpha_r = \frac{\mu_r}{\sum_s \beta_s E_s c_{rs}^{-\sigma^L}} \quad \text{and} \quad \beta_s = \frac{\lambda_s}{\sum_r \alpha_r M_r c_{rs}^{-\sigma^L}} \tag{28}$$

Now when the balancing factors have been obtained, κ_s can be determined from the second equation in (26). Since income, Y_r, and parameters associated with commodity consumption, δ^i , are already found in the previous calibration steps (we assume that the rate of expenditure to general goods consumption is given), both g_{r2} and Γ_r become functions of only δ_r^h . Therefore, the first equation of (26) can be used to calibrate δ_r^h . All these parameters are required to calculate housing demand d_r^h in (18b). Housing demand is consistently related to trip demand as well as employment in this way.

13.3.3 Solving the SCGE Model

In order to find market prices arising from improvements in transportation systems, the nonlinear equation systems (6) together with (9) should be solved under a given set of factor prices. The Newton-Raphson method is a convenient, widespread method for that purpose. Substituting prices $\{p_s^j, q_r^i\}$ into (10), we can calculate trade coefficients. Because income is given by (17), the utility level in region s is obtained from (14) by putting $Y_s = e_s$. Then the final demands are computed from (18a)-(18c), which are substituted into the I-O equation to obtain the production of commodity i in region r. Due to price changes, the demand for travel may shift. However, the changes in travel pattern must be consistent with the exogenously given demographic data. This will also be ensured by applying the balancing factor method so that only the balancing factors are adjusted. The value of X_r^i obtained from (22) can then be used in (21) to determine each E_r^j, K_r^j and L_r that yield the excess-factor-demand system,

$$\sum_{j=1}^{J+1} E_s^j = \overline{E}_s, \qquad s = 1, \ldots, S \tag{29a}$$

$$\sum_{j=1}^{J+1} K_s^j = \overline{K}_s, \qquad s = 1, \ldots, S \tag{29b}$$

$$L_s = \overline{L}_s, s = 1, \ldots, S \tag{29c}$$

Since $\{q_r^h\}$ are determined by solving cost-covering prices system, (6) and (9), and $\{c_{rs}\}$ are given, the unknown variables are only factor prices in (29). Thus, the number of equations coincides with that of the unknown variables. Owing to changes in factor prices, value-added coefficients should be revised by (5). Thus, an iterative calculation is repeated until a stable result is obtained. If a set of factor prices is found such that the excess factor demands, (21), in each region are zero, a complete general equilibrium has been found. Demand-supply equalities in commodity markets must prevail since factor demands are derived so as to meet commodity demands. At the same time, zero-profit conditions hold by finding cost-covering prices. Therefore, Walras's law holds and this implies that trade balance must hold for each region (Shoven and Whalley, 1992). The main information flow is depicted in Fig. 13.1.

In order to compare equilibria arising from different options, we need an aggregate measure of welfare. Kanemoto and Mera (1985) examined several economic issues associated with benefit evaluation of a large transportation project in a simple general-equilibrium framework. See Kanemoto and Mera (1985) for more details.

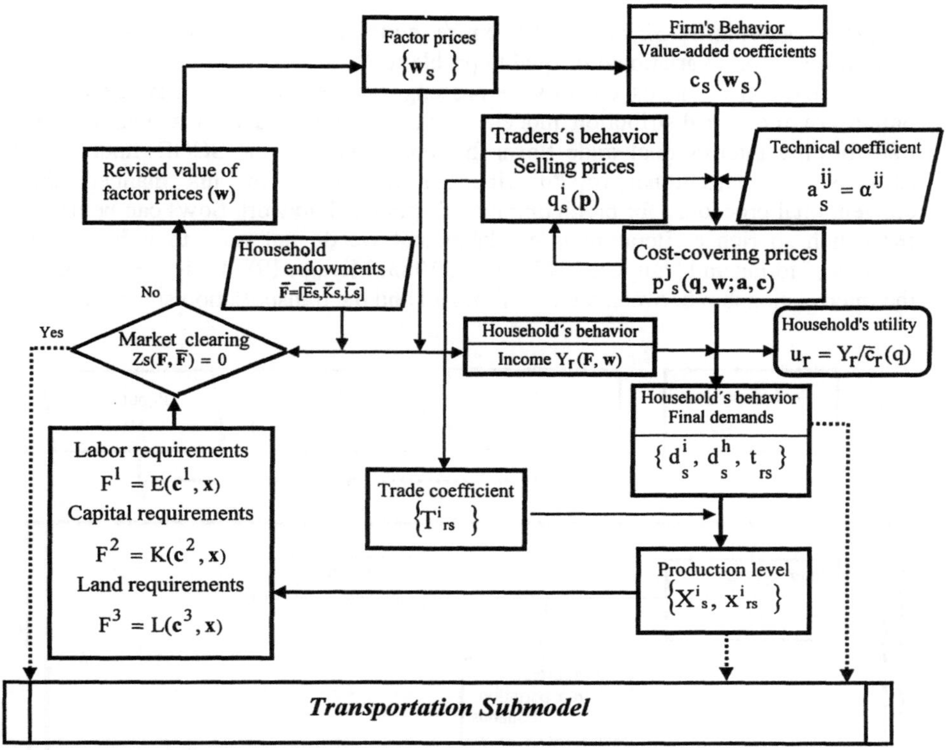

Fig. 13.1. Main information flow in SCGE model

13.4 Linking SCGE and Transportation Equilibrium

Transport services are provided by the so-called carriers who actually move goods and carry passengers between various origin-destination (OD) pairs. Thus, while behavior of shippers and households account for the demand side of the transportation market, the carriers enter into the transportation market as the supplier and determine the service levels of both freight and passenger transportation between each OD pair. In this section we propose a hierarchical model structure to treat the transportation market and discuss the interactions between supply and demand in the transportation market to some extent.

The transportation submodel is used to estimate both trips by each transportation mode and the flows and congestion levels on transportation networks. However, the present model assumes that the shippers already have information on congestion costs and make up an account based on those costs as delivered prices. Therefore, no information flow from the transportation submodel to shipper's behavior is conveyed. To take into account the interactions between network congestion and shipper's decision-making needs a more complicated framework that constitutes a reference scheme for the development of an integrated network/SCGE model. The entire model must be expressed as a fixed-point problem or a variational inequality problem.

Interregional demands consists of passenger travel demands that are direct outputs of the trip distribution model, (18c), and freight demands obtained by transforming interregional trade flows that are outputs of the SCGE model, i.e. (22). Since interregional trip distributions are given in this manner, the conventional procedure for predicting modal splits and network flows can be used, for which so rich a literature is available that we do not refer to it here (for example, Florian and Nguyen, 1978; Sheffi, 1985; Miyagi, 1989). Fig. 13.2 depicts the interrelationship between the SCGE model and the transportation submodel.

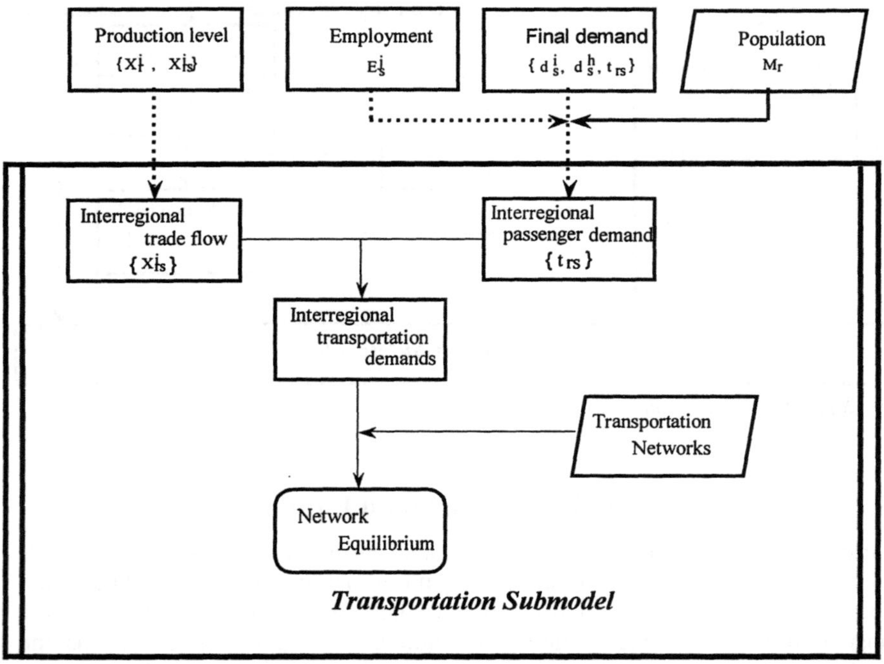

Fig. 13.2. Transportation submodel

An elaborate approach is proposed by Harker (1987), Harker and Friesz (1986a; 1986b) for analyzing the freight transportation market. As the demand side of the freight and passenger transportation has been already mentioned, we will turn to the supply side of the market. Suppose that there are several carriers who have their own transportation modes and routes, and who transport from region r to region s. Since the shipper must, in general, choose a sequence of carriers to move his good, transport cost depends on the shipper's decision on what type of combinations of the set of carriers to choose, and on the rate of transport offered by the carriers as well. We assume that each carrier behaves so as to maximize his profit. If the transportation market is regulated, then each carrier takes price as given and his optimal strategy is described by the traditional system optimal traffic assignment problem. More details about carrier behavior can be seen in Harker (1987) and no further attempt to explore the characterization of carrier behavior is done here.

13.5 Concluding Remarks

I have tried to show how the SCGE model is coupled with the land use/transportation model. The SCGE model developed here is different from the traditional land use/transportation models in that the SCGE approach can nest various models associated with land use and travel demands into one wisdom-box and provide a theoretically sophisticated model. In addition, the use of calibration reduces the efforts that are normally spent on statistical parameter estimation problems. A prominent feature of calibration is that no statistical test of the model specification is used, because a deterministic procedure of calculating parameter values from the equilibrium observation is adopted. This deterministic nature of parameter estimation encourages us to use the neural network technique as an alternative method for calibration of parameter values. However, the calibration technique may bring about another problem: since there is no method for testing the validity of calibrated parameters, we cannot evaluate equilibria arising from the use of different types of production and utility functions.

One of the greatest difficulties in applying the SCGE model arises from complicated computation work. In this paper, we have adopted a sequential method using the Newton-Raphson procedure, instead of a more rigorous, but intricate, fixed-point method. Unfortunately, the method presented here is not proved to converge to a unique equilibrium state. We can show the coincidence between the number of equations and unknowns; however, it is not very persuasive. Practical experience may be required to show the validity of the procedure.

It is obvious that other aspects associated with land use/transportation modeling are left open. The most important one among them seems to be how to handle externalities. The essence of cities is the presence of many people and firms in close quarters. Public services, traffic congestion and pollution all involve externalities. Another issue, which is a rather technical one, but important for SCGE applications, is how to generate "benchmark equilibrium data". If the region in question is wide enough to be able to utilize the existing I-O table, the difficulties are tractable. However, how is one to generate such a set of data for city-level applications? One possible way to obtain the input-output accounting for a city is to use a two-stage method where the estimated input-output accounting by the SCGE model application is further used as the reduced form of the benchmark equilibrium data. These problems should be addressed in further research.

References

Alonso, W. (1964), *Location and Land Use*, Harvard University Press, Cambridge, MA.

Asami,Y., Fujita, M. and Smith, T.E. (1987), 'On the foundations of land use theory', *Working Papers in Regional Science and Transportation'*, no. 112, University of Pennsylvania, Philadelphia, PA.

Bergh, J.C.J.M. van den, Nijkamp P. and Rietveld, P. (1996), 'Spatial equilibrium models: A survey with special emphasis on transportation', in Bergh, J.C.J.M. van den, Nijkamp, P. and Rietveld, P. (eds), *Recent Advances in Spatial Equilibrium Modelling*, Springer-Verlag, Berlin, pp. 48–76.

Bröcker, J. (1992), 'Operational spatial computable general equilibrium modeling', A paper presented at the International Seminar on Trade, Knowledge and the Network Economy, Mallacoota, Australia.

Chenery, H.B. (1953), 'Regional analysis', in Chenery, H.B., Clark, P.G. and Cao-Pinna, V. (eds), *The Structure and Growth of Italian Economy*, U.S. Mutual Security Agency, Rome.

Florian, M. and Nguyen, S. (1978), 'A combined trip distribution, modal split and trip assignment model', *Transportation Research*, vol. 12, pp. 241–246.

Friesz, T.L., Harker, P.T. and Tobin, R.L. (1984), 'Alternative algorithms for the general network spatial price equilibrium problem', *Journal of Regional Science*, vol. 24, pp. 475–507.

Fujita, M. (1989), *Urban Economic Theory*, Cambridge University Press, Cambridge, MA.

Greenhut, M.L., Norman G. and Hung, C. (1987), *The Economics of Imperfect Competition. A Spatial Approach*, Cambridge University Press, Cambridge, MA.

Harker, P.T. (1987), *Predicting Intercity Freight Flows*, VNU Science Press BV, Utrecht.

Harker, P.T. and Friesz, T.L. (1986a), 'Prediction of intercity freight flows, I: Theory', *Transportation Research -B*, vol. 20B, pp. 139–153.

Harker, P.T. and Friesz,T.L. (1986b), 'Prediction of intercity freight flows, II: Mathematical formulations', *Transportation Research -B*, vol. 20B, pp. 155–174.

Henderson, J. V. (1977), *Economic Theory and the Cities*, Academic Press, New York, NY.

Karmann, A.J. (1981), *Competitive Equilibria in Spatial Economies*, Oelgeschlager, Gunn & Hain, Cambridge, MA.

Kanemoto, Y. and Mera, K. (1985), 'General equilibrium analysis of the benefits of large transportation improvements', *Regional Science and Urban Economics*, vol. 15, pp. 343–363.

Mills, E.S. (1967), 'An aggregate model of resource allocation in a metropolitan area', *American Economic Review*, vol. 57, pp. 197–210.

Mills, E.S. (1972), *Studies in the Structure of the Urban Economy*, Johns Hopkins University Press, Baltimore, MD.

Miyagi, T. (1989), 'A combined residential-location and transportation network equilibrium model', *Proceedings of the 5th World Conference on Transport Research*, vol. 4, pp. 123–137.

Miyagi, T. and Honbu, K. (1993), 'Estimation of interregional trade flows based on the SCGE model', *Proceedings of Infrastructure Planning* (in Japanese), no. 16, pp. 879–886.

Miyagi, T., Ohashi, K. and Honbu, K. (1997), 'Measuring the impacts of the projected expressway networks in the Chyubu region: SCGE approach', *Planning Administration and Policy Science* (in Japanese), vol. 20, pp. 69–79.

Moses, L. (1955), 'The stability of interregional trading patterns and input-output analysis', *American Economic Review*, vol. 45, pp. 803–832.

Muth, R.F. (1969), *Cities and Housing*, University of Chicago Press, Chicago, IL.

Nijkamp, P., Rietveld, P. and Snickars, F. (1984), 'A survey of multiregional economic models', in Bahrenberg, G., Fischer, M.M. and Nijkamp, P. (eds), *Recent Developments in Spatial Data Analysis: Methodology, Measurement, Models*, Gower, Aldershot, pp. 93–115.

Nijkamp, P., Rietveld, P. and Snickars, F. (1987), 'Regional and multiregional economic models: A survey', in Nijkamp, P. (ed.), *Handbook of Regional and Urban Economics, Vol. I: Regional Economics*, North-Holland, Amsterdam, pp. 257–294.

Rietveld, P. (1989), 'Infrastructure and regional development: A survey of multiregional economic models', *The Annals of Regional Science*, vol. 23, pp. 255–274.

Roson, R. and Vianelli, P. (1993), 'Developing a multi-regional CGE model for freight transport analysis', *Proceedings of International Workshop on Transportation and Spatial CGE models*, Venice.

Roson, R. (1996), 'The macroeconomic impact of traffic congestion: A CGE analysis', in Bergh, J.C.J.M. van den, Nijkamp, P. and Rietveld, P. (eds), *Recent Advances in Spatial Equilibrium Modelling*, Springer-Verlag, Berlin, pp. 261–277.
Sato, K. (1967), 'A two-level constant-elasticity-substitution production function', *Review of Economic Studies*, vol. 34, pp. 201–218.
Schweizer, U., Varaiya, P. and Hartwick, J. (1976), 'General equilibrium and location theory', *Journal of Urban Economics*, vol. 3, pp. 285–303.
Sheffi, Y. (1985), *Urban Transportation Networks*, Prentice-Hall., Englewood Cliffs, NJ.
Shoven, J.B. and Whalley, J. (1974), 'On the computation of competitive equilibrium on international markets with tariffs', *Journal of International Economics*, vol. 4, pp. 341–354.
Shoven, J.B. and Whalley, J. (1984), 'Applied general-equilibrium models of taxation and international trade: An introduction and survey', *Journal of Economic Literature*, vol. 22, pp. 1007–1051.
Shoven, J.B. and Whalley, J. (1992), *Applying General Equilibrium*, Cambridge University Press, Cambridge, MA.
Solow, R.M. (1973), 'On equilibrium models of urban locations', in Parkin, J.M. (ed.), *Essays in Modern Economics*, Longman, London, pp. 2–16.
Varian, H.R. (1992), *Microeconomic Analysis* (3rd edition), Norton, New York, NY.
Webster, F.V., Bly, P.H. and Paulley, N.J. (eds) (1988), *Urban Land-Use and Transport Interaction*, Avebury, Newcastle-upon-Tyne.
Whalley, J. (1985), *Trade Liberalization among Major World Trading Areas*, The MIT Press, Cambridge, MA.
Wilson, A.G., Coelho, J.D., Macgill, S.M. and Williams, H.C.W.L. (1981), *Optimization in Locational and Transport Analysis*, John Wiley & Sons, New York, NY.

Part III

Integrated Analysis of Activity Location and Transportation in Urban and Regional Systems

14 Applied Models of Urban Land Use, Transport and Environment: State of the Art and Future Developments

Michael Wegener
Institute of Spatial Planning
University of Dortmund
D-44221 Dortmund, Germany

14.1 Introduction

The idea that computer models of urban land use and transport might contribute to more rational urban planning was born in the 1950s and culminated in the 1960s. The 'new tools for planning' (Harris, 1965) were thought to be a major technological breakthrough that would revolutionise the practice of urban policy making. However, the diffusion of urban models faltered soon after the pioneering phase, for a variety of reasons (see Batty, 1994; Harris, 1994). The most fundamental reason was probably that these models were linked to the rational planning paradigm dominant in most Western countries at that time. They were perhaps the most ambitious expression of the desire to 'understand' as thoroughly as possible the intricate mechanisms of urban development, and by virtue of this understanding to forecast and control the future of cities (Lee, 1973). Since then the attitude towards planning has departed from the ideal of synoptic rationalism and turned to a more modest, incrementalist interpretation of planning that has at least partly determined the failure of many ambitious large-scale modelling projects.

However, today the urgency of the environmental debate has renewed the interest in integrated models of urban land use and transport. There is growing consensus that the negative environmental impacts of transport cannot be reduced by transport policies alone but that they have to be complemented by measures to reduce the need for mobility by promoting higher-density, mixed-use urban forms more suitable for public transport. In the United States new legislation inspired by growing environmental awareness such as the Intermodal Surface Transportation Efficiency Act of 1991 requires that transport planning must consider the interaction between transport and land use in a consistent fashion – as it can be done only by land-use/transport models.

This new interest in land use models presents new challenges to land use modelling. A new generation of travel models such as activity-based travel demand models require more detailed information on household demographics and employment characteristics. New neighbourhood-scale transport planning policies to promote the use of public transport, walking and cycling require more detailed information on the precise location of activities. In addition, the models need to be able to predict not only economic but also environmental impacts of land-use and transport policies, and this requires small area forecasts of emissions from stationary and mobile sources as well as of immissions in terms of affected population.

Today there exist operational urban land-use/transport models that have the potential to respond to these challenges. There is a growing number of university laboratories, public agencies or private firms on five continents where research and development in urban and regional modelling is actively being conducted, and of urban/regional models of varying degrees of comprehensiveness and sophistication that have been and are being applied to real-life metropolitan regions for purposes of research and/or policy analysis. Rapid advances in information and computing technology have removed technical barriers besetting earlier generations of land-use/transport models. At the same time there exist exciting opportunities to incorporate new theoretical developments and methodologies into the field.

This chapter reviews the current state of the art of operational integrated urban models. In the first section an overview is given on existing land-use/transport (LT) models. The remaining sections explore how these models might be extended to become land-use/transport/environment (LTE) models and to what degree this has been achieved in pioneering modelling approaches.

14.2 Existing Land-Use/Transport Models

This section presents a brief overview on the state of the art in operational integrated urban land-use/transport models based on Wegener (1994). The term model is used here to indicate mathematical models implemented on a computer and designed to analyse and forecast the development of urban or regional land use systems. The models must be integrated, i.e. must incorporate the most essential processes of spatial development; this implies that they must include urban land use, where land use denotes a range of land uses such as residential, industrial and commercial. This excludes partial models addressing only one subsystem such as housing or retail. It is essential that the links from transport to land use are considered; transport itself may be modelled either endogenously or by an exogenous transport model. The models must be operational in the sense that they have been implemented, calibrated and used for policy analysis for at least one metropolitan region.

The number of real-world applications of models falling under the above definition has increased steadily over the last decade. There has been a continuous reflection of purpose, direction and theoretical basis of land-use/transport modelling as witnessed by volumes edited by Hutchinson et al. (1985), Hutchinson and Batty (1986) and Webster et al. (1988) and by reviews by Harris (1985), Wegener (1986b; 1987), Kain (1987), Boyce (1988), Berechman and Small (1988), Aoyama (1989), and Batty (1994), Harris (1994) and Wegener (1994).

To assess the current state of the art in urban modelling, in this section first a framework for the classification and evaluation of urban models is established. Then fourteen contemporary operational urban models are compared using as criteria comprehensiveness, overall structure, theoretical foundations, modelling techniques, dynamics, data requirements, calibration and validation, and operationality and applicability. This section is an updated summary of more detailed information presented in Wegener (1994).

14.2.1 A Model of Urban Models

For the evaluation of operational urban models, an idealised urban model is first sketched out as a benchmark by which existing models are classified and evaluated. Eight types of major urban subsystem are distinguished. They are ordered by the speed by which they change, from very slow to very fast (cf. Wegener et al., 1986):

- *Very slow change: networks, land use.* Urban transport, communications and utility *networks* are the most permanent elements of the physical structure of cities. Large infrastructure projects require a decade or more, and once in place are rarely abandoned. The *land use* distribution is equally stable; it changes only incrementally.
- *Slow changes: workplaces, housing.* Buildings have a life-span of up to one hundred years and take several years from planning to completion. *Workplaces* (non-residential buildings) such as factories, warehouses, shopping centres or offices, theatres or universities exist much longer than the firms or institutions that occupy them, just as *housing* exists longer than the households that live in it.
- *Fast change: employment, population.* Firms are established or closed down, expanded or relocated; this creates new jobs or makes workers redundant and so affects *employment*. Households are created, grow or decline and eventually are dissolved, and in each stage in their life-cycle adjust their housing consumption and location to their changing needs; this determines the distribution of *population.*
- *Immediate change: goods transport, travel.* The location of human activities in space gives rise to a demand for spatial interaction in the form of *goods transport* or *travel.* These interactions are the most volatile phenomena of spatial urban development; they adjust in minutes or hours to changes in congestion or fluctuations in demand.

There is a ninth subsystem, the *urban environment.* Its temporal behaviour is more complex. The direct impacts of human activities, such as transport noise and air pollution are immediate; other effects such as water or soil contamination build up incrementally over time, and still others such as long-term climate effects are so slow that they are hardly observable. All other eight subsystems affect the environment by energy and space consumption, air pollution and noise emission, whereas only locational choices of housing investors and households, firms and workers are co-determined by environmental quality, or lack of it. All nine subsystems are partly market-driven and partly subject to policy regulation.

14.2.2 Fourteen Urban Models

Fourteen models were selected for the comparison. The selection does not imply a judgement on the quality of the models, but was based on the availability of information. These are the fourteen models:

- *BOYCE*: the combined models of location and travel choice developed by Boyce (Boyce et al., 1983; 1985; Boyce, 1986; Boyce et al., 1992).
- *CUFM*: the California Urban Futures Model developed at the University of California at Berkeley (Landis, 1992; 1993; 1994).
- *MUS*: the '5-Stage Land-Use Transport Model' developed by Martinez for Santiago de Chile (1991; 1992a; 1992b).
- *HUDS*: the Harvard Urban Development Simulation developed by Kain and Apgar (1985).
- *IMREL*: the Integrated Model of Residential and Employment Location by Anderstig and Mattsson (1991; 1998).
- *IRPUD*: the model of the Dortmund region developed by Wegener (1985; 1986a; Wegener et al., 1991).
- *ITLUP*: the Integrated Transportation and Land Use Package developed by Putman (1983; 1991; 1998).
- *KIM*: the non-linear version of the urban equilibrium model developed by Kim (1989) and Rho and Kim (1989).
- *LILT*: the Leeds Integrated Land-Use/Transport model developed by Mackett (1983; 1990c; 1991a; 1991b).
- *MEPLAN*: the integrated modelling package developed by Marcial Echenique & Partners (Echenique et al., 1990; Hunt and Simmonds, 1993, Echenique, 1994; Williams, 1994; Hunt 1994).
- *METROSIM*: the microeconomic land-use and transport model developed by Anas (1992).
- *POLIS*: the Projective Optimization Land Use Information System developed by Prastacos for the Association of Bay Area Governments (Prastacos, 1986).
- *RURBAN*: the Random-Utility URBAN model developed by Miyamoto (Miyamoto et al., 1986; Miyamoto and Kitazume, 1989).
- *TRANUS*: the transport and land-use model developed by de la Barra (de la Barra et al., 1984; de la Barra, 1989; 1998).

These fourteen models are now compared with respect to the criteria listed above.

Comprehensiveness. All fourteen models are comprehensive in the sense that they address at least two of the eight subsystems identified above (the urban environment will be discussed later). Only MEPLAN and TRANUS encompass all eight subsystems. IRPUD, LILT and METROSIM address all subsystems except goods transport, KIM models goods movements but not physical stock and land use, HUDS has a housing supply submodel but does not model non-residential buildings. Half of the models make no distinction between activities (population and employment) and physical stock (housing and workplaces). Four models (CUFM, HUDS, POLIS and RURBAN) do not model transport and hence rely on input from exogenous transport models. Only HUDS, IRPUD and LILT model demographic change and household formation.

Model Structure. With respect to overall model structure, two groups can be distinguished. One group of models searches for a unifying principle for modelling and linking all subsystems; the others see the city as a hierarchical system of interconnected but structurally autonomous subsystems; The resulting model structure is either tightly integrated, 'all of one kind', or consists of loosely coupled submodels, each of which has its own independent internal structure. The former type of model is called 'unified', the latter 'composite' (Wegener et al., 1986). Five of the fourteen models (BOYCE, MUS, KIM, METROSIM and RURBAN) belong to the unified category, the remaining nine are composite. The distinction between unified and composite model designs has important implications for the modelling techniques applied and for the dynamic behaviour of the models (see below).

Theory. In the last twenty years great advances in theories to explain spatial choice behaviour and in techniques for calibrating spatial choice models have been made. Today there is a broad consensus about what constitutes a state-of-the-art land use model: Except for one (CUFM), all models rely on random utility or discrete choice theory to explain and forecast the behaviour of actors such as investors, households, firms or travellers. Random utility models predict choices between alternatives as a function of attributes of the alternatives, subject to stochastic dispersion constraints that take account of unobserved attributes of the alternatives, differences in taste between the decision makers, or uncertainty or lack of information (Domencich and McFadden, 1975). Anas (1983) showed that the multinomial logit model resulting from random utility maximisation is, at equal levels of aggregation, formally equivalent to the entropy-maximizing model proposed by Wilson (1967; 1970); he thus laid the foundation for the convergence and general acceptability of formerly separate strands of theory.

Underneath that uniformity, however, there are significant differences between the theoretical foundations of the models. Eight models (MUS, HUDS, IMREL, KIM, MEPLAN, METROSIM, RURBAN and TRANUS) represent the land (or floorspace or housing) market with endogenous prices and market clearing in each period; one (IRPUD) has endogenous land and housing prices with delayed price adjustment. These models are indebted to microeconomic theory, in particular to Alonso's (1964) theory of urban land markets or bid-rent theory. The six models without market equilibrium rely on random utility maximisation; however, two of the microeconomic models (MUS and RURBAN) are hybrids between bid-rent and random utility theory. All models with transport submodels use random utility or entropy theory for modelling destination and mode choice.

Only KIM and METROSIM determine a general equilibrium of transport and location with endogenous prices. The other models are equilibrium models of transport only (IRPUD, ITLUP), of transport and activity location separately

(IMREL, MEPLAN and TRANUS), or of transport and location combined, but without endogenous prices (BOYCE and LILT). Five models apply concepts of locational surplus (IMREL, POLIS), random utility (IRPUD and ITLUP) or profitability (CUFM) to locate activities. ITLUP may be brought to general equilibrium, but this is not normally done; METROSIM may produce a long-run equilibrium or converge to a steady state in annual increments.

Several other theoretical elements are built into some models. MEPLAN and TRANUS use export base theory to link population and non-basic employment to exogenous forecasts of export industries. HUDS, IRPUD and LILT apply standard probabilistic concepts of cohort survival analysis in their demographic and household formation submodels. IRPUD also utilises ideas from time geography, such as time and money budgets, to determine action spaces of travellers in its transport submodel.

Modelling Techniques. In all fourteen models, the urban region is represented as a set of discrete subareas or zones. Time is typically subdivided into discrete periods of between one and five years. This classifies all models except IMREL (which is static) as recursive simulation models.

In seven models (BOYCE, IMREL, KIM, LILT, MEPLAN, RURBAN and TRANUS) transport and location are simultaneously determined in spatial-interaction location models, in which activities are located as destinations of trips; in the remaining models (and in the employment location model of IMREL) transport influences location via accessibility indicators. In the ten models with network representation state-of-the-art modelling techniques are applied with network equilibrium the dominant trip assignment method despite its well-known weakness of collapsing to all-or-nothing assignment in the absence of congestion. Only ITLUP, MEPLAN and TRANUS have multiple-path assignment allowing for true route-choice dispersion.

For representing flows of goods, multiregional input-output methods are the standard method. KIM, MEPLAN and TRANUS use input-output coefficients or demand functions for determining intersectoral flows and random utility or entropy models for their spatial distribution. MEPLAN and TRANUS have generalised this to incorporate industries and households as consuming and producing 'factors' resulting in goods movements or travel.

With the exception of CUFM and HUDS, all models are aggregate at a meso level, i.e. all results are given for medium-sized zones and for aggregates of households and industries. CUFM and HUDS are disaggregate, i.e. apply microsimulation techniques. HUDS works on a sample of individual households in list form, whereas CUFM uses detailed land information in map form generated by a geographical information system. IRPUD starts with aggregate data but uses microsimulation techniques in its housing market submodel.

Dynamics. All but one of the fourteen models are recursive simulation models. Recursive simulation models are called quasi-dynamic because, although they model the development of a city over time, within one simulation period they are in fact cross-sectional. This is however only true for strictly unified models.

Composite models consist of several interlinked submodels that are processed sequentially or iteratively once or several times during a simulation period. This makes composite models well suited for taking account of time lags or delays due to the complex superposition of slow and fast processes of urban development (cf. Wegener et al., 1986). However, this feature is insufficiently used by most models, because the typical simulation period of five years has the effect of an implicit time lag – a too long time lag in most cases.

Data Requirements. The data collection for a model of a large metropolis has remained a major effort. However, in many cases the introduction of computers in local government has generated a pool of routinely collected and updated data that can be used as the information base for a model, in particular in the fields of population, housing, land use and transport. Another factor reducing the data-dependency of urban models is the significant progress made in urban theory in the last decades. The models of today are more parsimonious, i.e. can do with less data than previous models. Examples illustrating this are the techniques to generate regional input-output matrices from national input-output matrices and regional totals through biproportional scaling methods; or techniques to create artificial microdata as samples from multivariate aggregate data.

Calibration and Validation. All fourteen models of the sample have been (or could have been) calibrated using observed data, using readily available computer programs and following well-established methods and standards. In particular, maximum-likelihood estimation of the ubiquitous logit model has become routine. Yet, while calibration has become easier, the limits to calibrating a model with data of the past have become visible. Calibration of cross-sectional models, as it is practised today, provides the illusion of precision but does little to establish the credibility of models designed to look into the far future. There has been almost no progress in the methodology required to calibrate dynamic or quasi-dynamic models.

In the face of this dilemma, the insistence of some modellers on 'estimating' every model equation appears almost an obsession. It would probably be more effective to concentrate instead on model *validation*, i.e. the comparison of model results with observed data over a longer period. In the future, the only real test of a model's performance should be its ability to forecast the essential dynamics of the modelled system over a past period at least as long as the forecasting period. There are only two models in the sample following this philosophy, IRPUD and MEPLAN. These models are partly calibrated not by statistical estimation, but by manual fine-tuning in a long, interactive process.

Operationality. All the models in the sample are operational in the sense that they have been applied to real cities. However, only few models are on their way to becoming standard software for a wider market. Among these, TRANUS stands out as a particularly advanced and well documented software with an attractive user interface in Spanish or English. The time seems not far when any planning office will be able to buy a complex and versatile urban model with full documentation, default values and test data sets for less than a thousand dollars.

Applicability. If one considers the enormous range of planning problems facing a typical metropolitan area in industrialised countries today, the spectrum of problems actually addressed with the fourteen urban models in the sample is very narrow. The majority of applications answer traditional questions such as how land use regulations or housing programs would affect land use development and transport, or how transport improvements or changes in travel costs would shift the distribution of activities in an urban area. These are and will continue to be important questions – questions that can only be answered with the models discussed here. However, other issues are likely to become prominent in the future, and it will be essential that the models are able to contribute to their rational discussion.

14.3 Modelling the Urban Environment

The new interest in land-use models has its origin in the imperative to make cities more sustainable. Therefore future urban models need to be able to model the urban environment.

Ecological modelling has been an established field of scientific work long before the present debate about environmental sustainability. Important pioneering insights into the nature of complex dynamic systems originated in ecology (Lotka, 1920; Volterra, 1931; see Nijkamp and Reggiani, 1992).

Urban modellers have for a long time ignored ecological aspects of the processes simulated in their models and have only recently been prompted to redirect their attention from economic to environmental impacts of land use and transport policies. The main reason for this is the threat of long-term climate change due to production of greenhouse gases by the burning of fossil fuels for heating and transport. A major additional thrust to include environmental impacts into urban models has come from the United States Intermodal Surface Transportation Efficiency Act (ISTEA) which shifts the criteria for new transport investment from travel time savings to environmental benefits such as air quality or reduction of single-occupancy vehicle trips. To demonstrate these benefits requires different models.

Therefore all over the world urban modellers are turning their attention to the urban environment. Existing land-use/transport (LT) models are being augmented by environmental submodels to become land-use/transport/environment (LTE) models. However, today there exist no full-scale urban LTE models. In this section it will be attempted to identify the environmental aspects a model deserving that name would have to contain.

For this the field 'urban environment' will be classified into sub-fields using criteria such as policy relevance, quantifiability, data availability, availability of theory and relationship with other submodels. Table 14.1 summarises the interactions among the sub-fields and between them and land use and transport.

In particular the relationship to land use and transport can be used to exclude environmental aspects which may be important in their own right but are only little

Table 14.1. Cross-impact matrix: land use, transport and environment

			Resources						*Emissions*						*Immissions*		
Cause / *Effect*	Land use	Transport	Energy	Water	Land	Vegetation	Wildlife	Microclimate	CO_2 emission	Air pollution	Water quality	Soil contamination	Solid waste	Noise	Air dispersion	Noise propagation	Surface/ground water flows
Land use	●	●	●	●	●	●	●	●	●	●	○	●	○	●	●	●	●
Transport	●	●	●	·	●	*a*	*b*	*a*	●	●	○	○	·	●	●	●	○
Energy	○	○	●	·	·	·	·	*a*	●	●	·	·	·	·	●	·	·
Water	·	·	·	●	·	*c*	*c*	*c*	·	·	·	·	·	·	·	·	●
Land	●	·	·	●	●	●	●	●	·	·	·	·	·	·	·	○	●
Vegetation	○	·	·	○	·	●	●	●	●	·	·	·	·	·	·	●	·
Wildlife	·	·	·	·	·	·	●	·	·	·	·	·	·	·	·	·	·
Microclimate	○	·	·	·	·	○	○	●	·	·	·	·	·	·	·	·	·
CO_2 emission	·	·	·	·	·	·	·	·	●	·	·	·	·	·	·	·	·
Air pollution	*d*	·	·	·	·	*d*	*d*	●	·	●	*d*	·	·	·	●	·	*d*
Water quality	*c*	·	·	●	·	●	●	·	·	·	●	·	·	·	·	·	●
Soil contamination	●	·	·	·	●	●	●	·	·	·	·	●	·	·	·	·	·
Solid waste	○	·	·	○	○	·	·	·	·	·	·	●	●	·	·	·	●
Noise	*e*	·	·	·	·	·	*e*	·	·	·	·	·	·	●	·	●	·
Air dispersion	●	·	·	·	·	●	○	●	·	·	○	○	·	·	●	·	○
Noise propagation	●	·	·	·	·	·	●	·	·	·	·	·	·	·	·	●	·
Surface/ground water flows	·	·	·	●	·	●	●	●	·	·	·	·	·	·	·	·	●

· *no impact* ○ *weak impact* ● *strong impact*
a via air dispersion and land consumption
b via air dispersion, noise propagation and land consumption
c via surface/ground water flows
d via air dispersion
e via noise propagation

affected by the processes dealt with in the models from the analysis. It is useful to remember that the models discussed here are intended for forecasting the impacts of land use and transport policies. Therefore only those environmental impacts are relevant for them which result from changes in the spatial distribution of activities. For instance, the sewerage system and efficient waste water treatment are important for a sustainable city; however, while the cost of the sewerage system depends on the topography and physical layout of the city, the cost of the treatment plant probably does not. In Table 14.1 urban environmental impacts are classified under the headings of *resources*, *emissions* and *immissions*.

14.3.1 Resources

Most human activities consume resources. Some of them are global resources which are brought into the region such as energy, some are local resources such as water. Sustainable development aims at using non-renewable resources as little as possible in the interest of future generations. From the point of view of urban modelling the most important resources are energy, water and land:

- *Energy*. Energy is a global resource which is imported to the urban region in the form of non-renewable fossil fuel or electricity. Renewable kinds of energy such as solar or wind energy presently play a minimum role. Energy is consumed for process heat, for the heating of buildings and for transport. Energy use for heating is affected by type of building and density. Transport energy consumption is affected by modal choice, number and length of trips, travel speed, vehicle occupancy and energy efficiency of vehicles. Choice of building type and density as well as travel and shipping behaviour are influenced by energy cost, however, today this effect is diluted by underpriced fuel. Energy consumption of land use and transport are therefore candidates for being included in urban models; the relationships to be modelled are straightforward.
- *Water*. Water cannot be easily transported over great distances and is therefore consumed close to the source. Modern agriculture, manufacturing techniques and life styles all tend to higher water consumption. Water supply has therefore become a serious problem for many cities. There is a relationship between urban density and water consumption as suburban gardens and swimming pools tend to consume large quantities of water. There is no significant effect of urban transport on water consumption. However, both land use and transport affect water supply by sealing off land through buildings, pavings and roadways and so impeding rainfall from reaching the ground water. Because of this higher urban densities with smaller land coverage and less roads are environmentally preferable over disperse suburbs with one-storey buildings and a high percentage of paved road area. The effects of ground coverage on ground water supply can be modelled; so can the effect of policies to reduce water consumption by utilisation of rainfall.
- *Land*. Land is the ultimate resource of cities. With growing affluence and increasing substitution of (renewable) human labour by (non-renewable) mechanical energy, all human activities, from housing, manufacturing and services to transport

tend to consume more land. The amount of open space in and around cities is therefore continuously declining. This not only reduces ground water supply (see above) but has also negative effects on vegetation, wildlife and microclimate. Research on the impact of size, shape and interconnectedness of open spaces on the variety of plant and animal species and the microclimate in adjacent areas is still underway, but there are sufficient results available to include these effects in urban models.

14.3.2 Emissions

Most human activities give rise to metabolisms producing obnoxious emissions. Emissions are produced locally but have local, remote and global effects. From the point of view of urban modelling the most important emissions are gases, waste water, soil contamination, solid waste and noise:

- *Gases.* Most gaseous emissions originate from chemical or combustion processes in stationary or mobile sources. In cities stationary sources are chemical or manufacturing plants, power stations and residential areas; mobile sources are cars, lorries and buses. Pollutants such as CO, NO_x, HC, SO_2 and dust particles affect the well-being of humans at points of immission, whereas CO_2 is a greenhouse gas with global effects. It is therefore sufficient to predict CO_2 emissions for the whole urban region irrespective of where they occur, whereas for other pollutants air dispersion models calculating immissions from emissions are required (see below). To predict the magnitude and composition of industrial emissions requires information about the type and quantity of processes and the efficacy of emission abatement technology. Reasonable assumptions about the emissions by residential heating given a certain level of insulation and heating technology are possible. Gaseous emissions of cars are a well researched field; it is possible to predict them as a function of link traffic volumes, composition of flow, vehicle duty cycles and prevailing emission abatement technology with any desired detail. However, fuel consumption and emission characteristics of commercial vehicles still need more attention.
- *Water quality.* The amount of waste water produced in a city is a function of water consumption (see above), rainfall and irrigation; however, reasonable assumptions about waste water per capita or per worker by industry can be made. As indicated above, the amount of ground coverage through buildings and pavements affects the volume of rain water in the sewerage system and hence the required capacity of water treatment plants. Urban density affects the length and cost of sewerage networks. Intensive use of fertilisers or untreated industrial or domestic effluents lead to the degradation of ground water or streams and rivers and to the degeneration of aquatic habitats. However there is no causal relationship between type of land use or density and water quality. Therefore water quality is not a likely topic of urban models. Urban transport contributes to water pollution by oil and particles washed from roadways. This effect can be modelled, but is not likely to be significant compared with other sources of water contamination.

– *Soil.* Soil contamination through chemicals and obnoxious liquids by former manufacturing or extraction activities is a serious problem in old industrial or mining regions, but should soon become a thing of the past through stricter enforcement of emission standards. However, existing contamination can be a strong deterrent for potential residential or industrial investors.

– *Solid waste.* The generation of solid waste is not a function of land use type or density or urban transport, but of manufacturing and packaging technologies and personal life styles and recycling legislation. Solid waste disposal generates traffic and requires land for disposal sites or incineration plants and so affects urban transport and land use. These effects might be modelled but are likely to be relatively insignificant compared with those of other urban activities. Faulty disposal sites may be the cause of soil contamination, and incineration plants are suspected to emit dioxin, but these effects are not caused by land use or transport and are therefore not likely to be considered in urban models.

– *Noise.* Like air pollution, noise is emitted from stationary and mobile sources. Fixed sources are industrial processes and construction sites, mobile sources are vehicles. Noise from stationary sources (except construction sites) has been reduced by encapsulation of machinery and physical separation between industry and residences. Traffic noise, in contrast, is increasing and has become the most obnoxious and ubiquitous kind of emission in cities. Like industrial air pollution, industrial noise is difficult to predict without information about the processes at work. Traffic noise can be simulated as a function of traffic volume, composition of flow and speed. Noise propagation declines rapidly with distance, so calculation of noise emissions without calculation of noise immissions is not sufficient (see below).

14.3.3 Immissions

Air pollution, noise and water contamination are environmental impacts of which emission and immission points differ. As their effect is felt at immission points, calculation of immissions from emissions is critical for these kinds of impacts. Three types of emission-to-immission models are candidates for being included in urban models:

– *Air dispersion.* Air pollution can be carried over long distances, as the phenomenon of acid rain thousands of kilometres from the emission source has demonstrated. Also photochemical ozone-generating processes are known to occur far away from emission sources. Within urban areas air streams are important not only for the dispersal of pollutants but also as carriers of cool air from the countryside or mountains in the summer. Air dispersion models calculate immissions from emissions as a function of location and elevation of sources, topography and prevailing wind direction and speed. The use of these models can suggest which parts of the urban area should be left undeveloped as cold air ventilation corridors.

– *Noise propagation.* For the assessment of noise intrusion it is necessary to know the number of people affected by different noise levels. There exist several methods of calculating noise immissions from simple distance buffers around point or line sources to sophisticated sound propagation models taking account of multiple reflection of noise from roadways, topography, buildings and the effects of sound protection measures such as protective dams or walls. The latter methods, however, require spatially disaggregate information on topography, built form and distribution of population.

– *Surface/ground water flows.* Hydrological modelling includes surface water models such as rainfall-runoff or streamflow simulation models and groundwater models such as groundwater flow and groundwater contamination transport models. Hydrological modelling is a complex field requiring extensive information on rainfall probability, land cover and the geological formation and the river system of the urban region. However, with imminent depletion of water resources in many cities, this type of model may become more prominent in the future. Rainwater management policies designed to reduce water consumption and sewerage volumes require site-specific information on roof areas and ground coverage and appropriately spatially disaggregate models.

14.4 Implications for LTE Models

Modelling the urban environment in integrated urban models presents new challenges as new subsystems with different dynamics and spatial resolution need to be incorporated into the models.

14.4.1 Dynamics

Most operational urban land-use/transport models are quasi-dynamic in the sense that their transport or land-use submodels or both are cross-sectional equilibrium models. In addition they have relatively long simulation periods of five or more years. The rationale behind this is that adjustment processes in urban areas are slow.

Environmental processes, however, have a different time scale. Some processes such as air dispersion and noise propagation are very rapid and can be dealt with in cross-sectional submodels. However, some processes such as the impacts of development on water supply, vegetation, wildlife and water quality have very long response times between several years and one or more generations.

The problems arising from this for the temporal organisation of the models may be fundamental. The longer time perspective necessary for environmental analysis is likely to make equilibrium approaches less appropriate and to favour dynamic approaches allowing for a variety of different speeds of adjustment in different parts of the modelled system.

14.4.2 Spatial Resolution

Urban models have always been spatially aggregate with zones of varying size such as boroughs or statistical districts as units of spatial reference. As the internal distribution of activities and land uses within a zone is not known, a homogenous distribution across the area of the zones has to be assumed. However, even though the number of zones of some models has increased substantially in recent years, the spatial resolution of zone-based models is much too coarse to appropriately deal with environmental processes. In particular emission-immission algorithms such as air dispersion, noise propagation and surface and ground water flows, but also microclimate analysis, require a much higher spatial resolution:

- Air distribution models typically work with raster data of emission sources and topographic features such as elevation and surface characteristics such as green space, built-up area, high-rise buildings and the like.
- Noise propagation models require spatially disaggregate information on emission sources, topography and sound barriers such as dams, walls or buildings as well as the three-dimensional location of population.
- Surface and ground water flow models require spatially disaggregate data on the river system and geological information on ground water conditions in the region.
- Microclimate analysis depends on small-scale mapping of green spaces and built-up areas and their features.

In all four cases the information needed is configurational. This implies that not only the attributes of the components of the modelled system such as quantity or cost are of interest but also their physical location. This suggests a fundamentally new organisation of data of urban models.

A data organisation in which topological and other attributes of spatial systems are integrated is called a geographic information system (GIS). Geographic information systems, in particular raster-based GIS, therefore promise to have great importance for future integrated urban models. The tendency away from zonal to spatially disaggregate raster-based data structures suggested by environmental modelling is not only in line with the enormously increased memory and computing capacity of modern computers but also conforms with the trend to disaggregate activity-based models in urban transport planning and the current revival of microsimulation approaches in urban modelling (Wegener and Spiekermann, 1996).

14.5 LTE Models: A Survey

Because the increased attention of urban modellers for environmental aspects is a relatively recent phenomenon, it is difficult to get an overview of the state of the art in this rapidly developing field. Therefore a quick, ad-hoc mini survey among some of the authors of urban models was conducted. The survey does not attempt to provide a comprehensive inventory of urban LTE models existing in the world today. It can be assumed that in the United States under the impression of the

ISTEA legislation numerous new modelling activities are being launched by local governments of all sizes. However, the modellers surveyed belong to the small group of researchers who have developed operational urban land-use/transport models, so it is likely that they represent the forefront of urban LTE modelling.

Altogether 24 models or model versions were named as LTE models by their authors. Some of them have not yet been published, so no reference can be given; in some cases the references refer to the original models and not to their yet unpublished environmental extensions. The 24 models are, in alphabetical order, ARCTRAN-AIR (Kim et al.), CODMA (Lundqvist, 1989; 1996; 1998), IMREL (Anderstig and Mattsson, 1991; 1998), IRPUD (Wegener, 1996), ITLUP (Putman, 1983; 1991; 1998), LAND (Gu et al., 1992; Young and Gu, 1993; 1996), LET (Anjomani), LILT (Mackett, 1983; 1990b; 1990c; 1991a; 1991b), MAPLE (Hayashi and Tomita, 1989), MASTER (Mackett, 1990a), MEPLAN Edmonton (Hunt), MEPLAN Helsinki/London/Santiago/Vicenza (Echenique et al., 1990; Hunt and Simmonds, 1993, Echenique, 1994; Williams, 1994; Hunt, 1994), MOUSE (Diappi et al.), MUS (Martinez, 1996), PSS (Anjomani), RURBAN (Miyamoto and Udomsri, 1996), SALOC (Lundqvist, 1996; 1998), START/DSCMOD (Simmonds, 1995), SUSTAIN (Roy et al., 1996; 1998), TRANUS (de la Barra et al., 1984; de la Barra, 1989; 1998), TRANUS/CUFM (by de la Barra and Landis, adapted by Johnston).

Table 14.2 summarises the main results of the survey. It shows the environmental indicators presently being calculated in urban models or being considered for inclusion in the near future. There are clear priorities. Of the 24 models included in the survey fifteen calculate (or are considering to calculate) land consumption, as might be expected from land use models. Sixteen models calculate (or plan to calculate) energy consumption and CO_2 emission of transport. Air pollution of transport is modelled by thirteen models. All other indicators are listed much less frequently. Energy consumption and CO_2 emissions and air pollution of land use are considered by only seven models. Surprisingly, only four models calculate traffic noise. Only between one and three models deal with water supply, vegetation, wildlife, microclimate, waste water, soil contamination, solid waste and industrial noise. Only seven models have (or will shortly have) an air dispersion submodel. Other immissions are almost absent in present LTE models. Only one model deals with noise propagation and two with surface and ground water flows. Another question asked in the survey was whether the environmental indicators are calculated only as output for later exogenous evaluation or are fed back into the land use or transport parts of the models. The purpose was to find out whether the models recognise a two-way relationship between land use and environment, and transport and environment, respectively, in the same way as they take account of the two-way interdependency between land use and transport.

A look at the cross-impact matrix of Table 14.1 shows that in the real world the relationships between the environment and land use and transport are not symmetric. Land use and transport affect almost all environmental indicators but the reverse is not the case. Only land use changes, i.e. location decisions by

Table 14.2. Environmental impacts modelled by urban LTE models

Models *(Authors)*	*Resources*							*Emissions*									*Immissions*		
	Energy consumption by land use	Energy consumption by transport	Water supply	Land consumption	Vegetation	Wildlife	Microclimate	CO_2 emission by land use	CO_2 emission by transport	Air pollution by land use	Air pollution by transport	Water quality	Soil contamination	Solid waste	Industrial noise	Traffic noise	Air dispersion	Noise propagation	Surface/ground water flows
ARCTRAN-AIR	·	·	·	·	·	·	·	●	●	●	●	·	·	·	·	·	●	·	·
CODMA	·	·	·	·	·	·	·	·	●	·	●	·	·	·	·	●	●	·	·
IMREL	·	●	·	○	·	·	·	·	●	·	○	·	·	·	·	·	·	·	·
IRPUD	·	●	·	●	·	·	·	·	●	·	·	·	·	·	·	·	·	·	·
ITLUP	·	·	·	●	·	·	·	·	*a*	·	*a*	·	·	·	·	·	·	·	·
LAND	·	·	·	·	·	·	·	·	·	·	●	·	·	·	·	●	·	·	·
LET	·	●	●	●	●	○	·	·	·	·	·	·	●	·	·	·	·	·	·
LILT	·	●	·	●	·	·	·	·	·	·	·	·	·	·	·	·	·	·	·
MAPLE	·	●	·	·	·	·	·	·	●	●	·	·	·	·	·	○	○	·	·
MASTER	·	·	·	●	·	·	·	·	·	·	·	·	·	·	·	·	·	·	·
MEPLAN Edmonton	○	●	·	●	·	·	·	○	●	○	○	·	·	·	·	·	·	·	·
MEPLAN Helsinki	·	●	·	●	·	·	·	·	·	·	·	·	·	·	·	·	·	·	·
MEPLAN London	·	●	·	●	·	·	·	·	●	·	●	·	·	·	·	·	·	·	·
MEPLAN Santiago	·	●	●	●	●	·	·	·	●	·	●	·	·	·	·	·	·	·	·
MEPLAN Vicenza	·	●	·	·	·	·	·	·	●	·	●	·	·	·	·	·	●	·	·
MOUSE	○	●	·	●	·	·	·	○	●	·	·	·	·	·	·	·	○	○	·
MUS	●	●	·	·	·	·	·	●	●	●	●	·	·	·	·	·	●	·	·
PSS	·	·	●	●	●	○	·	·	·	·	·	·	●	·	·	·	·	·	·
RURBAN	●	·	·	·	·	·	·	○	·	·	·	●	·	●	·	·	●	·	●
SALOC	●	●	·	●	·	·	·	·	·	·	·	·	·	·	·	·	·	·	·
START/DSCMOD	·	●	·	·	·	·	·	·	●	·	·	·	·	·	·	·	·	·	·
SUSTAIN	·	○	·	●	·	·	·	·	○	·	○	·	·	·	·	·	·	·	·
TRANUS	●	●	·	●	·	·	·	○	○	○	○	·	·		○	○	·	·	·
TRANUS/CUFM	●	●	·	·	·	○	○	○	○	○	○	○	·		·	·	·	·	○

· *not modelled* ○ *under development or planned* ● *applied or operational*
a links to standard EPA emission models (MOBIL5)

firms and households, are strongly affected by land availability, soil contamination, air pollution and noise; all other feedbacks from the environment are weak or potentially strong only in the case of a major change in the decision framework such as a substantial change in energy cost. Transport decisions are not affected by environmental indicators at all, except potentially by rising fuel costs. Nevertheless, as a minimum, feedback from environment to land use, i.e. the impact of environmental indicators on location decisions, should be included in any LTE model. However, in only 10 of the 24 models environmental indicators enter the attractiveness functions of land use location decisions. In two models transport decisions are affected by environmental indicators, mainly by energy cost. Changes in trip generation through changes in location of activities listed as feedback from environment to transport by one author are in fact indirect via land use. Policies to reduce energy consumption and CO_2 emissions by transport planning can also not be counted as endogenous feedback.

In summary, most present urban models are still far from deserving the name land-use/transport/environment (LTE) models. Many environmental topics, which figure high on the list of controversial issues in contemporary cities, have not been taken up by the models even though there exist suitable methods and data. In the majority of cases the environmental indicators calculated are not fed back into the models and so have no impact on the behaviour of the model actors. This is particularly surprising in the case of land use as it is well known that environmental quality has become a more and more important component of locational attractiveness not only for households but also for services and even for manufacturing. The little feedback from the environment to travel behaviour, on the other hand, is realistic and reflects one of the main problems of planning for sustainability: that the negative impacts of the automotive society are felt by everybody but are not linked to individual behaviour: it does not pay to behave environmentally. It is one of the key tasks of planning for sustainability to link the environmental indicators, through incentives and penalties, to the daily travel decisions of each individual. It is to be hoped that future urban LTE models will be able to model that kind of feedback.

14.6 Conclusions

This paper has been an attempt to review the current state of the art of operational land-use/transport models in the light of the new challenges presented by the environmental debate. It has been shown that there have been immense achievements in land use and transport modelling during the last two decades. There exist operational land use and transport models which have been and are being used for real-life applications in cities all over the world. There is a growing number of active urban modelling centres on five continents in which new approaches are being developed and tested.

However, the review has also exposed deficiencies of current models. Many land-use/transport models are still too aggregate in space, time and substance for state-of-the-art environmental modelling. Most zone-based land-use/transport models lack the spatial resolution, required for modelling environmental impacts such as air dispersion or noise propagation, and surface and ground water flows. Some models have remained captive in the tradition of cross-sectional equilibrium poorly suited to cope with the complex temporal structure of environmental processes. Because of the limited number of environmental aspects addressed, only few current models qualify as full-scale land-use/transport/environment (LTE) models, though efforts to incorporate more environmental indicators in the models are increasing. Only very few models have yet implemented feedback from environment to land use.

These deficiencies suggest the agenda for modelling research in the next decade to make future land-use/transport models more responsive to environmental issues:

A first field of research will imply a new quantum leap in terms of disaggregation of variables – possibly down to the individual – and of spatial and temporal resolution. Fortunately, further increases in memory and speed of computers and the growing availability of spatially disaggregate data will make this feasible, even though the number and magnitude of conceptual problems still to be solved may be immense. The association, or even integration, of land-use/transport/environment models with geographic information systems will become standard practice, although, given the lack of flexibility of current GIS to be linked with other software, this may be a sizeable research program in its own right.

A second field of research will be to integrate the formerly separate traditions of transport, land use and environmental models. Transport models will have to be embedded into land use models (or vice versa) and environmental models into land-use/transport models. The current practice of feeding land-use and transport indicators off-line into exogenous environmental models will only be an interim solution as it negates feedback from environment to land use and transport. This also disqualifies feeding transport indicators into separate 'land use models'. The future urban/regional model will be an integrated land-use/transport/environment (LTE) model.

A third major task is to select environmental submodels suitable for integration into land use and transport models and adapt them to the new framework. Environmental submodels without doubt will further increase the data requirements of land-use/transport models, so careful consideration of what is essential is needed. For many standard indicators public-domain software routines ready to be interfaced with land-use/transport models might be provided by public agencies in order to avoid duplication of effort and to guarantee consistency and comparability of the indicators derived.

Other research needs apply to the way models are used and embedded into the decision making process. One important field of research will have to address problems of evaluation of policy impacts and issues of equity. Predominantly economic evaluation techniques such as cost-benefit analysis need to be complemented by multicriteria methods capable of measuring non-monetary aspects of mobility and neighbourhood and environmental quality and their distribution across privileged and disadvantaged socioeconomic and spatial groups of the population. The feasibility of such disaggregate evaluation will be greatly enhanced by the availability of disaggregate land use and population data required by activity-based transport models.

Finally, more efforts will be necessary to make land-use/transport/environment models a routine tool for a widening range of institutions and individuals, including non-experts. This must be supported by the development of attractive and efficient user interfaces for interactive model calibration, scenario formulation and inspection of results. The Windows-based user shell of TRANUS, Young's gaming simulation LAND (Gu et al., 1992; Young and Gu, 1993; 1996) and Batty's GIS-based model visualisation system (Batty, 1992) are leading the way in this direction.

The greatest challenge, however, seems to keep urban modelling open for new problems. Urban models have in the past been applied mainly to a very narrow set of planning problems, and have repeatedly failed to adapt to changing problem perceptions. The next decade will confront cities and regions in the developed world with complex new problems. Increasing social and spatial inequity, an ageing infrastructure and the need to significantly reduce energy consumption and CO_2 emission will require innovative solutions if social conflict is to be avoided. Only if the models prove that they are able to give meaningful answers to the urgent questions facing cities and regions can they establish for themselves a firm position in the planning process of the future.

References

Alonso, W. (1964), *Location and Land Use*, Harvard University Press, Cambridge, MA.

Anas, A. (1983), 'Discrete choice theory, information theory and the multinomial logit and gravity models', *Transportation Research*, vol. 17 B, pp. 13–23.

Anas, A. (1992), 'NYSIM (The New York Area Simulation Model): A model for cost-benefit analysis of transportation projects', Research Report, Regional Plan Association, New York, NY.

Anderstig, C. and Mattsson, L.-G., (1991), 'An integrated model of residential and employment location in a metropolitan region', *Papers in Regional Science*, vol. 70, pp. 167–84.

Anderstig, C. and Mattsson, L.-G. (1998), 'Modelling land-use and transport interaction: Policy analyses using the IMREL model', Chapter 17, this volume.

Aoyama, Y. (1989), 'A historical review of transport and land-use models in Japan', *Transportation Research*, vol. 23 A, pp. 53–61.

Batty, M. (1992), 'Urban modelling in computer-graphic and geographic information system environments', *Environment and Planning B: Planning and Design*, vol. 19, pp. 663–688.

Batty, M. (1994), 'A chronicle of scientific planning: The Anglo-American modeling experience', *Journal of the American Planning Association*, vol. 60, pp. 7–16.

Berechman, J. and Small, K.A. (1988), 'Research policy and review 25: Modeling land use and transportation: An interpretive review for growth areas', *Environment and Planning A*, vol. 20, pp. 1285–1309.

Boyce, D.E. (1986), 'Integration of supply and demand models in transportation and location: Problem formulation and research questions', *Environment and Planning A*, vol. 18, pp. 485–489.

Boyce, D.E. (1988), 'Renaissance of large-scale models', *Papers of the Regional Science Association*, vol. 65, pp. 1–10.

Boyce, D.E., Chon, K.S., Ferris, M.E., Lee, Y.J., Lin, K.T. and Eash, R.W. (1985), 'Implementation and evaluation of combined models of urban travel and location on a sketch planning network', Working Paper, Department of Civil Engineering, University of Illinois and Chicago Area Transportation Study, Urbana-Champaign/Chicago, IL.

Boyce, D.E., Chon, K.S., Lee, Y.J., Lin, K.T. and LeBlanc, L. (1983), 'Implementation and computational issues for combined models of location, destination, mode, and route choice', *Environment and Planning A*, vol. 15, pp. 1219–1230.

Boyce, D.E., Tatineni, M. and Zhang, Y. (1992), 'Scenario analyses of the Chicago Region with a sketch planning model of origin-destination mode and route choice', Final Report to Illinois Department of Transportation, Urban Transportation Center, University of Illinois at Chicago, IL.

de la Barra, T. (1989), *Integrated Land Use and Transport Modelling*, Cambridge University Press, Cambridge.

de la Barra, T. (1998), 'Improved logit formulations for integrated land use, transport and environmental models', Chapter 16, this volume.

de la Barra, T., Pérez, B. and Vera, N. (1984), 'TRANUS-J: Putting large models into small computers', *Environment and Planning B: Planning and Design*, vol. 11, pp. 87–101.

Domencich, T.A. and McFadden, D. (1975), *Urban Travel Demand: A Behavioral Analysis*, North-Holland Publishing Company, Amsterdam.

Echenique, M.H. (1994), 'Urban and regional studies at the Martin Centre: Its origins, its present, its future', *Environment and Planning B: Planning and Design,* vol. 21, pp. 517–533.

Echenique, M.H., Flowerdew, A.D.J., Hunt, J.D., Mayo, T.R., Skidmore, I.J. and Simmonds, D.C. (1990), 'The MEPLAN models of Bilbao, Leeds and Dortmund', *Transport Reviews,* vol. 10, pp. 309–322.

Gu, Q., Haines, A. and Young, W. (1992), 'The development of a land-use/transport interaction model', Report 2, Department of Civil Engineering, Monash University, Melbourne.

Harris, B. (1965), 'New tools for planning', *Journal of the American Institute of Planners*, vol. 31, pp. 90–95.

Harris, B. (1985), 'Urban simulation models in regional science', *Journal of Regional Science*, vol. 25, pp. 545–567.

Harris, B. (1994), 'Science in planning: Past, present, future', *Journal of the American Planning Association*, vol. 60, pp. 31–34.

Hayashi, Y. and Tomita, Y. (1989), 'A micro-analytic residential mobility model for assessing the effects of transport improvement', in *Transport Policy, Management and Technology – Towards 2001*, Selected Proceedings of the Fifth World Conference on Transport Research, Yokohama, Western Periodicals, Ventura, pp. 91–105.

Hunt, J.D. (1994), 'Calibrating the Naples land-use and transport model', *Environment and Planning B: Planning and Design*, vol. 21, pp. 569–590.

Hunt, J.D. and Simmonds, D.C. (1993), 'Theory and application of an integrated land-use and transport modelling framework', *Environment and Planning B: Planning and Design*, vol. 20, pp. 221–244.

Hutchinson, B. and Batty, M. (eds) (1986), *Advances in Urban Systems Modelling*, North-Holland Publishing Company, Amsterdam.

Hutchinson, B., Nijkamp, P. and Batty, M. (eds) (1985), *Optimization and Discrete Choice in Urban Systems*, Springer-Verlag, Berlin.

Kain, J.F. (1987), 'Computer simulation models of urban location', in Mills, E.S. (ed.), *Handbook of Regional and Urban Economics*, vol. II, Elsevier Publishers, Amsterdam, pp. 847–875.

Kain, J.F. and Apgar, W.C. Jr. (1985), *Housing and Neighborhood Dynamics: A Simulation Study*, Harvard University Press, Harvard, MA.

Kim, T.J. (1989), *Integrated Urban Systems Modeling: Theory and Applications*, Kluwer Academic Publishers, Dordrecht.

Lotka, A. (1920), 'Analytical notes on certain rhythmic relations in organic systems', *Proceedings of the U.S. National Academy of Sciences*, vol. 6, pp. 410–415.

Landis, J.D. (1992), 'BASS II: A new generation of metropolitan simulation models', Working Paper 573, Institute of Urban and Regional Development, University of California at Berkeley, CA.

Landis, J.D. (1993), 'CUF model simulation results: Alternative futures for the Greater Bay Area Region', Working Paper 592, Institute of Urban and Regional Development, University of California at Berkeley, CA.

Landis, J.D. (1994), 'The California Urban Futures Model: A new generation of metropolitan simulation models', *Environment and Planning B: Planning and Design*, vol. 21, pp. 399–422.

Lee, D.B. Jr. (1973), 'Requiem for large-scale models', *Journal of the American Institute of Planners*, vol. 39, pp. 163–178.

Lundqvist, L. (1989), 'A model system for metropolitan energy studies', in Lundqvist, L., Mattsson, L.-G. and Eriksson, A. (eds), *Spatial Energy Analysis*, Avebury, Aldershot, pp. 245–269.

Lundqvist, L. (1996), 'Using combined network equilibrium models for environmental assessments of land-use transportation scenarios', in Hayashi, Y. and Roy, J. (eds), *Transport, Land Use and the Environment*, Kluwer Academic Publishers, Dordrecht, pp. 359–381.

Lundqvist, L. (1998), 'A combined model for analysing network infrastructure, land-use/transportation interactions', Chapter 18, this volume.

Mackett, R.L. (1983), 'The Leeds Integrated Land-Use Transport Model (LILT)', Supplementary Report SR 805, Transport and Road Research Laboratory, Crowthorne,

Mackett, R.L. (1990a), 'MASTER Model (Micro-Analytical Simulation of Transport, Employment and Residence)', Report SR 237, Transport and Road Research Laboratory, Crowthorne.

Mackett, R.L. (1990b), 'Comparative analysis of modelling land-use transport interaction at the micro and macro levels', *Environment and Planning A*, vol. 22, pp. 459–475.

Mackett, R.L. (1990c), 'The systematic application of the LILT model to Dortmund, Leeds and Tokyo', *Transport Reviews*, vol. 10, pp. 323–338.

Mackett, R.L. (1991a), 'A model-based analysis of transport and land-use policies for Tokyo', *Transport Reviews*, vol. 11, pp. 1–18.

Mackett, R.L. (1991b), 'LILT and MEPLAN: A comparative analysis of land-use and transport policies for Leeds', *Transport Reviews*, vol. 11, pp. 131–154.

Martinez, F.J. (1991), 'Transport investments and land values interaction: The case of Santiago City', *Proceedings of the PTRC Summer Annual Meeting*, PTRC, London, pp. 45–58.

Martinez, F.J. (1992a), 'Towards the 5-stage land-use transport model', in *Land Use, Development and Globalisation*, Selected Proceedings of the Sixth World Conference on Transport Research, Lyon, Presse de l'Imprimerie Chirat, St.-Just-la-Pendue, pp. 79–90.

Martinez, F.J. (1992b), 'The bid-choice land-use model: An integrated economic framework', *Environment and Planning A*, vol. 24, pp. 871–885.

Martinez, F.J. (1996), 'Analysis of urban environmental policies assisted by behavioural modelling', in Hayashi, Y. and Roy, J. (eds), *Transport, Land Use and the Environment*, Kluwer Academic Publishers, Dordrecht, pp. 233–257.

Miyamoto, K. and Kitazume, K. (1989), 'A land-use model based on random utility/rent-bidding analysis (RURBAN)', in *Transport Policy, Management and Technology – Towards 2001*, Selected Proceedings of the Fifth World Conference on Transport Research, Yokohama, Western Periodicals, Ventura, vol. IV, pp. 107–121.

Miyamoto, K., Nakamura, H. and Shimizu, E. (1986), 'A land use model based on disaggregate behavioral analyses', in *Proceedings of the Fourth World Conference on Transport Research*, pp. 1535–1550.

Miyamoto, K. and Udomsri, R. (1996), 'An analysis system for integrated policy measures regarding land use, transport and the environment in a metropolis', in Hayashi, Y. and Roy, J. (eds), *Transport, Land Use and the Environment*, Kluwer Academic Publishers, Dordrecht, pp. 259–280.

Nijkamp, P. and Reggiani, A. (1992), 'Spatial competition and ecologically based socio-economic models', *Socio-Spatial Dynamics*, vol. 3, no. 2, pp. 89–109.

Prastacos, P. (1986), 'An integrated land-use-transportation model for the San Francisco region', *Environment and Planning A*, vol. 18, pp. 307–322 and 511–528.

Putman, S.H. (1983), *Integrated Urban Models: Policy Analysis of Transportation and Land Use*, Pion, London.

Putman, S.H. (1991), *Integrated Urban Models 2. New Research and Applications of Optimization and Dynamics*, Pion, London.

Putman, S.H. (1998), 'Results from implementation of integrated transportation and land use models in metropolitan regions', Chapter 15, this volume.

Rho, J.H. and Kim, T.J. (1989), 'Solving a three-dimensional urban activity model of land use intensity and transport congestion', *Journal of Regional Science*, vol. 29, pp. 595–613.

Roy, J.R., Marquez, L.O., Taylor, M.A.P. and Ueda, T. (1996), 'SUSTAIN – a model investigating Sustainable Urban STructure And Interaction Networks', in Hayashi, Y. and Roy, J. (eds), *Transport, Land Use and the Environment*, Kluwer Academic Publishers, Dordrecht, pp. 125–145.

Roy, J.R., Marquez, L.O., Taylor, M.A.P. and Ueda, T. (1998), 'Development of a compact urban simulation model', Chapter 19, this volume.

Simmonds, D.C. (1995), 'DSCMOD and DSCMOD+', Unpublished Report, David Simmonds Consultancy, Cambridge.

Volterra, V. (1931), *Leçons sur la théorie mathématique de la lutte pour la vie*, Gauthier-Villars, Paris.

Webster, F.V., Bly, P.H. and Paulley, N.J. (eds) (1988), *Urban Land-Use and Transport Interaction: Policies and Models*, Report of the International Study Group on Land-Use/Transport Interaction (ISGLUTI), Avebury, Aldershot.

Wegener, M. (1985), 'The Dortmund housing market model: A Monte Carlo simulation of a regional housing market', in Stahl, K. (ed.), *Microeconomic Models of Housing Markets*, Lecture Notes in Economic and Mathematical Systems 239, Springer-Verlag, Berlin, pp. 144–191.

Wegener, M. (1986a), 'Transport network equilibrium and regional deconcentration', *Environment and Planning A*, vol. 18, pp. 437–456.

Wegener, M. (1986b), 'Integrated forecasting models of urban and regional systems', in *Integrated Analysis of Regional Systems*, London Papers in Regional Science 15, pp. 9–24.

Wegener, M. (1987), 'Transport and location in integrated spatial models', in Nijkamp, P. and Reichman, S. (eds), *Transportation Planning in a Changing World*, Gower, Aldershot, pp. 208–225.

Wegener, M. (1994), 'Operational urban models: State of the art', *Journal of the American Planning Association*, vol. 60, pp. 17–29.

Wegener, M. (1996), 'Reduction of CO_2 emissions of transport by reorganisation of urban activities', in Hayashi, Y. and Roy, J. (eds), *Transport, Land Use and the Environment*, Kluwer Academic Publishers, Dordrecht, pp. 103–124.

Wegener, M., Gnad, F. and Vannahme, M. (1986), 'The time scale of urban change', in Hutchinson, B. and Batty, M. (eds), *Advances in Urban Systems Modelling*, North-Holland Publishing Company, Amsterdam, pp. 145–197.

Wegener, M., Mackett, R.L. and Simmonds, D.C. (1991), 'One city, three models: Comparison of land-use/transport policy simulation models for Dortmund', *Transport Reviews*, vol. 11, pp. 107–129.

Wegener, M. and Spiekermann, K. (1996), 'The potential of microsimulation for urban models', in Clarke, G. (ed.), *Microsimulation for Urban and Regional Policy Analysis*, European Research in Regional Science 6, Pion, London, pp. 146–163.

Williams, I.W. (1994), 'A model of London and the South East', *Environment and Planning B: Planning and Design*, vol. 21, pp. 535–553.

Wilson, A.G. (1967), 'A statistical theory of spatial distribution models', *Transportation Research*, vol. 1, pp. 253–269.

Wilson, A.G. (1970), *Entropy in Urban and Regional Modelling*, Pion, London.
Young, W. and Gu, K. (1993), 'Modelling the land use-transport-environment interaction', Occasional Paper 107, Bureau of Transport and Communications Economics, Australian Government Publishing Service, Canberra.
Young, W. and Gu, K. (1996), 'Educating planners in testing of alternative strategies for cities of different sizes', in Hayashi, Y. and Roy, J. (eds), *Transport, Land Use and the Environment*, Kluwer Academic Publishers, Dordrecht, pp. 147–169.

15 Results from Implementation of Integrated Transportation and Land Use Models in Metropolitan Regions

Stephen H. Putman
Department of City and Regional Planning
University of Pennsylvania
Philadelphia, PA 19104
USA

15.1 Introduction

The general notion that operating agencies should attempt to integrate transportation and land use planning has long been espoused by both practicing planners and scholars. Some of the earliest work on the methodological aspects of this goal, as related to computer modeling of the processes, was sponsored by the U.S. Federal Highway Administration in an attempt to resolve what they then labeled as "the problem of premature obsolescence of highway facilities". The earliest successful attempt to develop a modeling approach to these issues demonstrated that there were important interactions which most traditional planning methods overlooked (Putman,1973). At the same time, this early study was unable to resolve important conceptual and computational problems of how to fully implement such integrated, or combined, model approaches. Part of the problem had to do with then available theory, and part with then available computer technology. The general problem had been formulated earlier (Beckmann et al., 1956), but the relevance of that work was not known to the transportation and land use modeling community. In 1973 the cost, at academic computing rates, of just one trip assignment model run was in excess of $200 (US), and most computer runs had to be submitted for overnight processing.

In summarizing the development of my own work with the ITLUP model package ten years later, though it was possible to report on some experiments which clearly showed the importance of the close linking of transportation and land-use models, there were still problems of model system stability and mechanisms to assure convergence (Putman, 1983). Shortly thereafter the work of Evans (1976) and Florian (Florian et al., 1975) began to be tested on small data sets and later on larger ones. Even so, most planning agencies were only just beginning to consider the use of User Equilibrium (UE) traffic assignment algorithms. The notion of nonlinear mathematical programming formulations of a combined model of location, trip making, and trip assignment was entirely beyond the reach of most practitioners and many academics as well.

Through the 1980's numerous experiments were done, for the most part in academic settings, which began to establish the practical possibility of such model systems and their solutions. By the late 1980's it had become possible, at least in a laboratory setting, to demonstrate the feasibility and computational practicality of the solving of these integrated model systems for full scale regional planning data sets (Putman, 1991). Even so, it has only been more recently that there were any actual planning agency applications of this approach. The integration of a set of location and land-use models (DRAM and EMPAL) with a set of travel demand models and a trip assignment model (TRANPLAN) was first accomplished, in an agency setting, at the Southern California Association of Governments (SCAG) in Los Angeles, CA in 1993. A second implementation of the concept was completed, again with DRAM and EMPAL, but this time the trip assignment model was EMME/2, at the Metropolitan Service District (METRO) in Portland, OR in 1994. Since that time similar integrated model systems have been implemented for Colorado Springs, Detroit, and Kansas City, and additional development of similar systems is well along in Atlanta and Sacramento (S.H. Putman Associates, 1996). This paper discusses some of the issues involved in these integrated model applications and also suggests some topics for further research.

In the next section of this paper I will discuss the general model system configuration used in these experiments. This will be followed by a discussion of the solution methods, and then a description of some indicators of convergence and solution optimality. There will then be brief overviews of the two agency applications, at SCAG and METRO, followed by a concluding section of the paper with suggestions as to next steps.

15.2 Overall Model System Configuration

In the most general sense, what is being done here is to begin with a trial forecast of employment and household location in a region, based, along with other variables, on either free flow or exogenously estimated congested flow travel times through the region's transportation network(s). This is followed by the computation of travel demands, trip origins, destinations, and subsequently trip interchanges, and perhaps mode splits. The trips are then assigned to the transportation network(s), and the congested network travel times (and/or costs) are estimated. These estimated times, now beginning to become consistent with the estimated location and land-use pattern, are used to estimate a *second* trial forecast of employment and household location. The entire process is then repeated for some number of what we will call outer, or inter-model, iterations, solving for an overall static equilibrium between transportation, location, and land use. A simplified schematic of the process, as used in a single five year forecast step, is shown in Fig. 15.1.

The method used for solution of the inter-model equilibrium will be discussed following this section of the paper. Here I will give brief descriptions of the location models and traffic assignment models that were used.

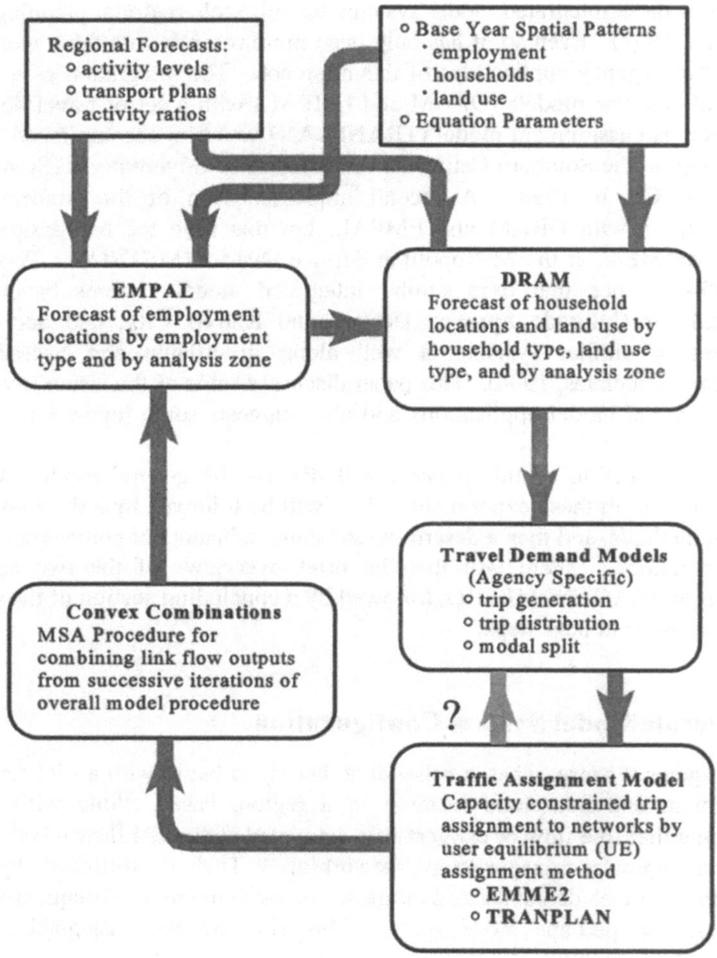

Fig. 15.1. Schematic of linked model system configuration

The models used for estimating the future location of employment and households were my own EMPAL model for the employment location, and DRAM for household location (Putman, 1983; 1991). I first began work on these models just over twenty years ago. After experimenting with several functional forms a spatial interaction structure (SI) was selected. This form was just emerging at that time as a mathematically consistent structure derived from entropy maximizing principles. In adapting the model for actual application I added the notion of a multivariate attractiveness function and a multiparametric travel cost deterrence function. This idea fell beyond the then current thinking about SI models, and also required the development of new methods of model calibration.

The SI approach did, at that time, provide a welcome organizing view of location models, but it lacked any appropriate grounding in terms of economic theories of activity location. Further, the use of the multivariate attractiveness term was clearly necessary for satisfactory model application, yet didn't seem to have a proper conceptual basis in the model derivations. It was not until the mid-to-late 1970's that work began appearing which would provide a satisfactory theoretical underpinning.

It is difficult to sort out which came first, as at that time there was a great flurry of work, and many informal paper drafts were in circulation. One very important paper was published by Cochrane (1975), in which a "location surplus" notion is developed. The derivation begins on the basis of an assertion that the trips which provide the trip-maker with the greatest *net* benefit are the trips that are chosen. The observed trip distribution pattern is thus indicative of the overall probability of trips being chosen on that basis. The approach taken to the subsequent derivation of a singly constrained SI model involves assuming that the probability of a particular trip-maker taking a trip from zone A to some other zone B, is the probability that a trip to zone B offers a surplus, or *net* benefit, greater than that which could be had from a trip to any other zone. It is then hypothesized that the probability that a trip to zone B for any individual trip-maker is the optimal trip (i.e. the trip which will actually be taken) increases with the number of opportunities for trip satisfaction in zone B, and decreases with trip cost between zones A and B, since the net benefit is reduced by greater trip cost.

In proceeding through the derivation of the functional form of the model it is assumed that the number of zones is large, say 100 or more. It is assumed that the underlying probability distribution is approximately exponential in the upper tail. It is shown that it is not necessary to know the actual number of trip possibilities represented by any trip attracting activity, but rather it is only necessary to assume that the number of trip possibilities is proportional to some measure of attraction.

The surplus then becomes the difference between the *probabilistic* utility u, which is the *gross* benefit of taking the trip, and a *deterministic* cost $c_{i,j}$ of actually taking the trip. Thus the *net* benefit, or surplus, from taking the trip is $s_{i,j} = u_{i,j} - c_{i,j}$, and the probability that the surplus will be of some particular value s, given all the preceding assumptions, is

$$\Phi_{i,j}(s) = \exp\left[-h A_j \exp\{-\omega(s - m + c_{i,j})\}\right] \qquad (1)$$

where $\Phi_{ij}(s)$ is the cumulative distribution function of the "location surplus" accruing from the *optimal* trip between zone i and zone j, and where h, ω, and m are constants, and A_j is a measure of the attractiveness of zone j.

With this, we maintain the assumption that the trip chosen by a trip-maker will be the trip that maximizes his/her personal surplus. The probability that the trip chosen will be a trip from origin i to a particular zone j is the probability that the maximum surplus offered by a trip terminating is j is greater than the maximum surplus offered by any other zone.

Continuing through the derivation, Cochrane winds up, given O_i trips originating from zone i, with the expected number of trips from zone i to zone j being given by

$$T_{i,j} = \frac{O_i A_j \exp(-\omega c_{i,j})}{\sum_j A_j \exp(-\omega c_{i,j})} \tag{2}$$

which is the usual equation of the singly constrained SI model. Note that ω is an empirically determined parameter. Further, the total surplus for all trips actually made is

$$S_T = \frac{1}{\omega} \sum_i O_i \left[0.577 + \ln \left(h\{\exp(\omega m)\} \sum_j A_j \exp\{-\omega c_{i,j}\} \right) \right] \tag{3}$$

Thus beginning with rather innocuous assumptions regarding a utility maximizing basis for trip-making behaviour, a robust economic underpinning can be developed for a SI model formulation. The relationship of this derivation, which is also called group-surplus maximization, to a derivation from random utility theory and probabilistic choice models is described in (Wilson et al., 1981).

The actual formulation of DRAM differs from equation (2) in that the attractiveness variable A_j is replaced by a multivariate formulation with a Cobb-Douglas function form, i.e., a product form with each term being an independent variable such as available land, or household income, raised to a parameter power. In addition, the simple exponential travel cost function used in equation (2) is replaced by the Tanner function which adds a power function of $c_{i,j}$ and requires the addition of the α and β parameters as well. Note that for DRAM, which is cast in terms of work-to-home trip making, e.g. the afternoon peak period, the $T_{i,j}$ are summed over j to get N_i, the number of persons (households) residing in zone i. This gives the following equation for DRAM:

$$N_i^n = \sum_j Q_j^n B_j^n W_i^n c_{i,j}^{\alpha^n} \exp(\beta^n c_{i,j}) \tag{4}$$

where

$$Q_j^n = \sum_k a_{k,n} E_j^k \tag{5}$$

and

$$B_j^n = \left[\sum_i W_i^n c_{i,j}^{\alpha^n} \exp(\beta^n c_{i,j})\right]^{-1} \tag{6}$$

and

$$W_i^n = (L_i^v)^{q^n} (X_i)^{r^n} (L_i^r)^{s^n} \prod_{n'} \left[\left(1 + \frac{N_i^{n'}}{\sum_n N_i^n}\right)^{b_{n'}^n}\right] \tag{7}$$

where

E_j^k = employment of type k (place-of-work) in zone j,

N_i^n = households of type n residing in zone i,

L_i^v = vacant developable land in zone i,

X_i = 1.0 plus the percentage of developable land already developed in zone i,

L_i^r = residential land in zone i,

$a_{k,n}$ = regional ratio of type n households per type k employee,

$c_{i,j}$ = impedance (travel time or cost) between zones i and j,

$\alpha^n, \beta^n, q^n, r^n, s^n$, and $b_{n'}^n$ = empirically derived parameters for DRAM.

In EMPAL the T_{ij} are summed over j to get E_i the number of persons working in i. Further, the actual formulation of EMPAL differed from that of DRAM principally in that an additive lag term is included. The current equation for EMPAL is as follows:

$$E_{j,t}^k = \lambda^k \sum_i N_{i,t-1}^T A_{i,t-1}^k W_{j,t-1}^k c_{i,j,t}^{\alpha^k} \exp(\beta^k c_{i,j,t}) + (1-\lambda^k) E_{j,t-1}^k \tag{8}$$

where

$$W_{j,t-1}^k = (E_{j,t-1}^k)^{a_k} (L_j)^{b_k} \tag{9}$$

and

$$A^k_{i,t-1} = \left[\sum_{\ell} (E^k_{\ell,t-1})^{a^k} (L_\ell)^{b^k} c^{\alpha^k}_{i,\ell,t} \exp(\beta^k c_{i,\ell,t}) \right]^{-1} \tag{10}$$

where

$E^k_{j,t-1}$ = employment (place-of-work) of type k in zone j at time t-1

$N^T_{i,t-1}$ = total households residing in zone i at time t-1,

L_j = total land area of zone j,

$c_{i,j,t}$ = impedance (travel time or cost) between zones i and j at time t,

$\alpha^k, \beta^k, \lambda^k, a^k$, and b^k = empirically derived parameters for EMPAL.

The usual practice is to link these two models so that the employment location forecast output from EMPAL becomes the input to DRAM. The household location forecast output from DRAM, in turn, becomes the input to a subsequent run of EMPAL. The two linked models take as input the base year locations of employment and households as well as land use inventories, and an estimate of zone-to-zone travel times and/or costs (in most cases travel time rather than travel cost is the available datum). The forecasts produced by these two models become the inputs to a second set of models having the function of estimating travel demand.

Countless different travel demand models have been developed over the years. Their purpose is to estimate 1) the trips originating from and terminating in each zone; 2) the destination zones of the trips originating from each zone, and the origin zones of the trips terminating in each zone; and 3) the numbers of trips traveling on each of the modes of transportation available between each origin-destination pair. These tasks may be accomplished sequentially, or with various degrees of simultaneity, dependent upon the model structure being implemented. The principal inputs to these travel demand models are the employment and household location outputs and land use patterns resulting from EMPAL and DRAM, and estimates of zone-to-zone travel times and/or costs.

The matrices of zone-to-zone trips which are produced by the travel models must be assigned to the various transportation networks (if a particular application includes the modeling of more than one network). This assignment process, too, may be done sequentially following the travel demand process, or simultaneously in some form of combined model structure (Boyce et al., 1993). Whichever way it is done, the trip assignment component of the process involves determining how the trips from each origin i to each destination j will travel over the links of the network. No traffic assignment model actually enumerates all the possible paths over the network from origin i to destination j. A few algorithms do enumerate selected paths, but most algorithms trace paths over networks by use of ingenious algorithms which obviate the need for path enumeration (Sheffi, 1985).

A central element of the traffic assignment component derives from the simple observation that as the number of trips assigned to (attempting to make use of) a network link increases, so too does the time taken to traverse the link. This congestion effect is embodied in the traffic assignment process by the assertion that link times and/or costs are a function of link flows (volumes). This is implemented in these models by use of volume/delay or volume/capacity functions which result in link times being calculated as a function of link volumes.

The most commonly used procedure for traffic assignment is the user equilibrium (UE) assignment algorithm. The underlying notion here is one of repeated assignment iterations with a mathematically appropriate procedure for moving the iterations towards an equilibrium solution. The definition of equilibrium follows Wardrop's principles (Wardrop, 1952). At equilibrium no trip-maker can shorten his/her trip over the network, from origin i to destination j, by changing their path. This implies that all paths in use by trip-makers traveling from origin i to destination j are of equal length, and that all unused paths are longer. Though the problem of UE assignment had been defined for some time (Beckmann et al., 1956), it was not until some years later that a simple approximation algorithm was developed for its solution (LeBlanc, 1973). The statement of the problem as a mathematical program is by now well known, and can simply be stated here, as minimizing the sum over all links of integrals, from 0 to the actual flow x_a, of the volume/delay function $t_a(x_a)$, as in:

$$\text{minimize} \quad Z(x)=\sum_{t} \int_{0}^{x_a} t_a(x_a)dx \tag{11}$$

subject to

$$\sum_{\rho} x_{\rho}^{rs} = q^{rs} \tag{12}$$

$$x_{\rho}^{rs} \geq 0 \tag{13}$$

$$x_a = \sum_{r,s,\rho} x_{\rho}^{rs} \delta_{a,\rho}^{rs} \tag{14}$$

where q^{rs} is a matrix of flows from each origin r to each destination s. Constraint equation (12) requires that the sum of the flows on all paths ρ, connecting origin r to destination s must equal the total flow from r to s. Constraint equation (13) prohibits negative path flows. Constraint equation (14) requires the total flow on any given link to equal the total of all path flows which use that link (specified in the form of the Kronecker delta, δ, which equals 1 for links on the specified path, and 0 for those which are not). The solution of the UE problem, is done by use of the Frank-Wolfe algorithm which works by solving a nonlinear objective function by use of successive linear approximations, and which has become more or less standard practice in solving the UE problem.

Other traffic assignment approaches are available, and in some small use by agencies. Laboratory tests of the linked transportation, location, land use model solution procedures, to be described in the next section of this paper, indicate that they work with all traffic assignment algorithms that might be used in the traffic model portion of a linked model set. The UE approach offers a rather clear means of describing the problem, and so is used in the following discussions, and was, of course, used in the model tests to be described.

15.3 Solution Methods for Systems of Linked Models

My early attempts to solve linked systems of transportation and land use models (Putman, 1973) were done in the absence of an understanding of their formal mathematical properties. Working strictly from a knowledge of computerized modelling systems, I tried numerous methods for, in effect, averaging the zone-to-zone travel times from successive iterations of linked models. None of these approaches were particularly successful, though they were quite informative, and laid the groundwork for further research on the problem (Putman, 1983).

Both the approach of Evans (1976) and that of Florian (Florian et al., 1975) were later evaluated as possible ways to solve the problem. Using small numerical examples of EMPAL and DRAM linked to a traffic assignment program, the Evans approach was shown to work reasonably well. A further experiment was then done involving a variation on the Evans approach wherein the Method of Successive Averages (MSA) was used as a substitute for some rather more complex calculations used by Evans (Putman, 1991). The essence of the procedure is simply this: In successive iterations of the linked models the traffic *volumes* on network links must be averaged, and the resulting averaged volumes used to recalculate the minimum (congested) paths through the network. The averaging must be done in a particular (MSA) way, and the new congested travel times are used to perform another model iteration. The need for this procedure, as distinct from simply averaging the congested network travel times, is a consequence of the nonlinearity of the volume delay functions used for calculation of congested link times, and the fact that this procedure involves, in effect, finding different paths through the network *after* the volumes have been averaged.

In order to ensure that the system will converge, the procedure for averaging the link volumes must be one of convex combinations. This is done by using the following averaging scheme, called MSA, the method of successive averages (Powell and Sheffi, 1982). The form is:

$$x_a^{n+1} = (1-\rho)x_a^n + \rho(y_a) \tag{15}$$

where x_a^{n+1} is the new combined (averaged) flow on link a at iteration n+1, x_a^n is the previously calculated flow on link a at iteration n, with y_a being the most recently calculated flow on link a. The value of ρ is calculated as 1/n, where n is the number of the iteration. In the laboratory tests I did with full size data sets for Washington, DC and Houston, TX, the full linked model system began to converge rather rapidly, with near convergence after 3-5 iterations, and full convergence, depending upon the tolerance specified, being obtained in 6-9 iterations (Putman, 1991). In those experiments, the definition of convergence was in terms of the mean absolute percent change (MAP), by locator type, by zone, from one model system iteration to the next. In the Washington experiments, the values of MAP for all locator types were below 1% after the third iteration, and below 0.1% after the fifth iteration.

Having once defined the procedure to be followed, the actual solving of these systems by use of this method is not difficult. A means must be developed to save the calculated link flows at the end of each traffic assignment run, and the MSA calculation must be put in place to produce the convex combinations of link flows. Then, the link congestion which results from the new combined volumes must be calculated, and the new minimum paths traced. In practice the major difficulties are those of file management, and in agency practice, of developing the necessary computer connections and staff cooperation. While my original tests were done (in 1990) on a mainframe computer, both the sets of actual agency results to be described below were done on workstations. More recently similar systems have been solved on 32 bit Pentium class microcomputers in a few hours or less. In the next section of this paper I will discuss the use of measures to indicate the extent to which convergence has been achieved, and what they might mean with respect to policy evaluation.

15.4 Indicators of Convergence: UE Objective Function, and Location Surplus

In the initial tests of the solution of these systems, the criterion for whether a computational equilibrium had been achieved was specified in terms of iteration-to-iteration change in specific variables (Putman, 1991). The iteration-to-iteration change in numbers of locators, e.g. employees by type and households by type, in each zone was expected to become very small, on the order of 0.5% or less. The mean zone-to-zone travel time was also expected to show little or no iteration-to-iteration change. These criteria were easily met in the test runs. A stopping point in a computational procedure is not necessarily an equilibrium solution to the equation system, so there was more to be done. In fact, since the first version of this paper in 1994, work has been completed which demonstrates the existence and uniqueness of the solution to this model structure (Shen, 1995).

It is one thing to examine the behaviour of a large system of models and by use of some measure, such as the MAP criterion mentioned above, decide whether the system has converged to an equilibrium solution. It is quite another thing to cast the set of models, or rather, their functional forms, in the form of a mathematical programming problem which, if the necessary conditions can be described and met, may be amenable to solution by use of an appropriate algorithm. The solution of the programming problem will involve the maximization (or minimization) of some objective function. It is possible to formulate all or portions of the linked (combined) transportation and land use model system as such a mathematical program, and then proceed to develop the appropriate algorithms for its solution. While this is a perfectly appropriate approach for research purposes, it has several problems with respect to actual agency application. In particular, it is almost universally true that planning agencies have very specific, and often extravagantly detailed requirements, particularly with respect to the capabilities of their transportation, traffic assignment, procedures. The inclusion of these requirements, often including thousands of network links and flows, into a formal mathematical programming problem formulation would be extremely difficult. This is not to mention that there is often a mismatch in level of detail, both sectoral and geographic, between the land use models and the transportation models. In practice this mismatch is dealt with by procedures for disaggregating the land use model outputs prior to their input to the travel demand and trip assignment models, and a subsequent aggregation of the travel time matrix output from trip assignment so that it may be used, at the appropriate level of detail, as input to the next iteration of the land use models. Given these problems, the only practical solution to implementation of linked model solutions was by use of the existing EMPAL and DRAM model software combined with each operating agency's current choice of travel demand model(s) and traffic assignment package. This approach seemed to work, but left unresolved the issue of whether the results were, in effect, equivalent to actually solving the mathematical programming formulation.

Recall that the derivation of the SI model given above proceeded from a measure of location surplus which each trip-maker is attempting to maximize. Several related approaches, including one involving maximization of group surplus, are developed in Wilson et al. (1981). In general these can be framed as mathematical programming problems involving the maximization of total surplus for all household locators. If the constant terms are scaled out of equation (3), the total surplus is

$$S_T = \frac{1}{\omega} \sum_i O_i \ln\left[\sum_j A_j \exp\{-\omega c_{i,j}\}\right] \tag{16}$$

If this is further modified by substituting the Tanner function of impedance, and by putting the attractiveness term in a form similar to that of DRAM, where here the different variables are merely by X's, and defining the location of place of residence as a function of both place of work and residential attractiveness, then one can state

$$\text{maximize} \quad S_T = \frac{1}{\beta}\sum_j E_j \ln\left[\sum_i X_{1i}^{\mu} X_{2i}^{\gamma} X_{3i}^{\delta} \ldots X_{ni}^{\eta} c_{i,j}^{\alpha} \exp\{-\beta c_{i,j}\}\right] \tag{17}$$

as the locational surplus maximization portion of the linked models' objective function, where E_j is employment at place of work in zone j. The traffic assignment portion of the linked model system is calculated as a UE problem as described in equations (11)-(14). Within the assignment portion of each linked model iteration, the UE objective function given in equation (11) will be minimized. The *net* benefit, or surplus, calculated as the net of locational surplus minus travel cost, is what we would expect to see being maximized as the model system converged to equilibrium. The experimental question was whether this would be the observed numerical result? Laboratory tests had shown that it should work (Lee, 1992). It should be noted that while the proposed solution procedure was known to be robust, there were so many departures from the strict mathematical programming formulation of the problem, that I had some concern as to whether the numerical "stopping point" for the procedure would correspond to a maximization of the net location surplus. It did. In the next sections of the paper I will discuss the results from two separate sets of runs done at two different regional planning agencies.

15.5 Computational Results: Southern California Association of Governments (SCAG)

The SCAG experiments were the first to be done. Preliminary tests of the linked model system were completed in July of 1993. The EMPAL and DRAM runs are done at a 772 zone level of detail, with four household types and eight employment types. These results are disaggregated, by a model developed at SCAG, to finer geographic detail, comprising about 1600 zones. SCAG's travel demand models work at this level of geographic detail, and produce trips for assignment to a network with a corresponding set of about 1600 load nodes, and about 30,000 one-way links. Note that in these preliminary experiments the mode-split step of the transportation models was omitted in order to reduce computational effort. In principle, this should have had no significant effect on the experiments, while it would, of course, have some effect on the specific numerical values obtained. The traffic assignment is done by TRANPLAN, and involves the use of the UE algorithm. At the completion of the traffic assignment run the 1600x1600 zone travel time matrix is collapsed to a 772x772 zone matrix for use in the next iteration as input to EMPAL and DRAM. The whole system is run on IBM RISC/6000 workstations.

In the preliminary tests five 'outer' iterations were done, cycling from EMPAL/DRAM to travel demand to TRANPLAN and back again. The initial travel time matrix input to EMPAL/DRAM was based on 1990 free-flow travel times. This caused EMPAL/DRAM to produce a dispersed location pattern with considerable trip-making. The consequent network congestion caused the next run of EMPAL/DRAM to produce an over concentrated spatial pattern. Even so, after the third iteration, and using the MSA procedure to do convex combinations of link volumes, the system had settled down considerably. The iteration-to-iteration percent change in activity types by zone was well under 1%, and the change in mean zone-to-zone travel time was 1.4%. To give a general notion of the apparent convergence of these first tests, Fig. 15.2 shows the values of what we call the SO objective function, the total vehicle-hours traveled on the network, along with the

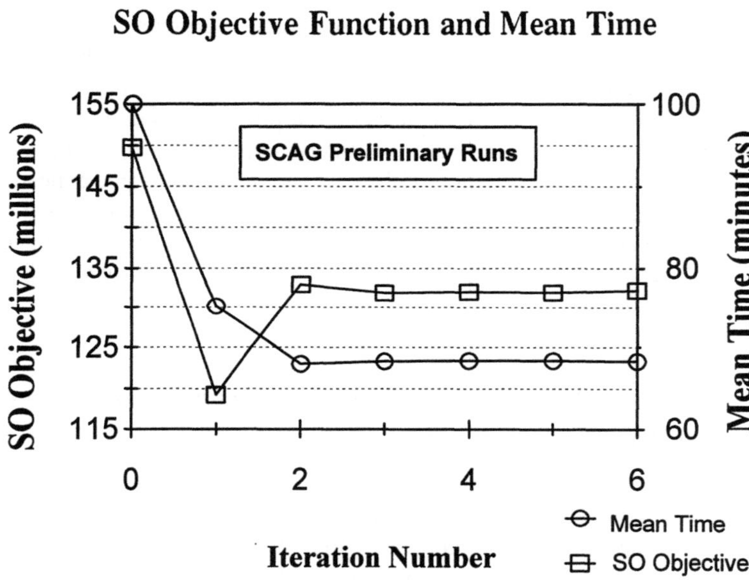

Fig. 15.2. Computational results: SCAG

mean value of travel time output from trip assignment, and then used as input to EMPAL and DRAM, for five iterations of the linked model system. Note again, that the Mean Travel Time was the value of the mean of the zone-to-zone travel times, either a trial value, or the output of a prior iteration of the travel models, which then were used as the input to the next iteration of EMPAL and DRAM. Their output was then used to produce the trips (only work trips in these first test runs) which, when assigned to the links of the network, result in the congested link times used to calculate the value of the SO objective function. In the subsequent, more carefully prepared runs of the linked model system, the results have been even more successful, with convergence coming more smoothly, and after fewer iterations. In part this was due to the use of a more appropriate set of starting impedances, rather than the completely uncongested 1990 free flow travel times which were used as starting values in the preliminary tests.

15.6 Computational Results: Metropolitan Service District (METRO)

In the Spring of 1994, just under a year after the preliminary runs done at SCAG, an extensive set of linked transportation and land use model test runs was begun for the Portland, OR region. Here the EMPAL/DRAM runs were done at a 100 zone level of detail, with five household types and four employment types. In the SCAG data set there was a mean of 6084 households per zone for 772 zones, while for METRO there was a mean of 5537 households per zone for 100 zones. We note, in passing, that in both cases, despite the substantial difference in number of zones being used in the analyses, the geography is probably too coarse for the best land use modeling results to be obtained. There are two levels of detail at which the METRO travel models and trip assignment procedures can be run. The *sketch* level of detail uses an aggregated network specification, and does both the travel demands and trip assignments using a network with 100 load nodes exactly corresponding to the EMPAL/DRAM zone centroids. The *detailed* level of detail uses a network with 1189 load nodes and 18,960 one-way links. When running tests with the networks at the detailed level it is necessary, as with the SCAG runs, to disaggregate the 100 zone output from EMPAL/DRAM to 1189 traffic zones for use as input to the travel demand models. The outputs of the travel demand models, at the 1189 traffic zone level of detail, are used as input to the traffic

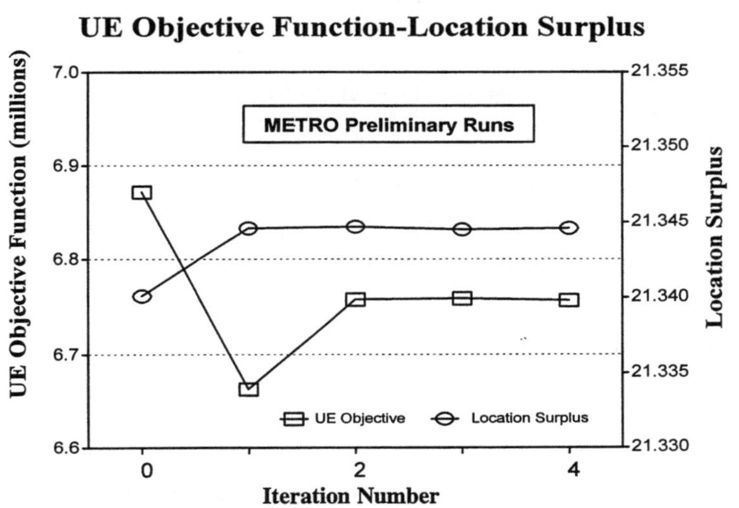

Fig. 15.3. Computational results: METRO

assignment step. Traffic assignment is done by EMME/2, and involves the use of the UE algorithm. At the completion of the traffic assignment run the 1189x1189 zone travel time matrix is collapsed to a 100x100 zone matrix for use in the next iteration as input to EMPAL and DRAM. The EMPAL/DRAM portion of the model system was run on a PC in the Philadelphia area, and the travel models and EMME/2 were run remotely from Philadelphia on METRO's Sun SPARCstation.

The first of the many test runs which were done with the METRO data was a simple run done at the *sketch* level of detail. We can see in Fig. 15.3, that for this run the full linked model system reached equilibrium very quickly. The household location surplus measure was at its maximum after the first iteration, while the UE objective function for trip assignment was, after an initial adjustment from the starting value, minimized after the second iteration (each ± a trivial amount). Within each traffic assignment procedure there were about 14 iterations. Overall, these results were quite satisfactory. Even so, while the linked model system had performed quite well with the METRO data, there were many further questions to be asked about numbers of iterations, both inner and outer, as well as in regard to what the implications of the results might be.

There are, of course, many questions of model system sensitivity to variations in model system configuration. One particular question is about the number of iterations of the travel demand models. In Fig. 15.1 there is an arrow going back

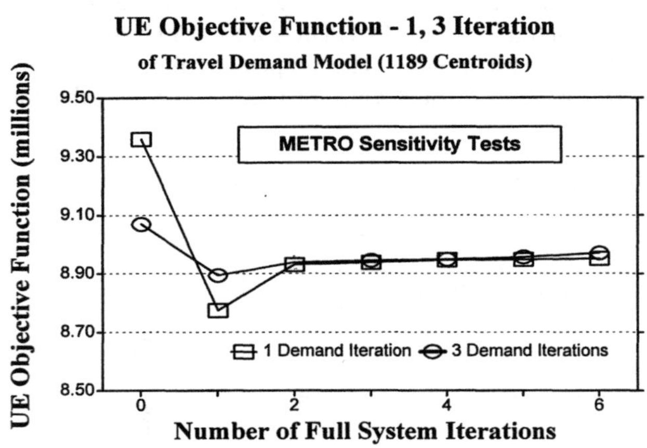

Fig. 15.4. Sensitivity to model system configurations

from "Traffic Assignment Model" to "Travel Demand Models". It is regular practice at METRO, though certainly not at all agencies, to do these iterations. These are an approximation to the notion of "combined models" of travel demand

and trip assignment. While there are still some issues to be resolved here, the results are significant. When three iterations are done between travel demand and trip assignment, the path to convergence taken by the complete linked model system is smoother, but the results are not significantly different, nor, as can be seen in Fig. 15.4, does it take any longer to achieve system wide equilibrium. A second question arose regarding the low level of congestion, system wide, on the Portland highway network. It was likely that with higher levels of congestion the linked models might not converge so readily. As a test of the sensitivity of these results to system wide congestion levels, the trip generation rates were arbitrarily doubled. This, of course, led to substantially more congestion on the network (though, of course, the results serve only as a numerical test). In this case, the linked models did again converge, but far more iterations were necessary. As can be seen in Fig. 15.5, there was some considerable oscillation of both the UE objective function and the location surplus measure during the first several iterations of the linked model system. Yet, the system did move to convergence despite the obviously excessive congestion level that had been artificially induced. At convergence, the UE objective function was minimized, and the location surplus measure was maximized.

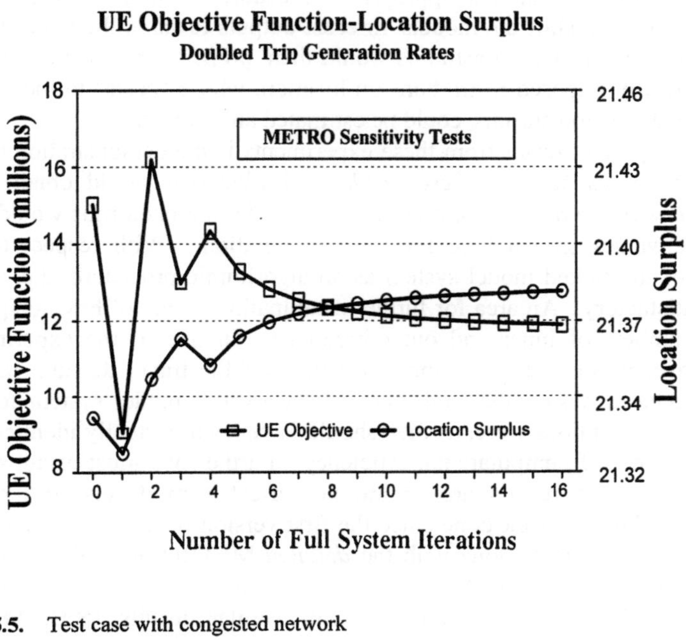

Fig. 15.5. Test case with congested network

15.7 Some Last Comments

With respect to the location surplus measure generally, it remains to examine its formulation in more detail. In the above experiments, though there wasn't the space here to discuss them, it was interesting to observe that the location surplus measures for the individual locating household types (DRAM typically operates with 4 to 8 household types, usually defined in terms of household income) while maximized, did follow different trajectories during the successive iterations of the model system. Of particular interest was the effect of imposing constraints on household location. DRAM permits several forms of absolute constraint and/or range constraint on household location, both by household type and by zone. In linked model runs where these constraints were imposed, the location surplus measure for constrained household types declined. In effect, when a restriction is placed on household location the sum of the net benefits to the locators decreases. While further research remains to be done here, one possible consequence of this result is that it will become possible to have a "measure" of the consequences of policy restrictions on household location. At present, some agency users of EMPAL and DRAM make use (perhaps excessively) of constraints to force the model outputs to conform to the expectations of their local technical advisory committees. This would compromise their ability to estimate policy consequences in terms of location surplus. The appropriate procedure would be to adjust zonal attractiveness levels to shift the models' forecast outputs towards the expectations of informed local planners. Constraints would then, properly, be used to actually represent politically imposed restrictions on location, whose "costs" to locators in terms of reduced location surplus, could be estimated and evaluated.

The most important result from these experiments is the demonstration of the *significant differences between these results* and what one would obtain *from model runs done in the absence of these linkages*. The unlinked runs would give as results the values shown here for the 0'th iteration. With respect to the performance of the linked model system as an algorithmic procedure, the results were quite satisfactory. An area for further examination is that of how to specify appropriate numbers of inner and outer iterations. In one of the experiments described above, it was clear that inner iterations of the travel demand models (METRO) produced a smoother trajectory of the overall model system toward equilibrium. The solutions were, as they should have been, virtually identical. A question exists here of computational efficiency. Further, we must establish the extent to which these results are or are not affected by levels of network congestion. In addition, work done since the first version of this paper confirms these results in experiments done with the *detailed* METRO network. Many of these experiments were repeated for four other urban areas, Colorado Springs, Detroit, Kansas City, and Sacramento, with similar results (S.H. Putman Associates, 1996).

We need also to examine some more general issues. The location models are calibrated with networks that are different, in terms of zone-to-zone travel times, from the equilibrium networks. Should there be some reconciliation? Would there be a significant difference in the results? Some recent experiments demonstrate that it is possible to calibrate the location models, run a linked model system to equilibrium, use the resulting zone-to-zone travel times to re-calibrate the location models, re-run the linked model system, etc. - until the whole "system" reaches an equilibrium of calibration and linked model runs (Im, 1993). This may well be an appropriate start to a sequence of linked model runs, and needs to be examined as well.

Finally, it is sometimes argued that *one* must use a combined equilibrium model system for forecasting purposes. I believe that an equilibrium solution to these model systems will not be a good forecast of the future. The issue is that of *rate of adjustment*. With respect to trip assignment, it seems reasonable to expect that trip-makers will shift paths rather readily in response to network congestion or traffic incidents. Solving the assignment problem for an equilibrium seems appropriate. The equilibrium solution is, subject to the possibility of some dispersion which could be described with a stochastic UE solution (SUE), a reasonable description of reality. Most users will manage to find their own "best" path and use it for most of their trips. With respect to location of employment or households, I continue to believe that such adjustments are *never* completed for all locators, much less so in a five year period. It is unreasonable to assert that all locators relocate in order to maximize their household utilities in each forecast period (usually of five years duration). Thus, at any instant, there is a frequency distribution of households which 1) have just relocated to maximize their utility, 2) are not at their maximum utility location, but are "close enough" so that the cost of relocation outweighs the increased utility to be gained, or 3) are just about to relocate. Further, it seems sensible to expect that as one locator changes location, other locators' location decisions will be affected. None of this takes place with the rapidity of trip reallocation on congested network links. By the time a significant number of locators have effected a location change, many other variables will have changed. *The whole system of locations will never be at equilibrium*. As I have said before (Putman, 1991) the appropriate approach seems to be along the lines of models of adjustment to disequilibrium. These models, known as PAMEQ (Partial Adjustment to Moving EQuilibrium) models have seen some investigation, but remain relatively undeveloped in the area of transportation and land use modeling (Putman, 1991). Knowing how to calculate a *potential*, or *target* equilibrium is an essential step, but it is not the end of the job. Next we must develop procedures for describing the adjustment processes for different locator types, and the appropriate methods for modeling them. After that we must tackle the equally important process of developing the data necessary to test these new models against reality.

Ten years ago there had been no full scale tests, much less applications at the levels of detail required for agency use, of linked location and transportation models in a system which was known to have the possibility of converging. In 1989-1990 when I was writing my most recent book, it was possible to demonstrate in a laboratory setting that such systems had computationally practicable solutions, but there were no agency applications even in process. Now there are perhaps four or five public agencies where these linked model systems are becoming a part of standard practice, and several more agencies where they are contemplating (no doubt with some trepidation) doing the same. In turn, most of these agencies have the intention of using the forecast outputs of these models as input to models of air quality. With the common use of advanced GIS systems in most planning agencies, it is only a matter of time until the data for testing PAMEQ models becomes available. What needs to be done first is sufficient investigation of their structure to permit them to be ready for examination and testing as the data become available to do so.

Acknowledgments

At the Southern California Association of Governments (SCAG) in Los Angeles, Terry Bills was in charge of the overall small area forecasting project. Shi-Cheng Fu did the actual model runs of the TRANPLAN computer package, developed by R. James W. Fennessy, Principal of The Urban Analysis Group. Jim Fennessy helped develop the computer code to link TRANPLAN with EMPAL and DRAM.

At the Metropolitan Service District (METRO) in Portland, OR, Keith Lawton, was in charge of the modeling work, and Scott Higgins did the computer runs of the EMME/2 computer package developed by Michael Florian of INRO Consultants Inc. Michael Florian helped develop the computer code to link EMME/2 with EMPAL and DRAM.

David Stiff of S.H. Putman Associates ran most of the combined model runs and helped in the statistical analyses on which some of the tables in the last part of this paper are based.

The work was funded by a contract to S.H. Putman Associates from the Federal Highway Administration.

References

Beckmann, M., McGuire, C. and Winsten, C. (1956), *Studies in the Economics of Transportation*, Yale University Press, New Haven, CT.

Boyce, D., Lupa, M. and Zhang, Y. (1993), 'Introducing 'feedback' into the four-step travel forecasting procedure vs. the equilibrium solution of a combined model', Paper prepared for presentation at the Transportation Research Board, Washington, DC.

Cochrane, R. (1975), 'A possible economic basis for the gravity model', *Journal of Transport Economics and Policy*, vol. 9, pp. 34-49.

Evans, S. (1976), 'Derivation and analysis of some models for combining trip distribution and assignment', *Transportation Research*, vol. 10, pp. 37-57.

Im, Y. (1993), 'A study on the system calibration of the integrated transportation and land use models', Ph.D. Dissertation in City and Regional Planning, University of Pennsylvania, Philadelphia, PA.

Florian, M., Nguyen, S. and Ferland, J. (1975), 'On the combined distribution-assignment of traffic', *Transportation Science*, vol. 9, pp. 43-53.

LeBlanc, L. (1973), 'Mathematical programming algorithms for large scale network equilibrium and network design problems', Ph.D. Thesis, Department of Industrial Engineering and Management Sciences, Northwestern University, Evanston, IL.

Lee, H. (1992), 'The investigation of equilibrium methods in spatial interaction models', Ph.D. Dissertation in City and Regional Planning, University of Pennsylvania, Philadelphia, PA.

Powell, W. and Sheffi, Y. (1982), 'The convergence of equilibrium algorithms with predetermined step sizes', *Transportation Science*, vol. 16, pp. 45-55.

Putman, S. (1973), 'The interrelationships of transportation development and land development', Report on Contract DOT-FH-11-7843 to the U.S. Department of Transportation, Washington, DC.

Putman, S. (1983), *Integrated Urban Models: Policy Analysis of Transportation and Land Use*, Pion Limited, London.

Putman, S. (1991), *Integrated Urban Models 2: New Research and Applications of Optimization and Dynamics,* Pion Limited, London.

Sheffi, Y. (1985), *Urban Transportation Networks*, Prentice-Hall, Englewood Cliffs, NJ.

Shen, P.-N. (1995), 'Optimized network equilibrium models of combined travel and residential location choices', Ph.D. Dissertation, Department of City and Regional Planning, University of Pennsylvania, Philadelphia, PA.

S.H. Putman Associates, Inc. (1996), *Integrated Transportation and Land Use Forecasting: Sensitivity Tests of Alternative Model System Configurations,* Report on work done under Contract No. 903-042 for METRO, Portland, OR, and Contract DTFH61-93-C-00216 with the Federal Highway Administration, Washington, DC.

Wardrop, J. (1952), 'Some theoretical aspects of road traffic research', *Proceedings of the Institute of Civil Engineers*, Part II, vol. 1, pp. 325-378.

Wilson, A., Coelho, J., Macgill, S. and Williams, H. (1981), *Optimization in Location and Transport Analysis*, Wiley, Chichester.

16 Improved Logit Formulations for Integrated Land Use, Transport and Environmental Models

Tomás de la Barra
Modelistica
PO Box 47709
Caracas 1041-A, Venezuela

16.1 Introduction

Integrated land use and transport models have been a matter of interest for a long time. It is almost universally accepted that the interaction between activities is the driving force behind transport demand, and in turn, accessibilities *shape* land uses and prices. In spite of this, such integrated models have developed slowly, and transport policies are, by and large, still evaluated in their own terms.

Now this situation is changing rapidly, mainly due to environmental concerns. After all, the transport sector is responsible for a substantial share of our present ecological and energy related problems. When the main interest of a study is to estimate short-term demand, such as traffic expected on a proposed highway or modal split along a particular corridor, it is possible to apply a disaggregated model to the affected area. By contrast, when the interest is to evaluate the overall effects of a set of policies on energy consumption and the environment, large scale comprehensive and aggregated models are required. For instance, road pricing may be viewed as a way to rapidly increase throughput in congested areas, thus reducing emissions and energy consumption. This may be true in the short run, but in the long term it could be that central areas reduce their attraction in favor of less efficient urban sprawl. Hence, integrated models capable of representing the land use-transport interaction and assess their long-term implications are clearly required. A number of models of this kind exist (see Wegener, 1994), but most of them are at a research level.

From a theoretical point of view, nested multinomial logit (NMNL) models represent a reasonable solution to simulate the highly integrated and complex decision chains and hierarchies needed to build such comprehensive models, as suggested by Ben-Akiva and Lerman (1985). A modeling framework of this kind is developed by de la Barra (1989). His strategy is to build several models to represent the various links along the decision tree, such as residential location, retail, industry, trip generation, mode choice and route choice. In this scheme, each link conditions the probabilities of the next decision level. At each level, choice probabilities generate a composite utility that, in turn, affects the previous, higher level decisions.

The author and colleagues have implemented a specific modeling package based on this idea (TRANUS), including a logit-based assignment procedure, and have put it into practice in a large number of applications over the last ten years. As a result of this experience, it has become evident that the standard nested logit model shows some important weaknesses that may have been overlooked when dealing with specific transport related topics, such as modal split. They become apparent, though, when applied to larger and complex systems, as the number of links in the chain increases. They also emerge in mixed level applications, such as large metropolitan areas, in which short trips and interactions mix with longer ones. Another important implication is that when dealing with large scale models of this type, the disaggregate approach to modeling is no longer useful, and an aggregate representation seems the only practical solution. This fact raises additional theoretical questions regarding the general random utility theory. Furthermore, the way we currently use logit models and composite costs may generate unreasonable results in evaluation. The issue of evaluation, that is, the estimation of benefits to land use and transport consumers, is the real objective of modeling when broad and long-term policies are assessed from an economic and environmental point of view. Ideally, we should develop and implement policies that reduce energy consumption and improve the quality of the environment and at the same time produce benefits to consumers and suppliers of the various *urban goods*. This calls for models that not only simulate the behavior of the main elements of the urban system, but that can also estimate the benefits derived from policies in a consistent way.

This paper concentrates on two problems related to the standard NMNL model. The first one, termed *constant perception of utility*, becomes evident when different decision levels are dealt with simultaneously. The second problem refers to the unreasonable property of the composite cost indicator of producing negative results under certain conditions, which is inconsistent with economic principles. In this paper, a solution to these problems is proposed, together with a discussion of its main properties. The paper begins by presenting the standard logit and nested logit models, mainly as a reference, with a digression on the case of aggregation. Next, the problem of the constant perception of utility, implicit in the standard logit formulation, is presented; a possible solution, based on the assumption of marginally decreasing perception of utility is proposed, leading to a *scaled utilities* form. The discussion then turns to the composite cost indicator, and its unreasonable negativity property. A solution to this is also proposed, which satisfies the main conditions that such an indicator should ideally have. A final section presents some conclusions and raises a few questions for future research.

16.2 The Standard Logit Model

This section presents the classical derivation of the logit model, leading to the MNL and NMNL forms. This description is only for the sake of consistency with later sections, and adds nothing new to this well-covered subject. The derivation draws mainly from Ben-Akiva and Lerman (1985), in turn based on Domencich and McFadden (1975) and Williams (1977a; 1977b). Of the many possible interpretations, the logit model is viewed here as a demand model, relating quantities of goods or services with perceived costs. This point of view affects the interpretation of the concepts, but not the derivation itself.

Consider first an individual that must choose one option out of a set of alternatives C available to him. According to consumer theory, the individual is assumed to have consistent and transitive preferences that determine a ranking. If U_i is a utility index associated with each alternative $i \in C$, it is assumed that a particular alternative i will be chosen if:

$$U_i > U_j, \ \forall j, i \neq j \,. \tag{1}$$

Using the concept of indirect utility, the utility function U_i may be defined in terms of its attributes, such that:

$$U_i = U(z_i)\,, \tag{2}$$

where z_i is a vector of the attributes of alternative i from the point of view of the decision maker.

Observation of reality reveals that the above framework does not always hold. Apparent inconsistencies may be observed, in which individuals of similar characteristics are seen to choose different options. Random utility theory assumes that such differences are the result of observational deficiencies on the part of the analyst. In order to be consistent with the theory of consumption, the individual is always assumed to behave in a rational way, i.e., will always select the option with the highest utility. Because utilities are not known to the analyst with certainty, a better approach is to treat them as random variables. From this point of view, the choice probability that option i is selected is equal to the probability that the utility of such option is greater than that of all other options, that is:

$$P(i \mid C) = \Pr\left[U_i > U_j, \forall j, i \neq j\right], \tag{3}$$

assuming that ties do not occur. There are several possible sources of randomness in this probability, and the literature usually includes elements such as unobserved attributes, unobserved taste variations, imperfect information, and the use of instrumental or proxy variables. An important assumption that follows logically from the identification of the sources of variation is that it is possible to derive a probabilistic model assuming a joint probability distribution for the set of random utilities. It is possible to express the random utility of an alternative as a sum of observable and unobservable components as follows:

$$U_i = V_i + \varepsilon_i, \quad (4)$$

where V_i and ε_i are the observable and unobservable elements. If the term V_i represents the perceived costs for the consumer, expression (3) may be rewritten as:

$$P(i \mid C) = \Pr\left[V_i + \varepsilon_i < \min_{j \neq i}(V_j + \varepsilon_j)\right]. \quad (5)$$

To derive a specific model, an assumption must be made about the form of the joint probability distribution of the set of disturbances ε. It is usually assumed that the disturbances are *independently and identically distributed* (IID). If it is further assumed that the negative of the disturbances are Gumbel-distributed, the multinomial logit model can be derived. The adoption of a Gumbel distribution can be defended as a reasonable approximation to the normal for convenience. Important properties of this assumption for the binary case are:

a) if ε_1 and ε_2 are independent Gumbel-distributed with parameters (η_1, μ), (η_2, μ), then $\varepsilon^* = \varepsilon_1 - \varepsilon_2$ is logistically distributed:

$$F(\varepsilon^*) = \frac{1}{1 + \exp(\mu(\eta_1 - \eta_2 - \varepsilon^*))}; \quad (6)$$

b) the $\max(\varepsilon_1, \varepsilon_2)$ is itself Gumbel-distributed with parameters:

$$\left(\frac{1}{\mu}\ln[\exp(\mu\eta_1) + \exp(\mu\eta_2)], \mu\right). \quad (7)$$

To derive the multinomial logit model, define the *minimum achieved disutility* of all options other than the chosen i as:

$$U^* = \min_{j \neq i}(V_j + \varepsilon_j) \text{ where } -\varepsilon_j \text{ is Gumbel-distributed with parameters } (0, \mu). \quad (8)$$

Then $-U^*$ is Gumbel-distributed with parameters[1]:

$$\left(\frac{1}{\mu}\ln\sum_{j \neq i}\exp(-\mu V_j), \mu\right). \quad (9)$$

[1] The negative signs in (9) and many of the equations that follow are due to the fact that V_j represents perceived costs. The notation $exp(-V_j\mu)$ would be more consistent with the interpretation of μ as a parameter in the Gumbel distribution. However, this notation has not been adopted in keeping with existing literature.

We can write $U^* = V^* + \varepsilon^*$, where $-\varepsilon^*$ is Gumbel-distributed with parameters $(0, \mu)$, where:

$$V^* = -\frac{1}{\mu} \ln \sum_{j \neq i} \exp(-\mu V_j).$$

Since the probability that option i is chosen is:

$$P(i) = \Pr(V_i + \varepsilon_i < V^* + \varepsilon^*)$$
$$P(i) = \Pr\left[(V^* + \varepsilon^*) - (V_i + \varepsilon_i) > 0\right], \tag{10}$$

by property (6) we have:

$$P(i) = \frac{1}{1 + \exp(-\mu(V^* - V_i))} = \frac{\exp(-\mu V_i)}{\exp(-\mu V_i) + \exp(-\mu V^*)}$$

$$P(i) = \frac{\exp(-\mu V_i)}{\exp(-\mu V_i) + \exp[\ln \sum_{j \neq i} \exp(-\mu V_j)]} = \frac{\exp(-\mu V_i)}{\sum_{j \in C} \exp(-\mu V_j)}. \tag{11}$$

The minimum expected disutility over all options is then:

$$\tilde{V} = -\frac{1}{\mu} \ln \sum_{j \in C} \exp(-\mu V_j), \tag{12}$$

minus a constant (γ/μ). This value is usually taken as a *composite cost* indicator or as an accessibility indicator. Williams (1977a) claims that (12) is an *exact* measure of benefit and that there exists no freedom in the selection of a composite cost function. In his argument, if (11) is the model, then (12) must be used to estimate consumer surplus as a result of the introduction of a policy. This argument is discussed in a later section.

An important property of the multinomial logit model is that it satisfies the axiom of *independence of irrelevant alternatives* (IIA) that states that, in any choice set, the ratio of one probability over another is unaffected by the presence or absence of any additional alternatives. This is a direct consequence of adopting homoscedastic distributions (of equal variance); it can be shown to hold in the logit model as follows:

$$\frac{P(i)}{P(k)} = \frac{\exp(-\mu V_i) / \sum_{j \in C} \exp(-\mu V_j)}{\exp(-\mu V_k) / \sum_{j \in C} \exp(-\mu V_j)} = \frac{\exp(-\mu V_i)}{\exp(-\mu V_k)} = \exp(-\mu(V_i - V_k)), \tag{13}$$

a result that shows that the relative odds of option i versus k are independent of any other options present in the choice set. This property has been widely discussed and can lead to unreasonable results if some of the alternatives are highly correlated, such as in the case of the well-known red-bus-blue-bus paradox. The invariance property is thoroughly discussed in a paper by Lindberg et al. (1995), in which they argue that, contrary to the opinion of many other authors, this is not necessarily a drawback, but a benefit of models of this kind. These authors claim that, although invariance may not hold in practice in many situations, "*it provides a yardstick for measuring the effects of heterogeneity. This has no counterpart in models not exhibiting invariance, such as [multinomial probit] models.*"

There are two main situations in which the assumption of homoscedasticity may produce unrealistic results: a) if the population is sufficiently heterogeneous to justify segmentation of the model, and b) if there are strong correlations between some of the options.

The nested logit model is usually presented as a way of treating choice situations in which the unobserved attributes of a subset of options are highly correlated. In such a case the subset is grouped as a single option at a first level, and treated as separate options at a second, conditional, level. This scheme, however, can be very powerful to represent general models involving conditional decision chains, as argued by Ben-Akiva and Lerman (1985). This is precisely the main use given to the nested logit model in large scale modeling, the central concern of this paper. In an urban case, for instance, a residential location choice may be linked to a retail location choice, assuming that people first decide where to live and then decide where to go shopping. This, in turn, may condition the choice of transport mode to make the shopping trip, and in turn condition the route choice. In sum, the nested logit is not only useful to group similar options and treat them as sub-choices, but as a way to build larger choice constructs in the form of decision chains.

The general nested logit model may be represented formally as follows. Consider two related choice sets C_1 and C_2, in which, for example, C_1 represents the set of modes available to the trip-making population, such as private and public, and C_2 represents the set of transit routes within the public mode. In this case, the probability P_{kr} of choosing the public mode k from C_1 and a specific route r from C_2 may be written as:

$$P_{kr} = \frac{\exp(-\mu_1 \widetilde{V}_k)}{\sum_{j \in C_1} \exp(-\mu_1 V_j)} \cdot \frac{\exp(-\mu_2 V_r)}{\sum_{l \in C_2} \exp(-\mu_2 V_l)}, r \in C_2, k \in C_1, \tag{14}$$

where $\widetilde{V}_k$ denotes the composite cost of option k (the public mode) aggregated over all options (transit routes) in C_2:

$$\widetilde{V}_k = -\frac{1}{\mu_2}\ln\left(\sum_{l\in C_2}\exp(-\mu_2 V_l)\right). \qquad (15)$$

The invariance property in this case is not maintained across levels, because in principle $\mu_1 \neq \mu_2$. Furthermore, as shown in Ben-Akiva and Lerman (1985), the ratio μ_1/μ_2 must be positive and less than or equal to 1 and therefore must satisfy the condition $\mu_1 \leq \mu_2$. The nested form ensures that any change in the attributes of one or more options in the set C_2 has an effect on the probability of choosing among options in C_1.

16.3 The Aggregate Case

We now turn briefly to the question of aggregation, because of its importance in large scale environmental modeling. As pointed out in the introduction, when simulating a metropolitan area with the purpose of evaluating the overall impact of policies from economic and environmental points of view, an aggregated model is required for practical reasons. There are several procedures that have been proposed in the past to derive an aggregated model. Ben-Akiva and Lerman (1985) reviewed several methods. The approach taken here follows the same lines suggested by de la Barra (1989): derive a random utility model, assuming that aggregation is simply another source of variability in the model construction.

Aggregation not only affects demand entities, i.e., decision makers, but also supply options. In the case of demand, we are interested in populations of individuals taking decisions such as where to live or shop, or in which way to travel. In the case of supply, it is convenient to group, say, housing supply in a particular neighborhood, or a set of transport facilities made up of many different services. In this context, one of the main sources of aggregation is the spatial division of the study area into discrete zones, and the representation of the transport system in the form of a simplified network. These concepts, widely used in practice, introduce a considerable source of variability into the decision process.

Consider the example of Fig. 16.1 in which a population travels from an origin zone to a destination zone. The aggregation process implies that, even if the origins and destinations might be scattered all over both zones, for analytical and practical purposes it is assumed that all travelers originate at centroid i and end at centroid j, both relatively arbitrary point locations in space. Furthermore, a simplified transport network must represent all real supply options; in this example, only three travel options with engineering costs c_1, c_2, and c_3 are represented. The observed distribution of trips will probably show demand distributing among options, even if we assume that all travelers are identical from all other points of view. It is clear from this simple example that spatial aggregation is a source of variation of considerable importance. Even if the analyst has calculated that c_1 is the best option, a proportion of travelers will choose c_2 and c_3, not because they are not making a rational decision, but because for them c_2 and c_3 represent better options than c_1. For example, travelers from point a to b in the diagram might agree that option c_1 is the best option, while those traveling from d to e might rightly perceive c_2 as preferable.

It is clear, then, that the disturbance element ε in expression (4) must absorb all sources of variation due to the aggregation process. This means that, in the logit model of equation (11), as more aggregation is introduced, e.g. there are fewer zones and fewer links, variability increases, causing μ to decrease, consequently reducing the capacity of the model to capture net benefits resulting from proposed improvements. For calibration purposes, this means that the best estimate of the value of μ for, say, a modal split model, will vary depending on the number of zones and the degree of detail in the network. This, in turn, may question the practice of calibrating such models with disaggregated data. The problem of aggregation will be taken up later in further sections.

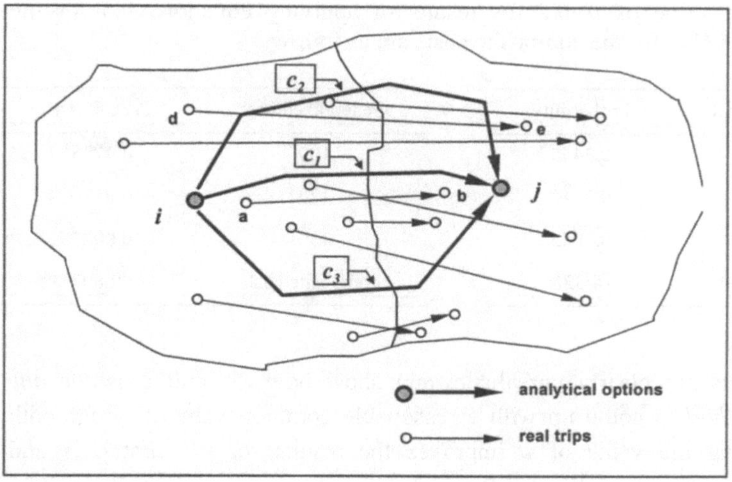

Fig. 16.1. Spatial aggregation of demand and options

16.4 Marginally Decreasing Perception of Utility

A first problem that becomes apparent when applying the standard logit formulation to large scale modeling is that it is very difficult to calibrate when decisions at different *scales* have to be dealt with simultaneously. This occurs because the logit model, as it stands, only perceives the absolute differences in utilities among options, as represented in (13), a property that may lead to unrealistic results. Accepting this property is equivalent to considering that a \$5 difference in generalized cost in a flight from San Francisco to Sydney has the same effect in demand as a \$5 difference in a trip from San Francisco to Berkeley, assuming that the same trip-makers are involved. This is an extreme example, and fortunately we are not usually being asked to simulate both cases in the same model; separate models with different dispersion parameters will adequately deal with this problem. However, in the case of large metropolitan areas or regions, or national scale modeling, which is the kind of scope integrated models must address, there are considerable differences in the scale of the decisions that must be represented simultaneously in the same model.

When a certain decision, such as mode choice, is being simulated simultaneously at largely different scales, the invariance assumption involving a single dispersion parameter, may cause serious problems. Consider the example shown in Fig. 16.2, which is very common in large scale applications. There is one origin and two destinations; destination 2 is hundred times farther away than 1. In each case there are three alternative modes with generalized costs as indicated. Intuitively, one would expect modal shares to be similar.

Assuming a value of $\mu=0.2$ the results of applying equation (11) for the probabilities and (12) for the composite costs are as follows:

Generalized cost	Probability
1.0	0.3445
1.2	0.3310
1.3	0.3245
composite cost	-4.328

Generalized cost	Probability
100	0.9796
120	0.0179
130	0.0025
composite cost	99.897

These results are clearly unrealistic, and show how difficult it is *for any calibration method* to come up with a reasonable common value of μ for both cases. Increasing the value of μ improves the results for the first case and eventually turns the composite cost positive, but the probabilities for the second case quickly become *1.0, 0.0* and *0.0*.

Practitioners have faced this problem in several ways. The most obvious one is to adopt different values of μ for each origin-destination pair. This is consistent with the suggestion by many authors that $\mu \approx 1/\text{mean trip cost}$. In this example, if $\mu=0.2$ is a reasonable value for destination 1, then a comparable value for destination 2 would be $\mu=0.002$, making the results very similar and far more reasonable. The practical difficulty with this approach is that it might involve several thousand calibrations, in the order of half the square of the number of zones. A reasonable approximation is to *pack* origins and destinations into groups of similar scale, and use a dozen or so parameters. This problem, however, becomes much more acute in the case of a location model, in which the different scales are part of the same choice situation. In the case of a model simulating the purchase of agricultural products from manufacturing industries, or the location of households from places of work, the same decision-making group will apply the concept of scale in evaluating the options. In the case of residential location, for example, households considering options costing, say, 3, 8, 20 and 25 will tend to perceive the difference between 8-3=5 as much more important in their decision than the difference between 25-20=5. Intuitively, one would assume that households would rank the option costing 3 as very close, 8 as intermediate, and both 20 and 25 as similarly distant. In a similar fashion, if a constant $50 tax is added to the fare of all flights in the USA, it is unrealistic to assume that this would have the same effect in all destinations, as the current form of the logit model would predict; probably short internal trips costing $60 to $100 would be very hard hit, while long international flights costing $ 1,500 or more would hardly show any significant change.

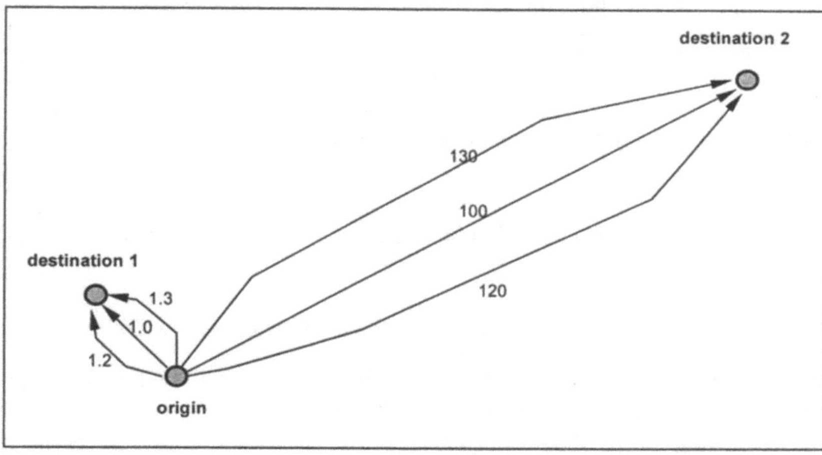

Fig. 16.2. Decisions at various scales

What lies at the core of this problem is that, in the logit model, the perception of utility is assumed to be constant and independent of the scale at which the decision process is taking place. An alternative, more realistic assumption, is that perception of utility *decreases marginally* as the scale of utility increases. This is equivalent to assuming that decision makers perceive utilities in relative terms, and this is the reason why demand curves are not straight lines but curves of diminishing slope.

There are many possible ways to represent diminishing perception of utility in the logit formulation. A straightforward approach is to assume that the marginal perception of utility decreases linearly. As a result, the strict utility term in the logit model can be replaced by the ratio of the utility of each option over a *scale parameter*. In this way, the perceived cost of each option is scaled to a common value. There are several values that might serve as a scale parameter, such as the average cost over all options, or the cost of the minimum cost option. If the latter is adopted, the utility term (4) may be rewritten as follows:

$$U_i = \frac{V_i}{\min\limits_{j \in C}(V_j)} + \varepsilon_i , \tag{16}$$

where $\min\limits_{j \in C}(V_j)$ represents the utility of the option with least perceived cost within the option set C. In this way, utilities have been *scaled*. Consequently, the logit model of equation (11) becomes:

$$P(i) = \frac{\exp(-\mu(V_i / \min\limits_{j \in C}(V_j))}{\sum\limits_{j \in C} \exp(-\mu(V_j / \min\limits_{j \in C}(V_j))} . \tag{17}$$

For consistency, the composite cost becomes:

$$\hat{V} = -\frac{1}{\mu} \ln \sum_{j \in C} \exp\left[-\mu(V_j / \min_{j \in C}(V_j))\right] \bullet \min_{j \in C}(V_j) . \tag{18}$$

Multiplying by $\min\limits_{j \in C}(V_j)$ *unscales* back the composite cost, without losing the property that utility is being perceived in relative terms. As will be seen below, this is important because composite costs may form part of a choice set at a higher level, competing with other composite costs; thus they must all be at the appropriate *utility scale*.

By contrast to the example of Fig. 16.2, in which very different probabilities were obtained in two cases where intuitively one would expect similar results, model (17) yields identical probabilities in both cases. Because the costs have the same relative differences in the two cases, the resulting probabilities will be the same as in the first case in the example above. This example is also useful to show why the composite costs must be unscaled back; if they are not unscaled, the composite cost calculation would yield the same result in both cases, making little sense for, say, a residential location model further up the decision chain.

The proposed model has some additional advantages for calibration. On the one hand, μ becomes independent of the units in which the generalized cost is measured, so that transferring values from one context to another gives, at least, a first approximation. On the other hand, model (17) requires smaller samples than (11) for the same degree of confidence. As was mentioned, the only practical solution to the estimation of the standard logit when different scales are involved is to segment the choice set, estimating separate parameters for each segment, based on random samples for each segment. In the case of the scaled utilities logit, because there is only one parameter to estimate, the random sample may be drawn from a single set, and only one calibration process is required.

A point of considerable debate concerning this new specification of the utility function with scaling is that it might be viewed as a way of relating the attributes of the different options, making the set of disturbances ε in (4) interdependent. In this case the utility function adopts a complex error structure and the model ceases to be a logit. This may also imply that the proposed form might not satisfy the IIA axiom. It is also a matter of debate whether it is important that the IIA property should be preserved. The IIA property, in turn, relates to the homoscedastic distribution of the error terms, and if this property is dropped, the error structure would, indeed, become very complex.

However, if the scaled utility is specified as (16) the $-\varepsilon$'s can be assumed to be IID Gumbel-distributed, and the derivation of the model remains unchanged, leading to a logit type model with well-known properties. It is also possible to prove that the scaled utility model (17) is still consistent with the IIA axiom *for any given choice set*. The proof is similar to (13), and the result is:

$$\frac{P(i)}{P(k)} = \exp(-\mu(\frac{V_i}{\min\limits_{j \in C}(V_j)} - \frac{V_k}{\min\limits_{j \in C}(V_j)})) = \exp(-\mu\frac{V_i - V_k}{\min\limits_{j \in C}(V_j)}), \qquad (19)$$

which is only valid if the minimum cost term is interpreted as an arbitrary constant to *set the scale* of the decision. If, however, the choice set itself changes for any reason, particularly if the perceived cost of the best option is further reduced, or if a new option with a cost smaller than that of the previous minimum cost is introduced, then the IIA property no longer holds. In other words, the argument is that the IIA property is kept only within a given choice set, but not across choice sets, or if the choice set itself changes in any way, particularly with respect to scale.

A common argument in favor of IIA is that it makes it easy to estimate the *new mode* problem. In this respect, the scaled utilities model provides a new insight into the problem, because it states that the introduction of a new option might alter the scale of the choice situation, or in other words, might alter the relative benefits perceived from each of the original options. The argument behind equation (17) is that if users are confronted with options costing, say, 20 and 25, they would perceive the first option as being considerably better than the second one; if a new option costing only 5 is introduced, people would change their views about the previous options and think that options 20 and 25 become almost equally unattractive. For example, the large variety of liner options that existed between London and New York practically disappeared when air services were introduced.

The main argument against IIA is that it produces unrealistic results if two or more options are highly correlated within a given choice set, as exemplified by the well-known red bus-blue bus paradox. It is not within the scope of this paper to discuss the problem of attribute correlation among alternatives, but clearly the scaled utilities model does not solve it. This could be taken as a further proof that homoscedasticity is preserved in the proposed scaled utilities model.

16.5 An Improved Composite Cost Indicator

Replacing the constant perception of utility with the marginally decreasing assumption improves the calculation of choice probabilities and facilitates the estimation of the dispersion parameter. However, it does not solve the negativity problem of the composite utility indicator. In the example shown in Fig. 16.2, the results of the composite utilities were unrealistic, but if equation (18) is applied, the results are even worse: -4.328 and -432.8 !

Let us first examine the conditions that a composite cost indicator $\widetilde{V}$ must satisfy. Williams (1977a) proposes, among others, two basic conditions:

$$\widetilde{V} < \min_{j \in C}(V_j)\text{, and} \tag{20}$$

$$\lim_{\mu \to \infty}(\widetilde{V}) = \min_{j \in C}(V_j)\,. \tag{21}$$

To these, the non-negativity condition can be added as:

$$\lim_{\mu \to 0}(\widetilde{V}) = 0\,. \tag{22}$$

Condition (20) states that the composite cost must be less than the measurable generalized cost of the best option, a direct consequence of the rationality assumption. When a proportion of a decision-making population faced with two options with costs, say 5 and 10, choose the *apparently* bad option, they do so because for them it is the best option. However small this proportion may be, for them their perceived cost is less than that of the apparently best option, i.e., less than 5.

Condition (21) implies that, as the value of μ increases, the composite cost tends towards the cost of the best option. This is because if μ is large, dispersion is very low and the probability of choosing the best option nears 1.0, approaching the composite cost towards the cost of the best option. This means that all other options were not the best option for a large majority of the population. This condition sets the upper bound for the composite utility indicator.

Condition (22) sets the lower bound for the value of the composite cost by stating that such value cannot be negative. This is consistent with consumer theory, because a demand curve *only exists in the positive price-quantity quadrant*: neither prices nor quantities may become negative.

The original composite cost indicator (12) satisfies conditions (20) and (21), but certainly not (22). In fact, as $\mu \to 0$, $\tilde{V} \to -\infty$, which is clearly unrealistic. In practice, the composite cost quickly becomes negative in normal conditions, making the calculation of consumer surplus an economic absurdity. As will be seen later below, this may also imply that differences in consumers' surplus when improvements are introduced, are bound to be overestimated.

The non-negativity condition is also a matter of debate. It may be accepted that users will always perceive positive costs or prices when evaluating options and responding to them. But the composite cost, or consumers' surplus, is often viewed as a mere indicator of utility, not a price, and hence, can be positive or negative. The only relevance of the composite cost, in this view, is to serve as an estimator of the change in utility when two alternative choice situations are being compared, so that it is the difference in surplus that matters; and in this sense, the composite cost will always show results with the correct sign. However, in a nested logit model, the composite cost is used as a price or as a perceived cost in choice sets higher up in the hierarchy, competing with other options that may or may not have been evaluated with a comparable composite cost measure. As a consequence, the composite cost must have the conditions of any price, including non-negativity. Fisk and Boyce (1984) address this problem and claim that the fact that the composite cost function can assume negative values is inconsistent with its interpretation as an *average cost* measure.

The composite cost indicator (12) was rigorously derived, yet it leads to inconsistent and unrealistic results. One possible explanation for this anomaly is that the choice of the mean value of the distribution of utilities as the composite cost indicator might not be appropriate. William's (1977b) argument that *'the mean value of the distribution of minimum costs ... may be formally identified as the appropriate (and unique) composite cost function'* is not necessarily true. It is certainly better than any measure of the average cost, so popular in practical applications, but it must be considered as one of several options available. This is particularly valid for aggregated models, in which the process of aggregation itself introduces a major cause of variability in the perception of utilities. One must bare in mind that the values of V_j, that is, the perceived cost associated with each option, only represent the result of the analyst's best effort to estimate the average cost of each option, but he or she cannot support this with certainty. Furthermore, there is no guarantee that all real options are being considered in the set C. Options and their associated utilities are merely convenient analytical devices that we hope behave numerically in a way similar to that which people would perceive in reality. The same criteria applies to the choice of the probabilistic model and, of course, to the composite cost indicator.

A possible alternative proposed here is that the composite cost indicator can be constructed from the probabilistic model, taking as point of departure the only element that we can be sure of: the proportions in which people choose among our analytical options. To do this, first define the following term for convenience:

$$G_i = \exp(-\mu(V_i / \min_{j \in C}(V_j)), \tag{23}$$

so that the model of equation (17) becomes:

$$P(i) = \frac{G_i}{\sum_{j \in C} G_j}. \tag{24}$$

Next define a series of the following form:

$$G^* = G_1 + (1 - G_1)G_2 + (1 - G_1)(1 - G_2)G_3 + \ldots \tag{25}$$

Finally calculate the composite cost as:

$$\breve{V} = -\frac{\ln G^*}{\mu} \min_{j \in C}(V_j). \tag{26}$$

It can be shown that this formulation satisfies the three conditions stated above. If there is only one option, the series (25) collapses to $G^* = G_1$ and the composite utility in (26) becomes $\breve{V} = V_i$. Thereafter, as more options are considered, the value of G^* increases, thus reducing the result of the logarithm in equation (26), making $\breve{V}$ decrease. Thus, the composite cost will always be less than the cost of the best option, as stated in condition (20).

In the limit, when $\mu \rightarrow \infty$, $G^* \rightarrow \exp(-\mu)$, making $\breve{V} = 1 \bullet \min_{j \in C}(V_j)$, condition (21) is satisfied. Finally, when $\mu \rightarrow 0$, $G^* \rightarrow 1.0$, making $\breve{V} = 0$ in equation (26), condition (22) is satisfied.

The following should be noted: (a) each additional choice when entered into the series reduces the composite cost in proportion to the probability that they are selected; (b) the order in which options are entered into the series is irrelevant, a property that makes it very easy to compute; (c) the series (25) is valid both for the standard and scaled utilities model forms.

Fig. 16.3 plots three alternative composite cost functions for different values of μ: simple average, the log-sum of equation (12), and the proposed indicator of equation (26). In this example a case of three options is considered with costs 4, 8 and 12. It can be seen that the original log-sum decays from the cost of the minimum option to $-\infty$. The simple average varies from the cost of the minimum option to an intermediate value between the three options. Finally, the proposed solution varies from the cost of the minimum option to zero, thus satisfying the required conditions. To make these plots comparable, in all cases utilities were not scaled.

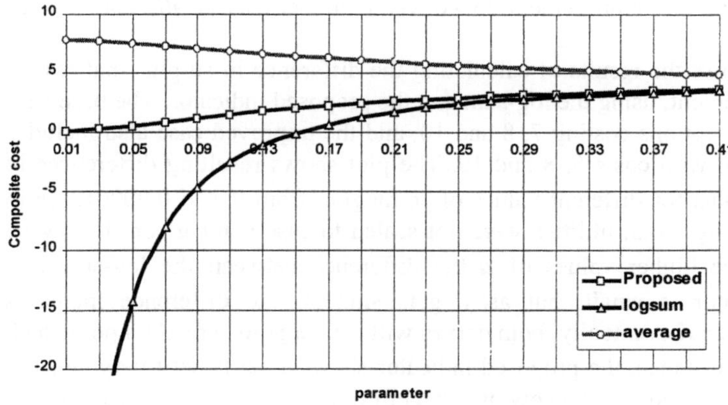

Fig. 16.3. Composite utility functions for different values of μ

In the table below, the example of Fig. 16.2 has been recomputed with the proposed formulation and the same value of $\mu=0.2$:

Generalized cost	Probability
1.0	0.3445
1.2	0.3310
1.3	0.3245
composite cost	0.0445

Generalized cost	Probability
100	0.3445
120	0.3310
130	0.3245
composite cost	4.4447

More realistic results are obtained with a higher value of $\mu=2.0$:

Generalized cost	Probability
1.0	0.4506
1.2	0.3021
1.3	0.2473
composite cost	0.6506

Generalized cost	Probability
100	0.4506
120	0.3021
130	0.2473
composite cost	65.0661

A final consideration is the effect of the proposed composite cost indicator on evaluation. In this respect, if the dispersion parameter μ is close to zero, it means that the measurable attributes included in the utility function V_i are of little relevance for the decision makers; hence, any change in the values of V_i, such as a reduction in cost, cannot generate benefits of significance. In this sense, both the standard composite cost indicator (12) and the proposed indicator (26) will cause the benefits from an improvement to converge to zero as $\mu \to 0$, but at very different rates.

Fig. 16.4 shows the results of calculating the difference in surplus that results from an improvement, using the log-sum and the proposed indicator. The base case consists of three options costing 7, 8 and 10, and the improved case also includes three options but with costs 5, 8 and 10. The plot shows resulting differences in consumers' surplus for different values of μ, ranging from 0.5 to 0.00001. As in the example of Fig. 16.3, utilities were not scaled to ease comparison. It may be seen that for the higher values of μ the difference between the log-sum and proposed indicator is small, but as μ gets smaller, the difference increases considerably, even if eventually, both curves will converge to zero. At a point such as $\mu=0.001$, for instance, the proposed indicator shows a net benefit very close to zero, while the log sum is still showing a benefit of about 0.67; at this same point, however, choice probabilities are virtually indistinguishable[2]. What these results are showing is that, according to the log-sum indicator, users would still perceive considerable benefits even if demand remains unchanged after the improvement has been introduced. This unreasonable result is not obtained with the proposed indicator because consistency between choice probabilities and surplus is guaranteed in its formulation.

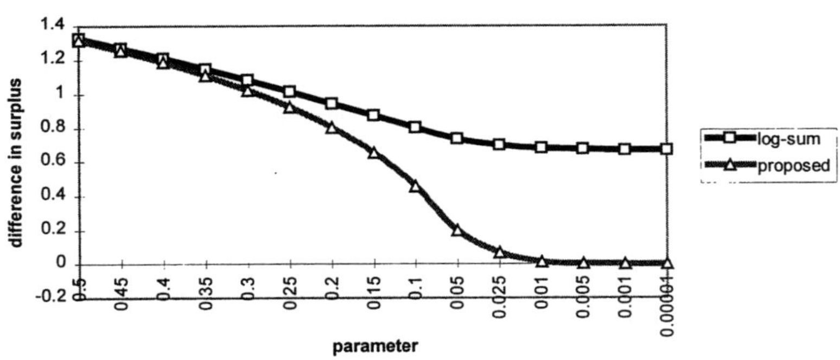

Fig. 16.4. Effect of the log-sum and proposed indicators on surplus

16.6 Conclusions and Further Work

Two problems related to the multinomial logit model are discussed in this paper. The first problem is the validity of the assumption that decision makers, when faced with several options, evaluate these options as a function of the difference in utilities. This implies that options are evaluated in absolute terms, with a constant

[2] With $\mu=0.001$, the probability of choosing the best option changes by 0.0004.

level of utility independent of the *scale* on which the decision is being made. Here, it is argued that the scale of the decision affects the way in which people evaluate each option, and hence, that the assumption of constant utility is not realistic. This, in turn, makes the standard logit model difficult to calibrate in practice, particularly if large scale modeling is involved. An alternative assumption, that of *marginally decreasing perception of utility,* was introduced. If it is further assumed that perception of utility decays linearly, a *scaled utilities* model results, in which the perceived cost of each option is divided by a scale parameter. In the proposed formulation, the cost of the least cost option was taken as the scale parameter. If a single IID of the error terms is maintained in the derivation, it was shown that the resulting model is a logit and complies with the IIA axiom for a given choice set. These properties, however, are not maintained across choice sets or if any of the attributes of a single set changes the scale of the decision.

It was argued that the proposed scaled utilities model is more realistic and easier to calibrate than the constant utility model. This discussion must be settled by empirical research, and it may turn out that reality lies somewhere in between these two extremes. A more general form of the scaled utilities model is obtained if the minimum cost term that sets the scale in the utility function is raised to the power of a second parameter. In this case the form of the utility function in (16) is transformed into:

$$U_j = V_j / (\min_{j \in C}(V_j))^{\alpha} + \varepsilon_j, \tag{27}$$

where α is a parameter that regulates the degree of scaling, $\alpha=0 \to 1$. If $\alpha=1$, a fully scaled model results, identical to (16); if $\alpha=0$, a fully unscaled model results, identical to the standard logit model in (4), making the standard logit a special case. This more general model form would have the advantage of being able to represent various degrees of scaling. It may turn out that in experiments in small urban areas, scaling is less noticeable than in large urban areas, and that scaling is more important for regional, state-wide and national models than for urban studies, or that scaling is more relevant to passengers than to commodities.

This is certainly an interesting piece of research that lies ahead. So far, practical results with the proposed fully scaled form are encouraging, although examples with real data were not presented in this paper for reasons of brevity.

The second problem discussed here was that of deriving an improved measure of composite utility. This is also of relevance to environmental modeling involving long, hierarchical decision chains. It is also important to evaluation, an inseparable part of environmental analysis. The literature is almost unanimously in favor of the assumption that the mean value of the random distribution of utilities, known as the log-sum in logit models, is the appropriate measure of composite cost. In this paper such an assumption was questioned on the grounds that the results are inconsistent with basic economic principles. In particular, it was argued that the fact that the log-sum may become negative depending on the value of the dispersion parameter, is unreasonable in nested logit models, is inconsistent with choice probabilities, and may overestimate users' benefits in the evaluation of alternative policies. Instead, it was argued that the economics underlying discrete choice models suggest that a composite cost measure must be a non-negative function that decays from the cost of the best option to zero. Furthermore, to be consistent with the logit model, the rate at which such indicator declines must be in the same proportion that decision makers choose each option. The proposed solution is based on the log of a series, and it was shown that the results satisfy three basic conditions that were set out.

It must be stressed that in both cases the proposed solutions are only some of many possible formulations. The standard logit model has been available for almost 20 years and has allowed a vast area of research to develop; researchers and practitioners, however, have had to live with its shortcomings. It is about time, then, that improved formulations be proposed and tested. Now that there is a growing interest in large-scale comprehensive models, time is ripe for such research to produce fruitful results. The task that lies ahead, then, is to pursue the analysis of alternative forms and their formal derivations, and perform the necessary experiments with real data to prove the validity of the assumptions.

As was mentioned, the proposed model has been incorporated into the programs of the TRANUS integrated activities-transport model, and used in practice in studies in the U.S., Europe and Latin America, involving both urban and regional levels. Results so far show an improvement in the realism of the simulations and easier calibration of the models. These results, however, have not yet been organized in a systematic way to show the potential benefits of the proposed model forms.

Acknowledgments

The author would like to acknowledge the contributions of Juancarlo Añez, Beatriz Pérez and Francisco Martínez that were fundamental to this paper. Of special interest were the comments made at the Smådalarö Conference by the discussant Tony Smith, and later by the editors and reviewers.

References

Ben-Akiva, M. and Lerman, S.R. (1985), *Discrete Choice Analysis: Theory and Application to Travel Demand*, The MIT Press, Cambridge, MA.

Cochrane, R.A. (1975), 'A possible economic basis for the gravity model', *Journal of the Royal Statistical Society*, Series A, vol. 114, pp. 490–496.

de la Barra, T. (1989), *Integrated Land Use and Transport Modelling*, Cambridge University Press, Cambridge.

Domencich, T. and McFadden, D. (1975), *Urban Travel Demand: A Behavioral Analysis,* North-Holland, Amsterdam.

Fisk, C.S. and Boyce, D.E. (1984), 'A modified composite cost measure for probabilistic choice modeling', *Environment and Planning A*, vol. 16, pp. 241–248.

Lindberg. P.O., Eriksson, E.A. and Mattsson, L.-G. (1995), 'Invariance of achieved utility in random utility models', *Environment and Planning A*, vol. 27, pp. 121–142.

Wegener, M. (1994), 'Operational urban models: State of the art', *Journal of the American Planning Association*, vol. 60, pp. 17–29.
Williams, H.C.W.L. (1977a), 'On the formation of travel demand models and economic evaluation measures of user benefit', *Environment and Planning A,* vol. 9, pp. 285–344.
Williams, H.C.W.L. (1977b), 'Generation of consistent travel demand models and user benefit measures', in Bonsall, P., Dalvi, W. and Hills, P. (eds), *Urban Transportation Planning,* Abacus Press, London.

17 Modelling Land-Use and Transport Interaction: Policy Analyses Using the IMREL Model

Christer Anderstig
Lars-Göran Mattsson
Department of Infrastructure and Planning
Royal Institute of Technology
S-100 44 Stockholm, Sweden

17.1 Introduction

In some respects there is a renaissance in large-scale urban models – maybe somewhat unexpectedly. There are many reasons for this, both on the supply and the demand side. A brief retrospection could help to get this into perspective.

When large-scale urban and regional models emerged on the scene in the sixties, a lot of expectations were raised about a paradigmatic change in planning practice, from an intuitive style to one based on scientific principles. A decade later a common view was that these efforts had essentially ended in failure (Lee, 1973; see also Lee, 1994). The models were accused of being too data hungry, non-transparent, and in need of extensive and costly computer hardware, just to mention a few points of criticism. Lee also argued that the traditional planning process, as well as the land-use models that were developed, fitted into a top-down or command-and-control approach that supported centralised planning aimed at a detailed specification of land uses and behaviour.

Large-scale urban land use modelling did not die, however, it survived and developed as an academic activity rather than as a regular instrument in comprehensive urban and regional planning. As a matter of fact, the following period turned out to be quite successful from an academic point of view, as rapid progress occurred in many areas of importance for building land use models.

Firstly, advances in random utility theory, discrete choice modelling, entropy maximisation, and network equilibrium theory have been of importance for the present understanding and sometimes also for the unification of different modelling approaches. Secondly, sophisticated applications of optimisation principles have become standard ingredients in formulating models consistent with rational behaviour, in solving different kinds of equilibrium problems, and in estimating model parameters efficiently. Thirdly, more recently, concepts from the theory of dynamical systems have enriched the repertoire.

On the whole this progress has been helpful in combating the black box syndrome and should in the long run increase the transparency of applied urban and regional models. Furthermore, computer speed has continued to increase as if without any saturation. It is less known that many computational algorithms have been improved at approximately the same rate. This has facilitated the application of large-scale models. Today data availability rather than computer capacity is the main restriction.

The environment in which urban and regional planning take place has also changed. Now it is neither acceptable nor possible for planning authorities to dictate land uses in detailed plans. No one believes that planning can be a strictly scientific activity, even if planners can become better in making use of scientific findings, methods, and tools. The focus of planning has changed from prescribing solutions to providing problem analyses and decision support. Often the planners have the role of finding good compromises among different conflicting interests. Negotiation planning has become an established form of planning.

These changes in planning practice do not imply that computer-based modelling instruments will be less important. On the contrary the demand for detailed forecasts, precision and flexibility will increase, as it becomes established practice to support decision-makers and the public with broad impact analyses. This is now increasingly the case in the field of large-scale infrastructure investment. These new demands often go in the direction of more detailed analyses, which may require more elaborate and perhaps also more complicated models, and may counteract the endeavour to make transparent and simple models.

The criticism of the command-and-control character of planning, and of urban and regional models, has sometimes taken the form of an attack on the rational mode of planning. Here a swing back can be noticed. The environmental issues, which are now so urgent, call for more rationalism, not less. Consider for example the issue of promoting economic development without sacrificing the opportunities of future generations, i.e. the aim of a sustainable urban development. Coping with such problems necessitates a comprehensive understanding of the urban system, and this is what operational urban and regional models attempt to provide.

Proposals of large infrastructure investment raise the question of the interaction between transportation system, land use structure, and the environment. To what extent do new roads and improved public transport systems contribute to excessive urban commuting? Will such investments lead to a more dispersed land use pattern, and hence further impair the market for public transport? Can various planning measures counteract such tendencies? What restrictions may a sustainable urban and regional development place on the freedom of mobility?

To investigate such strategic long-term issues, carefully designed operational land use and transportation models could be very useful. The models could help the planners to convince the decision-makers that the (often unpleasant) recommendations they propose will have the stated consequences. In many situations, when practical experiments are too costly or would take too long time until reliable conclusions could be drawn, they are perhaps the only realistic alternatives (Harris, 1994). To be useful such strategic models must be built on a systems view of the urban economy, reflecting a comprehensive understanding of the driving mechanisms. They must be able to simulate the urban housing, labour, and transportation markets realistically, to give useful evaluations of cost changes, different investment strategies, regulation policies, economic incentives, preferably within a common framework.

The impacts of major infrastructure projects in metropolitan areas are discussed intensely in Sweden today, as in many other countries. Better tools for analysing long-range impacts on the settlement structure have explicitly been demanded (Governmental Bill, 1994). Yet such analyses have not the same legislative support as in the U.S., where the Clean Air Act Amendments of 1990 and the Intermodal Surface Transportation Efficiency Act of 1991 explicitly require such analyses (Lakshmanan, 1998). However, in the Swedish follow-up of Agenda 21 of the 1992 UN conference in Rio de Janeiro, the municipalities were demanded to investigate the long-term relationship between the transportation system, the land use structure, and the environment. In particular they were supposed to consider to what extent a sensible regional planning policy would be capable of linking the long-term development in compliance with a sustainable development.

These and similar observations in other countries indicate that we may very well be on the eve of a breakthrough in the application of operational urban and regional models. For this to succeed it will be necessary to pay close attention to the problems and needs of the planning organisations, for which the models are meant to be useful. A closer interaction, co-operation and exchange of ideas between academic researchers and urban planners would definitely be necessary, and could also stimulate new research. Perhaps the final step from academic journals to user-friendly computer systems (and one should not underestimate the investment needed) is best performed in privatised forms as has been the case with the fortunate market penetration of some urban transportation planning software packages.

In this paper we report on the development and application of one particular urban land use modelling system, for which the authors are responsible: IMREL, Integrated Model of Residential and Employment Location. This system has been developed in close connection with a regional planning agency.[1] Although we have made use of some of the theoretical advances referred to above, it has also been necessary to restrict the modelling efforts, and the data requirements, in order to achieve an operational model system.

The paper is organised as follows: Section 17.2 gives a short presentation of the intentions behind the development of IMREL and a non-technical description of how the model works. The validation problem is touched upon in Section 17.3. Then Section 17.4 reports on three recent policy applications concerning transportation and land use interaction. Section 17.5, finally, gives some summarising comments and concluding remarks.

17.2 IMREL

17.2.1 Outlook

The ultimate aim of urban modelling spells applicability. This view may not concern *theoretical* urban models, but it implies that *operational* urban models should be appropriate to address urgent planning problems, qualified for application in real practice, and well-suited to deal with issues raised by public agencies. It is our belief that any modelling effort guided by this principle places strong restrictions on the complexity of the model and its adaptation to available data. In the development of IMREL, it has been natural to follow a "piecemeal modelling strategy",

[1] Office of Regional Planning and Urban Transportation, Stockholm County Council.

to be able to address specific planning and policy problems.[2] Before giving a description of the model, we will briefly declare some of our basic notions.

First, in the matter of dynamics: According to Wegener (1994) the typical operational urban model is quasi-dynamic, i.e., a recursive simulation model, subdividing time into discrete periods. This feature is "especially suited for taking account of time lags or delays due to the complex superposition of slow and fast processes of urban development" (p. 25).

Though the idea is theoretically sound, an operational quasi-dynamic model runs into problems as regards the definition of proper time lags and data requirement. In contrast to this, IMREL works in a static equilibrium framework. The approach may be justified by the main purpose, which is not to predict the evolution of the urban system, but analysing partial effects of various changes in the transportation system.

The second matter concerns feasible behavioural equations. In an appropriate model the spatial pattern is derived from conditions of supply and demand, represented by behavioural equations, usually based on profit and utility maximisation. However, whereas the location behaviour of residents in IMREL is based on utility maximisation, that of employment (workplaces) is not based on profit maximisation. This asymmetry is basically explained by a serious specification problem: Profit functions requiring data on wages, input and output prices are ruled out, since reliable and accurate data of this kind is lacking.[3]

17.2.2 Model Description

In the following we will give a fairly detailed non-technical description of the IMREL model system.[4] A general framework for integrated land use and transport modelling is presented in Fig. 17.1, which is adapted from de la Barra (1989). The way IMREL has been designed and linked to a network equilibrium model for traffic assignment suits well into that scheme.

At the regional level total volumes of employment (workplaces) and residents (housing units) are determined for a planning year, either exogenously as a scenario, or by separate demographic and employment sub-models. At the urban activity level these totals are distributed across a system of zones in the region under study, through a process of iterations between a residential (RES) and an employment (EMP) location sub-model. These sub-models use as input data, among other things, travel times and travel costs between zones by available modes of transport

---

[2] The model development has been motivated by a general need for a better understanding of the interaction between residential and employment location, and the transportation system. This need became manifest in relation to a large-scale proposed investment programme for transportation infrastructure in the Stockholm region, commonly referred to as the "Dennis Agreement" (Johansson and Mattsson, 1995). The analysis of this programme has served as a test case as regards the applicability and usefulness of the model, and has also inspired modifications and elaboration.

[3] In our opinion, a profit function that is feasible for operational models must rely on data mainly concerning accessibility and land rent. This may give a possibility for elaboration of IMREL that has not yet been explored.

[4] For precise definitions and mathematical details we refer to Anderstig and Mattsson (1991) and Mattsson (1987).

as calculated by a traffic assignment module of a linked travel demand model. The RES sub-model determines the households' combined choices of residential location and travel mode, assuming that the employment location is given by EMP. This means that RES also produces complete matrices for work trips subdivided by mode.

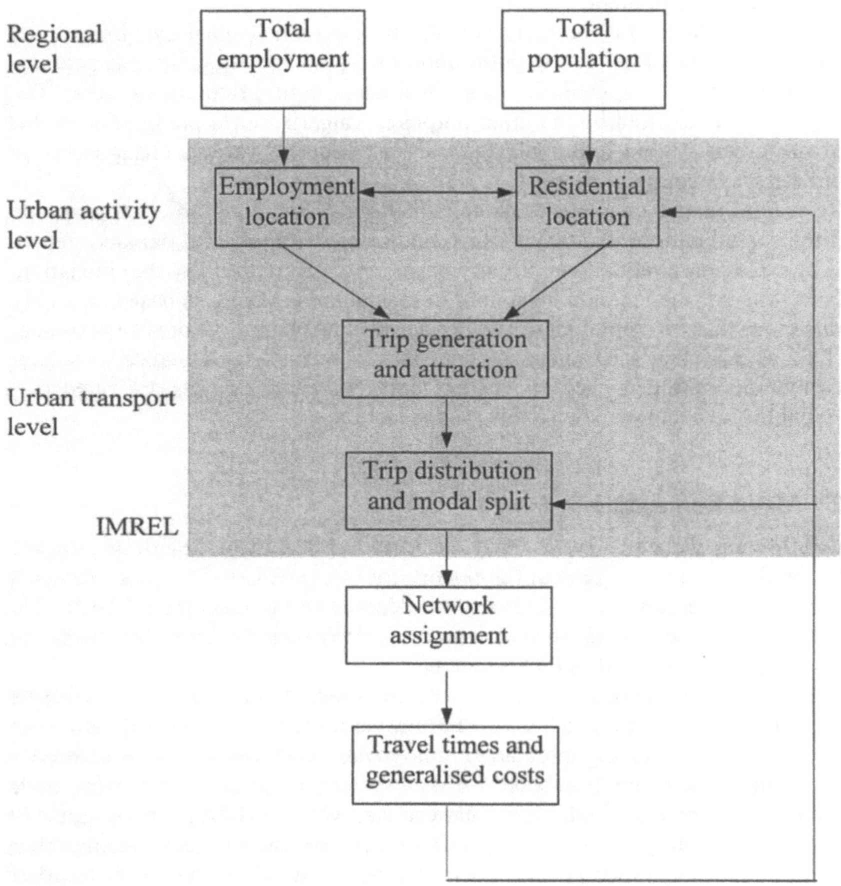

Fig. 17.1. A conceptual framework for an integrated land use and transport model

In early applications that were carried out (as exemplified in Section 17.4.2), input travel times and travel costs by zones and modes were exogenously estimated by a traffic assignment model, which in turn was part of a complete four-step travel demand model system, working at a detailed spatial level. There was no feedback to the travel demand model from IMREL's estimates of future location, and thus no check whether the changes in land use were consistent with the input car travel times and costs.

In recent applications, however, the predicted land use changes have been fed back into the travel demand model, to make times and costs for the car mode endogenous (see Section 17.4.1).[5] In this case a very detailed model system, T/RIM, built around a number of nested multinomial logit models, has been used (Karlsson and Svalgård, 1993; Svalgård, 1994). This system is implemented by means of the network equilibrium software package EMME/2 (INRO, 1993) and employs its user equilibrium traffic assignment module. T/RIM works at a finer spatial resolution than IMREL and handles three other trip purposes in addition to work trips. In the experiments that have been carried out so far, only a few iterations have been performed between the two model systems. It is therefore to early to draw any conclusions concerning the convergence of the iteration procedure.

Fig. 17.2 gives a more detailed description of IMREL, which corresponds better to the actual modules of the computer code. In particular the interaction between the residential (RES) and the employment (EMP) location sub-model is illustrated.

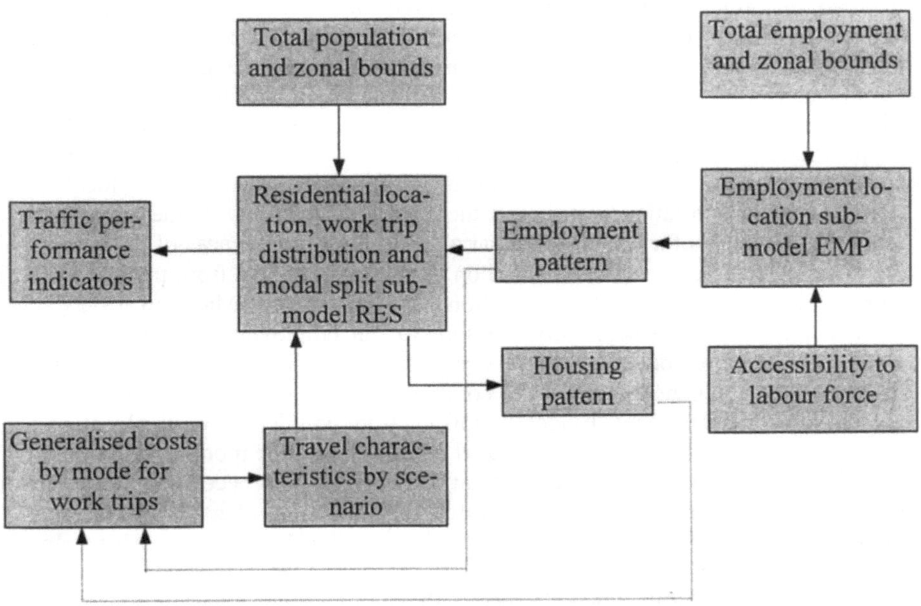

Fig. 17.2. The IMREL model system

RES is a normative model that allocates the total regional population (or housing) to the zones, so as to maximise a welfare measure based on locational (con-

[5] The approach is in fact very similar to Putman's (1998). As an alternative way to make travel times and costs by car endogenous, Boyce and Mattsson (1997) investigate how a simplified version of IMREL could be formulated and solved as a network equilibrium model, without relying on iterations with a separate travel demand model. This theoretical analysis could provide a basis for more efficient solution algorithms in the future.

sumer) surplus. In order to handle physical or other restrictions, or to enforce a predetermined land use, it is possible to impose upper and lower bounds on the number of residents (housing units) to be located in each zone. RES assumes that the employment location is given by EMP and that travel times and travel costs by available modes are exogenously provided (typically, as discussed above, after aggregation from a linked travel demand model). RES also determines the distribution and modal split of the work trips so that the distribution will be consistent with the (resulting) housing pattern and with the (input) employment pattern.

EMP is a positive model that simulates how employment location depends on accessibility to labour force and on other indicators of zonal attractiveness. The accessibility, in turn, depends on residential location, as generated by RES, and on travel times and travel costs exogenously provided in the same way as for RES. Again it is possible to impose zonal bounds on the location pattern. For given travel characteristics iterations between RES and EMP are carried out till the residential and employment patterns stabilise (which normally requires less than ten iterations).

EMP makes no distinction between employment and its physical representation in form of workplaces. Typically, two categories of employment are considered: local services and other workplaces. For each category, the probability that a workplace is allocated to a certain zone is modelled by a multinomial logit model. The attractiveness, or the representative utility, of a zone for a potential employer is specified as a weighted sum of a number of variables: labour force accessibility, supply of land for workplaces and indicators related to the presence of local stations in the transportation system and the degree of centrality of the zone in the region. The weights for the different variables in the attractiveness of the employment zones have been estimated based on cross-sectional data from travel surveys (for the most recent Stockholm application, see Anderstig and Mattsson, 1992).

The labour force accessibility measure for an employment zone essentially sums the number of potential workers residing in the different residential zones, discounted by the generalised cost of commuting to the employment zone, aggregated over available modes. Evidently, this measure will depend on the location of households and on travel times and travel costs by available modes, which in turn will depend on the supply of transportation services. The accessibility measure can, within a framework of random utility maximisation, alternatively be interpreted as the expected attractiveness of the best available worker in each employment zone.

In addition to the variables discussed above, a (positive or negative) shadow price (rent differential) may have to be added to the attractiveness of each employment zone to ensure that imposed bounds on the number of workplaces will not be violated. These shadow prices are determined through a simple iterative procedure each time the EMP sub-model is rerun.

The RES sub-model is, with Wegener's (1994) term, a *unified* model of residential location and travel choice. It combines a normative approach to the location of housing with a positive approach to the modelling of the households' behaviour on the housing and travel markets. From a mathematical and computational point of view it is a more sophisticated model than the employment location sub-model. Here we will only give a brief description of RES.

RES can be described as divided into three levels (see Fig. 17.3). At the lowest (household) level, RES models the choice behaviour of the individual household. Each household demands one unit of housing and consists of one worker and the average number of dependent persons. This is of course a restrictive assumption, not the least since all households are modelled as if they could choose the place of living conditional on the place of work for only one household member. However, it is a necessary simplification at the present stage of model development. The household decides about where to live and which mode of transport to use for the journey to work, conditional on its place of work, which is assumed to be given (by EMP).

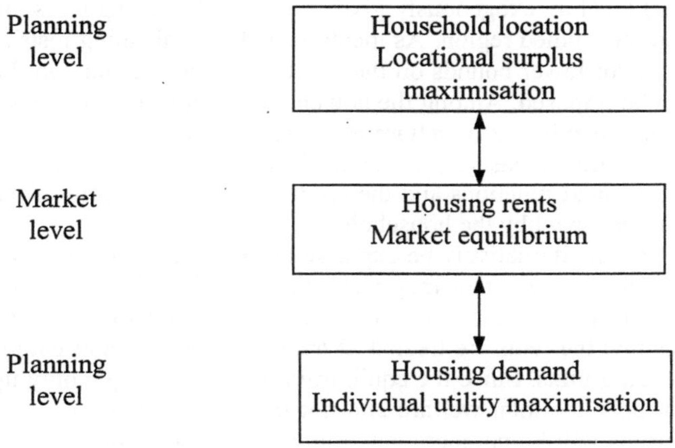

Fig. 17.3. Decision levels in the RES-sub-model

The representative utility of a combined mode and residential zone alternative is composed of the housing density, the (equilibrium) rent level and the generalised cost of commuting. In addition, there is a random utility term associated with each mode-zone alternative, accounting for individual taste variations. Assuming that the random utilities are generalised extreme value distributed (with a certain structure of statistical dependence), the combined location, origin-destination and mode choices of utility maximising households can be represented by nested multinomial logit models. The expected utility the households obtain from these choices can then be calculated as well as the expected demand for housing in each zone.

At the middle (market) level, housing rents that clear the local (zonal) submarkets are determined. In all applications so far it has been assumed that the total (regional) number of housing units exactly matches the total demand of the population, i.e. that the housing market is balanced at the regional level. This assumption is possible to relax, however. Given a spatial allocation of the housing units, i.e. housing supply specified by zone, and hence also a specification of the housing density by zone, the market clearing rent is determined for each zone so that the *expected* demand according to the household demand model equals housing supply in each zone – *stochastic equilibrium* (Anas, 1982).

It can be shown that in our setting such equilibrium rents always exist and that they are unique up to an additive constant (Mattsson, 1987). This means that the rent differentials, i.e. differences between rents in different zones, are uniquely determined. To achieve unique absolute rents one may arbitrarily fix the rent level in one zone. The actual computation of the equilibrium rents is not performed as a separate step. As the optimisation procedure is constructed, the equilibrium rents are automatically computed when the welfare maximising housing allocation is determined.

At the top (planning) level an exogenously given number of housing units are allocated to the zones in the studied region. As mentioned above, planning restrictions in form of upper and/or lower bounds on the number of housing units in the different zones can then be imposed. Among the housing allocations that are feasible, the one is selected that maximises a welfare measure based on locational surplus. More precisely, the welfare measure is the sum of the consumer surplus associated with the housing demand functions and the producer surplus in form of the summarised equilibrium rents paid by the households.

The consumer surplus can alternatively be expressed as the aggregate *expected* utility the households achieve, given that they maximise their individual random utilities. The consumer surplus includes as costs the equilibrium rents paid by the households. This means that the rents are treated as transfers from the households to the owners of the housing units. Since the equilibrium rents are unique only up to an additive constant, only the total welfare and not the division between consumer and producer surplus will be uniquely determined in this modelling setting (Mattsson, 1987).

The situation modelled by RES may be conceived as a bi-level optimisation problem or as a bi-level game. At the lower level each household maximises its own individual utility. The housing location, the densities and the travel costs are known to the households as well as the (equilibrium) rents.

At the top level the "planner" decides about the housing location within given constraints. The planner knows the representative utilities of the households and the equilibrium rents associated with a particular housing allocation. He or she locates the housing so as to maximise the locational surplus associated with the demand models that describe the households' behaviour on the housing and travel markets. The normative criterion used for assessing alternative housing allocations is thus consistent with the positive models employed for describing the behaviour of the households.

The optimisation problem implied by RES is not solved in the way it has been described. One reason is that the equilibrium rents are defined only as implicit functions of the housing allocation. However, by the help of duality theory the welfare maximisation problem of RES can be reformulated as an explicit unconstrained convex minimisation problem with one dual variable associated with each zone (see Mattsson, 1987). Any efficient non-linear optimisation algorithm can then be used to find the solution. We have used the MINOS optimisation package in our implementations (Murtagh and Saunders, 1987).

The nested multinomial logit models describing the residential location and travel choices of the households have all been estimated based on cross-sectional data from the same travel surveys as for the EMP sub-model, using the full information maximum likelihood method (Lundqvist and Mattsson, 1992).

17.3 Internal and External Validation

A properly validated model should have a proved capacity for predicting the development of the study area for some future or historical period. Few, if any, operational urban or regional models are validated in this respect, and IMREL is no exception. There are of course several reasons why modellers have paid little attention to this kind of "external" validation. A model is always a simplification of reality and can at best capture some fundamental relationships in the system that is studied. Discrepancies between model predictions and actual development can often be attributed to changes in external factors that were not included in the model, or that were considered exogenous, rather than to shortcomings in the model specification.

Therefore, to make reliable predictions we need not only good models but also reliable forecasts for exogenous factors that are inputs to the models. To improve the predictive power it would be necessary to increase the size of the models or to make them more sophisticated mathematically. This is likely to come in conflict with the endeavour to design simple and transparent models. There is a trade-off here. Often the analysts lack necessary data and support to carry out detailed comparisons of the performance of a model. When models are applied in actual decision-making situations there is an immediate interest in the predictions generated. At the time when it would be possible to compare predictions with outcome, it can be difficult to get necessary resources.[6] These remarks should not be conceived as excuses but rather as attempts to explain the present situation of poorly validated models. There is no doubt that much more effort must be put int validation issues in the future.

But apart from this external validation, there is another issue, concerning the "internal" validation of a model. Whether a specific model is validated in this sense can be judged reasonably well from the theoretical basis, the applied estimation techniques and the use of sensitivity analysis. Let us briefly touch on these three requirements.

First, as has been argued above, the theoretical improvements during last decades have been important for the theoretical underpinning of the present generation of operational urban and regional models. A sign of the maturity of the field is that there is now a certain consensus about the kind of modelling elements that are used in building large-scale models. Second, parameters in behavioural and other sub-models are now usually estimated with reliable techniques that exploit the data sources efficiently. Maximum likelihood estimation of logit models is but one example. The statistical procedures also provide information on the significance of different variables and on how well a particular model can reproduce observed

[6] However, see Fournier and Bjurklo (1994) for a careful evaluation of an implementation of the MEPLAN model.

data.[7] Third, systematic sensitivity analysis is perhaps the best way of creating credibility for a model (see e.g. Webster et al., 1988, for a thorough sensitivity analysis of a selection of operational urban models). One of the strongest arguments for the use of operational urban models in policy-making is their ability to give a comprehensive understanding of the urban system. Sensitivity analysis is a way of learning how the model behaves and, hopefully, also how the system functions.

The extent to which a model is able to provide an intuitive and consistent framework for communicating the common knowledge and view of a system in an organisation is then important. The capability of summarising knowledge, the usefulness as a decision support tool, and not least the acceptance by the planning professionals, must hence also be considered part of the validation of operational models. We argue that the IMREL modelling system is reasonably well validated in this "internal" sense, as it has been theoretically formulated, specified, estimated and applied.

17.4 Policy Applications

We will report on three applications where IMREL is being used for various kinds of policy analysis. All studies involve analysis of both land use and transport policies, but the main problems are somewhat varying, and the three cases refer to different regional settings.

17.4.1 Sketch Tool in the Stockholm Regional Planning Process

The first case concerns the use of IMREL as a sketch tool in the regional planning process in the Stockholm region. The main focus is about the interaction between infrastructure and environment, or put in policy terms, the tension between trends towards increased mobility and the requirements of a sustainable development. This issue is one of the most challenging for long-term infrastructure planning. In the Stockholm region it has become highly accentuated by the recently proposed transport investment programme (referred to as the "Dennis Agreement"), and in particular by the road projects therein.

We can expect that the consequences for transport demand and air emissions will be a focal point of the next regional plan, the preparation of which has recently started. What policy options, with respect to the future transport system, are available to reduce the negative impacts on the environment? Which trade-offs should be made between further development of the region and the protection of ecologically sensitive and valuable areas? To what extent can a sensible policy concerning the spatial arrangement of activities affect the travel pattern? The Information Technology (IT) revolution will certainly affect travel behaviour, but will total travelling increase or decrease? These are examples of the questions that have been raised and illustrated within the first study.

[7] Unfortunately, this does not help us much in deciding about appropriate values for model parameters in long-term policy applications. Here we are in need of a better methodology.

In the first phase of the preparation of a new regional plan for the Stockholm area, the responsible planning agency has initiated a project on using IMREL to evaluate various land use and transport scenarios. The analyses are intentionally meant to be at an aggregate level, to highlight the most important relationships between the transport system, the land use pattern and travel demand – in interaction with abatement technologies and traffic management instruments. Sensitivity analyses have been performed with respect to cost parameters and other uncertain parameters of significance for location and travel choices.

The main purpose is to investigate to what extent it is possible to affect travel demand, and car use in particular, by changes in the transport system, by land use measures, and by using road pricing mechanisms and other control instruments on the travel market – with the overall objective to reduce air emissions and other negative consequences of vehicle use. In a second phase the results of the analyses have been used in designing new or modified scenarios, subject to more detailed analyses and for which extensive predictions of the effects on travel demand, accidents, emissions etc. have been made. IMREL is thus utilised as an easy-to-use sketch tool at an early stage of the regional planning process.

Scenarios for the Year 2020. Three transport network scenarios have been designed: one according to the present regional plan, *RP91*, another "green" alternative with a strong emphasis on public transport, *PT Plus*, and a third network, *Road Minus*, defined as *PT Plus* exclusive of some controversial road projects.[8] Five land use patterns have been constructed/defined:

- *RP91* — According to the present regional plan
- *DENSE ("Subway")* — Expansion concentrated to the central zones in the region
- *TRACK ("Railway")* — Expansion in zones with dense population, especially around railway stations
- *SCATTERED ("Car")* — Expansion primarily in peripheral zones
- *MODEL (IMREL)* — Land use according to IMREL simulation results.

In the construction of *RP91*, *DENSE*, *TRACK* and *SCATTERED*, location of residential population is exogenous (= zonal bounds), whereas location of employment (workplaces) is modelled by use of IMREL. In *MODEL* also residential population is modelled by IMREL.

IMREL Analyses. First, the model system T/RIM has been used to generate travel times and travel costs for the three transport scenarios. These data have been aggregated to the spatial level of IMREL. Then the purpose of using IMREL has

[8] The road investments in the present regional plan are identical with those in the Dennis Agreement, comprising the completion of a Ring Road around inner Stockholm, and a north-south Outer Cross Route. The Ring has a radius of 3–4 km and includes a controversial section under the sea east of the city centre, called the Eastern Route. The Outer Cross Route also includes a controversial section, the Western Route, crossing Lake Mälaren in a semi-circular route at a distance of 10-12 km west of the city centre.

been to achieve a proper picture of the main features of the various land use and transportation scenarios. This means, e.g., illustrating how the choice of transport scenario might affect residential and employment location under different technology and cost assumptions. One point of special interest has been to investigate potential conflicts between proposed land use scenarios, and the land use patterns according to IMREL (*MODEL*). By comparing the locational surplus of different scenarios we also get an indication of the value that must be attributed to a particular scenario to justify its enforcement. Finally, the effects of the transport and land use scenarios on travel behaviour are of primary interest. In particular, we are interested in assessing to what extent it is possible to affect commuting distances and car use by land use measures.

Comparing the results from different scenarios gives information on a number of "key indicators": car share (%), public transport share (%), car travel time (min), public transport travel time (min), total car-km, and total distance travelled (km all modes). Fig. 17.4 illustrates the results with respect to one of these indicators, total daily distance travelled per person.

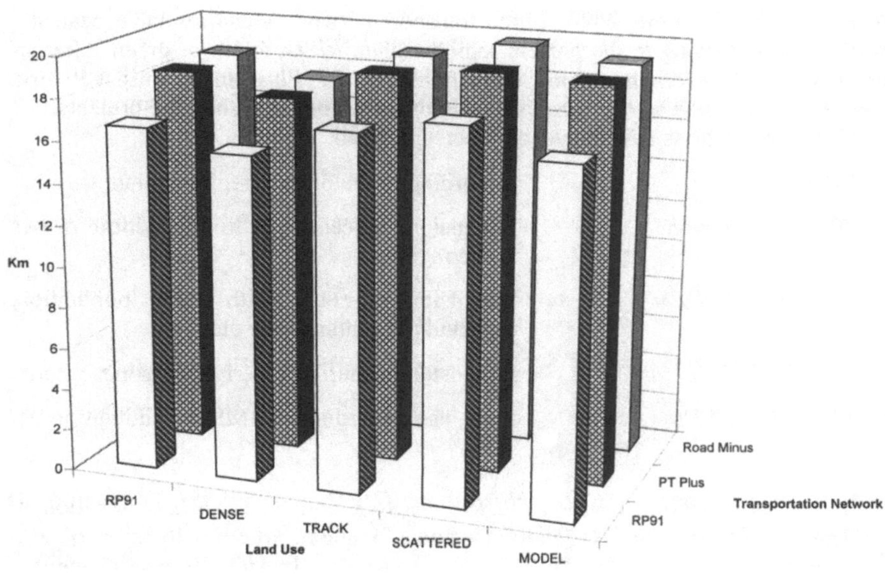

Fig. 17.4. Total daily distance travelled per person

The results of IMREL analyses, for all 15 combinations of land use and transport network scenarios, were presented for the regional planning agency. After discussions with the planners, and guided by this initial study, a limited number of scenarios were selected for further analysis. (In this process also other planning agencies at the regional level were consulted).

Detailed traffic analyses were performed for five transport network and land use combinations, those shaded in Fig. 17.5. These analyses were based on interaction between the detailed traffic model system T/RIM and IMREL, as described in Section 17.2.2 above.

Some Results and Conclusions. According to the result of this analysis some scenarios were considered to be of interest for further investigation. The features of the *DENSE* land use structure, combined with transportation network according to regional plan, *RP91*, was judged the most interesting: Less car traffic than for other combinations, and also smaller total travel. This combination also produces a relatively low average travel time, without being burdened with the high operating costs of an expansion in the public transport system.

	Land Use Scenario				
Transport Netvork	RP91	DENSE	RAIL	SCATTERED	MODEL
RP91	X	X	X	X	X
PT Plus	X	X	X	X	X
Road Minus	X	X	X	X	X

Fig. 17.5. Transport network and land use combinations. (Detailed analyses performed for shaded combinations)

The land use structure SCATTERED combined with *Road Minus*, i.e. the network where the two controversial road projects are excluded, was considered the least attractive: Car traffic as well as travel times and total travel at a much higher level than for other combinations.

As for the aim of reducing the car traffic, various sensitivity tests seem to call the efficacy of land use policies into question: The factors mainly affecting car traffic are assumptions about economic development and operating costs for car, i.e. petrol price and road pricing (tolls). Further, it is interesting to note that the effect on car traffic of doubling the public transport capacity is only marginal: a small reduction in the car modal share, fewer walking or biking, whereas the operating costs for public transport are heavily increased and the new capacity poorly utilised.

Effects of Telecommuting on Land Use and Traffic. When considering recent trends in commuting alternatives, a growing interest has been directed towards the effects of an increase in tele-commuting. According to the U.S. Department of Transportation (1994), the number of workers working at home increased from 2.2 million in 1980 to 3.4 million in 1990, and the work at home share increased from 2.3 percent in 1980 to 3.0 percent in 1990. Estimates for Sweden indicate a similar development. This probably reflects opportunities made available through the so-called IT revolution.

From a regional planning perspective this gives rise to several interesting questions. One issue is concerning the impact of tele-commuting on total travelling. Also in this case IMREL has been used as a sketch tool, illustrating the effects of an increase in tele-commuting on travel pattern and location behaviour. In the same way as above, the analyses have been based on interaction between IMREL and the detailed traffic model system T/RIM.

Of course, this kind of analysis claims several critical assumptions. First, the most strategic assumption is that the information technology will not basically change the preferences as represented in the parameters of the currently estimated models that describe the travel and location behaviour. A second assumption concerns the share of potential tele-commuters in a forecast/scenario year (2020).[9] Third, a crucial assumption relates to the expected use of this potential. Somewhat arbitrarily it has been assumed that the frequency of tele-commuters can be represented as a function of distance to workplace.[10] Finally, assuming that tele-commuting does not affect the generalised cost per trip, the total average travel cost for commuting will be reduced in the same proportion as the number of trips. Obviously, analysis based on these assumptions can only give rough indicators of possible effects of tele-commuting. In summary, the results indicate that:

- total number of trips will decrease
- travel times and distances will increase
- total number of km will not decrease
- the land use pattern will be more dispersed

These calculated effects are derived from a comparison with a scenario RP91 without tele-commuting, illustrated in Fig. 17.6.

17.4.2 Modelling Land Use Effects of "Trains in the Mälar Valley"

The second case refers to the application of IMREL in a larger regional setting, the Mälar Valley, defined by five counties around Lake Mälaren, where Stockholm is the central region. In this case the main focus is long-term relocation and land use effects of large-scale investments in the transportation network. The specific interest relates to the expected effects of introducing regional trains, "Trains in the Mälar Valley" (TIM). The regional political bodies consider TIM a way of supporting the expected continued integration by infrastructural investments, intended to benefit both the Stockholm region ("diverging excessive growth") and settlements suffering from problems of restructuring industries and declining population. The crucial question in this study is to illustrate to what extent TIM may contribute to a long-term change in the location pattern, especially with respect to cross-county relocation.

Municipalities and county boards in the Stockholm metropolitan region and surrounding counties have established "The Mälar Valley Council", a representa-

[9] It has been estimated that 45 % of all commuters in the Stockholm region will be potential tele-commuters by 2020, judging from forecasts of the number of employed persons in pertinent occupations and industries (no tele-commuting in, e.g., personal services). By the same criterion the tele-commuting potential today is about 30 %.

[10] The applied function means that, among the 45 % potential tele-commuters, those residing at distances shorter than 5 km from the workplace make three work trips per week, those residing at distances 5 – 20 km make two trips per week, and those at distances longer than 20 km make one trip per week. It is assumed that the remaining 55 % make five trips per week.

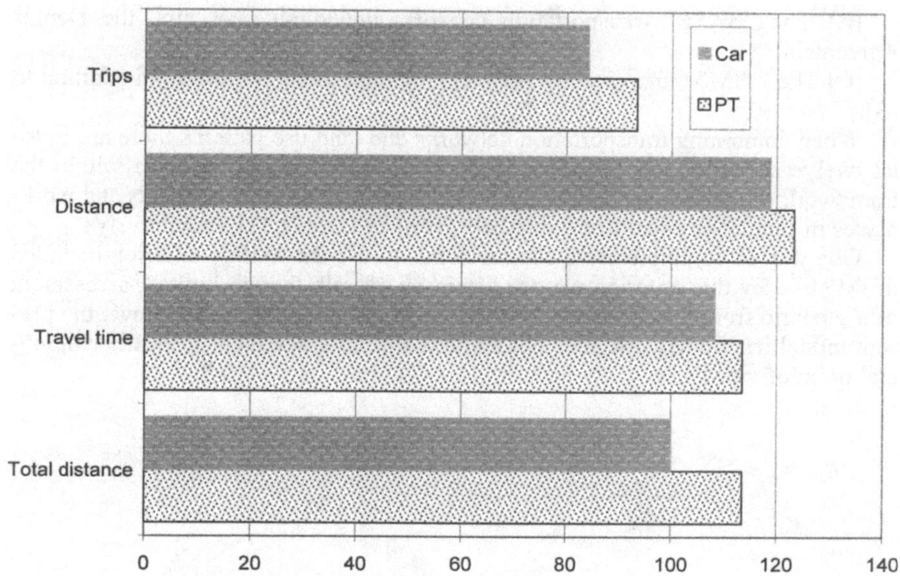

Fig. 17.6. Effects of tele-commuting for the year 2020. (Index RP91 = 100)

tive body for analysis and joint efforts in several areas, including infrastructure, housing and employment. As a part of their work, the Mälar Valley Council has begun to produce "futures" (scenarios), concerning various aims in the development of the Mälar Valley region. Within the area of transportation and land use, the state of things is still far from a common idea, or vision. So far three outlines have been constructed, with the following main features as to land use:

[1] The "Single Core" development, where the expansion of population and employment is concentrated to the Stockholm metropolitan region;

[2] The "Few Core" development, where the expansion mainly is located in six cities of medium size, in all surrounding counties;

[3] The "Multi Core" development, where the expansion is distributed to a large number of villages in various parts of the Mälar Valley region.

Being the outcome of a joint effort this urban/rural set-up of futures contains far more details than indicated by the brief description. It is, e.g., assumed that the transportation system in [2] and [3] includes regional trains, that [2] includes a new international airport in the western part of the region, etc. However, since the futures have been produced by ad hoc methods, the question has been raised whether these land use patterns are in compliance with, or in contrast to, a land use pattern in accordance with the expected behaviour of households and firms. Therefore, on commission by the Mälar Valley Council, IMREL has been applied to simulate such market oriented futures (year 2020) for the Mälar Valley region, with respect to three transportation network scenarios:

[A] The "Do Nothing" transportation network, i.e. 1992 transportation network;

[B] The "BASE" transportation network, including, inter alia, the Dennis Agreement;

[C] The "TIM" transportation network, regional trains operating, in addition to [B].

When combining transportation networks and land use patterns there are in total twelve scenarios to be compared. Such comparisons can be made within the framework of IMREL, by setting the bounds on the number of residents and workplaces in each zone equal to the numbers given by [1], [2], [3], respectively.

One way of comprehensive comparison is to use the welfare criterion included in IMREL. By this criterion we can obtain an overall, though limited, assessment of a scenario from the residents' point of view. As being discussed above, the present modelling framework does not admit a similar assessment from the employers' point of view.[11]

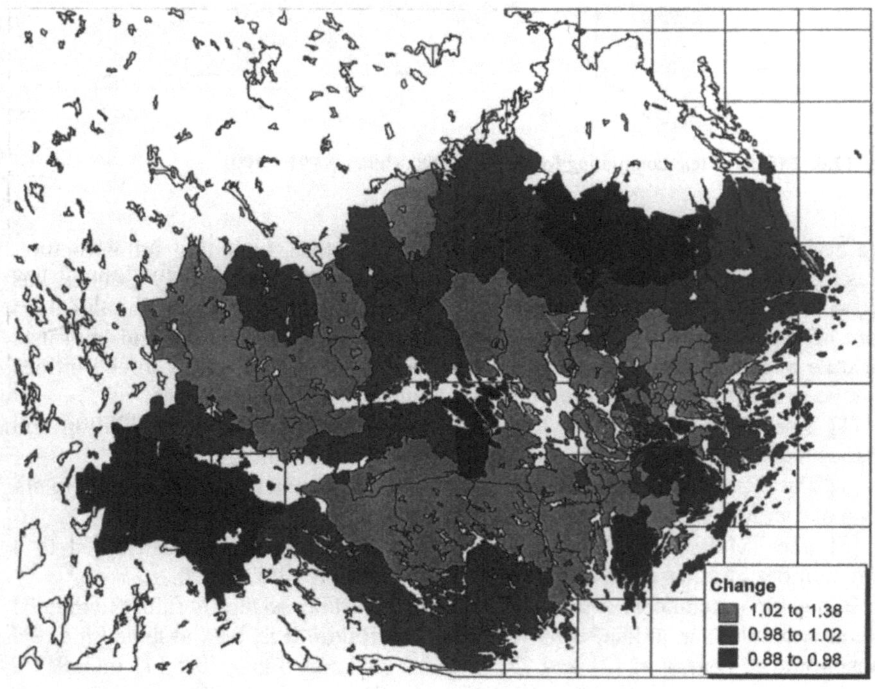

Fig. 17.7. Population change in 214 zones in the Mälar Valley region: "TIM" compared to "Do Nothing". (Index, No change = 1.0)

[11] The limitation implies that only those factors embraced by the choice models of the residents are taken into account, i. e. changes in travel time and cost for the journey to work including effects on mode choice, and changes in accessibility on the housing and labour markets including induced impacts on land use. Accidents and environmental effects are left outside the evaluation.

Comparing different scenarios in terms of the resulting consumer surplus shows, e.g., that [1, A] gives a higher surplus than [3, C], i.e., the "Do Nothing" transportation network and the "Single Core" land use give a higher consumer surplus than "TIM" and "Multi Core" land use.

Turning to the simulated land use for different transportation networks, Fig. 17.7 shows the population change between "TIM" and "Do Nothing".

In the year 2020 the population forecast assumes a total population of 3.1 million in the whole Mälar Valley region. According to the IMREL simulations, the impact of TIM implies that slightly more than 130 000 persons (4 %) would relocate between the 214 zones, comparing "TIM" with "Do Nothing". The relocation of workplaces is similar. Whereas there is a significant relocation effect between zones – in the county of Stockholm a decentralisation from inner to outer zones (see Johansson and Mattsson, 1994) – the relocation among counties is marginal or less than 1 %. The expectations that the introduction of the regional trains would lead to a considerable relocation from the Stockholm county to other counties in the Mälar Valley region are evidently not confirmed by these results.

17.4.3 Modelling Long-Term Induced Traffic of the Øresund Fixed Link

In the third and most recent case, IMREL is applied to the Øresund region, as one tool within a traffic model system aimed at a comprehensive traffic forecasting model for the Øresund link. At present the cities of Copenhagen and Malmö, in Denmark and Sweden respectively, are separated by the Øresund Straits, which connect the Baltic to the North Sea. The Swedish and Danish governments are in the process of constructing a sixteen-kilometre fixed link (part tunnel, part bridge) to provide a new road and rail alternative to existing ferry services.

In this study the central question is to estimate the potential for long-term induced traffic on the fixed link. The induced traffic is expected as a result of changes in the location pattern within the Øresund region, caused by the change in accessibility that the fixed link will bring about.

In contrast to the previous cases, where IMREL is used mainly as a sketch tool, this application is oriented towards direct decision support for the company operating the Øresund link. The objective is to provide a land use and traffic forecasting model capable of exploring the effects of different pricing strategies and variations in land use and exogenous variables, such as levels of economic growth.

The whole model system includes short-distance and long-distance passenger and freight models, where IMREL is used for long term forecasting purposes.[12] Given the specific locational context, linking two cities in different countries, the modelling involves some new issues, e.g. dealing with "barrier effects". (In this context a barrier effect denotes, in general terms, the disutility not included in the trip cost, i.e. travel time, headway etc., of crossing the Øresund Straits). The model system has been delivered to the operating company during 1997.

[12] The model system is developed by COMVIN, a joint venture between COWI (Denmark), the MVA Consultancy (United Kingdom), and INREGIA (Sweden).

17.5 Concluding Remarks

The changing role of transportation policy-making has created a renewed demand for carefully designed operational land-use and transportation models. The current issues go far beyond the traditional infrastructure expansion planning. Today the models are also expected to yield decision support for short- and long-term transportation demand management, including the locational control of traffic generating activities and other land use measures. In particular, environmental assessment of transportation and land use policies has received increasing interest, stimulated by strategic discussions on how a sustainable urban development could be realised.

Applicability is central if the models should ever be used by those they are intended to be tools for. This requires, in our opinion, genuine collaboration between model developers and users. This has been an underlying idea behind the present development of the IMREL model. As a consequence the model structure has intentionally been kept fairly simple and adapted to the kind of data that is available. Efforts have been made to produce a model responsive to the issues that the planners find urgent today. This approach has proved to be fruitful when it comes to application in real practice in various contexts: sketching, simulating land use effects and recently, forecasting induced traffic. We are convinced that the model exercises have improved the comprehensive understanding among the planners of how land-use and transport affect each other, and also created a better understanding of and deeper interest in land-use and transportation models for decision support. This will facilitate the future introduction of more advanced models.

Within the present model system there are some potential improvements that should be fairly easy to implement. One example is the segmentation of residents with respect to income or education, and of workplaces with respect to industry. Another would be to link output data concerning land-uses and traffic volumes to environmental models for calculation of energy use and emissions in a spatially disaggregated way (Ramjerdi et al., 1995).

Yet, it must be admitted that IMREL represents a simplified approach when it comes to the theoretical underpinning of location behaviour. For example, the residential location is unaffected by accessibility to other trip ends than work, and trip generation is not affected by accessibility[13]. These limitations are at the centre of attraction in an ongoing research project, where a new more versatile model, TILT, is being developed (Eliasson and Mattsson, 1997). This new approach includes several theoretical innovations, which offer a more realistic representation of the interaction between location and travel decisions. However, a lot of estimation, calibration and validation work remains before the applicability of the new approach in planning practice can be assessed.

Finally, the need for developing the methodology of validating and applying land-use and transportation models in long-term planning must be stressed. By nature, model validation is an ingenious issue in this context. Without good validation studies, however, we will simply not know whether we should trust the models or not.

---

[13] The telecommuting application in Section 17.4.1 illustrates a simple attempt to overcome this shortcoming.

References

Anas, A. (1982), *Residential Location Markets and Urban Transportation*, Academic Press, New York, NY.

Anderstig, C. and Mattsson, L.-G. (1991), 'An integrated model of residential and employment location in a metropolitan region', *Papers in Regional Science*, vol. 70, pp. 167–184.

Anderstig, C. and Mattsson, L.-G. (1992), 'Policy applications of an integrated land-use transport model in the Stockholm region', Paper presented at the 6th World Conference on Transport Research, Lyon.

de la Barra, T. (1989), *Integrated Land Use and Transport Modelling: Decision Chains and Hierarchies*, Cambridge University Press, Cambridge.

Boyce, D.E. and Mattsson, L.-G. (1997), 'Modelling residential location in relation to housing location and road tolls on congested urban highway networks', Submitted to *Transportation Research B*.

Eliasson, J. and Mattsson, L.-G. (1997), 'TILT – A model for integrated analysis of household location and travel choices', Paper presented at the Western RSA Conference, Hawaii.

Fournier, S.F. and Bjurklo, L.-G. (1994), ' "Purpose-specific" measures and assessment of flows in the Vägverket implementation of MEPLAN', Paper presented at the conference on Network Infrastructure and the Urban Environment, Smådalarö Gård, Stockholm.

Governmental Bill (1994), 'Finansiering av vissa väginvesteringar i Stockholms län m. m.' (Financing of some road investments in the county of Stockholm etc.), *Regeringens proposition 1993/94:86*.

Harris, B. (1994), 'The real issues concerning Lee's "Requiem"', *Journal of the American Planning Association*, vol. 60, pp. 31–34.

INRO Consultants (1993), 'EMME/2 user's manual', Montréal.

Johansson, B. and Mattsson, L.-G. (1995), 'From theory and policy analysis to the implementation of road pricing: The Stockholm region in the 1990's', in Johansson, B. and Mattsson, L.-G. (eds), *Road Pricing: Theory, Empirical Assessment and Policy*, Kluwer Academic Publishers, Boston, pp. 181–204.

Karlsson, P. and Svalgård, S. (1993), 'Synergetic – a new travel demand and land-use model for the Stockholm region, integrated with EMME/2', Paper presented at the 21st PTRC European Highways and Planning Annual Meeting, Manchester.

Lakshmanan, T.R. (1998), 'The changing context of transportation modeling: Implications of the new economy, intermodalism and the drive for environmental quality', Chapter 3, this volume.

Lee, D.B. (1973), 'Requiem for large-scale models', *Journal of the American Institute of Planners*, vol. 39, pp. 163–178.

Lee, D.B. (1994), 'Retrospective on large-scale models', *Journal of the American Planning Association*, vol. 60, pp. 35–40.

Lundqvist, L. and Mattsson, L.-G. (1992), 'Modelling travel demand in an extended metropolitan region', Paper presented at the 32nd European Congress of Regional Science Association International, Brussels.

Mattsson, L.-G. (1987), 'Urban welfare maximization and housing market equilibrium in a random utility setting', *Environment and Planning* A, vol. 19, pp. 247–261.

Murtagh, B.A. and Saunders, M.A. (1987), 'MINOS 5.1 user's guide', *Technical Report SOL 83–20R*, Department of Operations Research, Stanford University, CA.

Putman, S.H. (1998), 'Results from implementation of integrated transportation and land use models in metropolitan regions', Chapter 15, this volume.

Ramjerdi, F., Jensen, T. and Rand, L. (1995), 'Integrated land use, transport and environment models: State-of-the-art and a proposal for research', *TØI rapport 304/1995*, Institute of Transport Economics, Oslo.

Svalgård, S. (1994), 'The implementation of an integrated transport and land-use model, using new features of EMME/2 Release 7.0', Paper presented at the 3rd European EMME/2 Users Meeting, Stockholm.

U.S. Department of Transportation (1994), 'Commuting alternatives in the United States: Recent trends and a look to the future', Office of University Research and Education, Washington, DC.

Webster, F.V., Bly, P. H. and Paulley, N. J. (eds) (1988), *Urban Land-Use and Transport Interaction: Policies and Models*, Avebury, Aldershot.

Wegener, M. (1994), 'Operational urban models: State of the art', *Journal of the American Planning Association*, vol. 60, pp. 17–29.

18 A Combined Model for Analysing Network Infrastructure and Land-Use/Transportation Interactions

Lars Lundqvist
Department of Infrastructure and Planning
Division of Transport and Location Analysis
Royal Institute of Technology
S-100 44 Stockholm, Sweden

18.1 Sustainable Development and Land-Use/Transportation – an Introduction

Urban areas accommodate a rapidly growing share of the global population. The major problems of current and future societies, therefore, to a large extent need solutions on the urban scene. These problems include poverty, resource consumption and pollution. The concept *sustainable development* (emanating from the report of the Brundtland commission, WCED (1987)) summarises the joint emphasis on development *and* sustainability: "Sustainable development is development that meets the needs of the present without compromising the ability of future generations to meet their own needs". The development goal (standard of living adequate for health and well-being) has to be reconciled with the sustainability goal ("the planetary capital" in terms of natural systems, renewable and non-renewable resources should not be depleted).

Cities are *open systems* with flows of people, goods, energy, pollution and waste across the urban boundaries. This has to be taken into account when judging urban strategies for sustainable development. Land-use/transportation interactions and the organisation of other kinds of urban infrastructure (energy, water & sewage, waste disposal, telecommunications) are crucial for the attainment of welfare and environmental goals. This is because of the important structuring role of *urban network infrastructure* (UNI): UNI to a very large degree determines the patterns of land-use (corridors, mix of land-uses, densities) and transportation, which are vital for urban welfare as well as for urban environmental impacts.

In developed economies the role of *land-use/transportation interactions* (LUTI) is of major importance for development of the settlement structure, which in turn affects the options and costs for other kinds of UNI. In developing countries with very rapid migration to cities, the availability of land and the supply of basic water, sanitation and garbage services are most important for the development of the settlement structure. However, all over the world the human settlement and transportation systems account for major shares of the use of energy resources and of the emissions of environmental pollutants. Consequently, urban network infrastructure and land-use/transportation systems have crucial roles in the formation of urban strategies for sustainable development. Integrated models for analysis of interdependencies between UNI and LUTI constitute important starting points for development of more complete land-use/transport/environment modelling approaches, see Wegener (1998).

In this paper we will discuss the potential and limitations of *combined network equilibrium models* for analysing sustainable development of urban systems. Section 18.2 contains a brief review of combined network equilibrium models for integrated land-use/transportation analysis. One particular combined network equilibrium model is presented in Section 18.3. It is our intention to test this model in comparative studies of urban regions with very different character. The results of a first implementation in Chicago and the status a current application to Stockholm are reported in Section 18.4, and some features of the data sets and the computer implementations are compared. Finally, Section 18.5 attempts to evaluate the strengths and weaknesses of the present prototype model for analysing sustainable development in the urban context. Further developments of the model to increase its potential in this respect are outlined.

18.2 Combined Network Equilibrium Models for Integrated Land-Use/Transportation Analysis

Combined models of transportation market equilibria have a long tradition in research. As early as 1956, Beckmann et al. formulated a theoretically consistent equivalent optimisation model of the combined trip generation, trip distribution and route choice problem in a congested network context. The optimum conditions of the model ensured that demand (trip generation, trip distribution) was consistent with travel costs obtained from link performance functions at equilibrium vehicle flows and that user optimal route choices according to Wardrop's equilibrium conditions prevailed. Only the modal choice step (and the transit assignment step) of the traditional four-step transportation planning procedure was missing for the formulation of Beckmann et al. (1956) to be a fully integrated combined network equilibrium model of the transportation market.

No solution algorithm for the equivalent optimisation problem was available in 1956 and not until the first half of the 1970s were theoretically convergent algorithms devised for the fixed demand, congested route choice problem (Leblanc et al., 1975) and for combined network equilibrium models of trip distribution and route choice (Evans, 1976). As long as separable demand and link cost functions are employed, the solutions of combined network equilibrium models (trip distribution – route choice, modal split – route choice, trip distribution – modal split – route choice) are easily computed by using modified versions of the fixed demand, congested route choice algorithm of Leblanc et al. and following the partial linearisation technique suggested by Evans (1976). Given the efficient solution techniques of separable combined network equilibrium models, the slow introduction of these models in commercial transportation planning packages and planning practice is not easily explained (cf. the discussion in Boyce and Zhang (1998)). The EMME/2 package introduced the combined modal split - route choice model in 1989 and most packages still only provide equilibrium trip assignment (route choice). With only small increases in the computational burden and important gains in consistency, a more widespread practical use of combined network equilibrium models might have been expected.

Models allowing more general cost and demand functions were developed during the 1980s in terms of fixed point, non-linear complementarity or variational inequality problems. In particular, these formulations have been important for investigating existence and uniqueness of solutions to combined network equilibrium models, see Nagurney and Zhang (1998).

Large scale combined network equilibrium models covering all the four steps of the traditional transportation planning procedure were shown to be feasible in the late 1980s. Wegener (1986) used such a model as part of a broader sequential urban development model. The transportation model includes car ownership and trip generation, travel demand for four trip purposes, four income groups and three modes, and equilibrium assignment. Safwat and Magnanti (1988) presented a model including trip generation (related to accessibility), trip distribution and equilibrium choice of mode-route combinations. The model has been applied to networks with up to 520 zones, 2137 nodes and 7096 links.

The feasibility of extending the combined network equilibrium model to include land-use also has been demonstrated by Kim and his colleagues during the 1980s (Kim, 1989). These models can be described as integrated versions of a five-step land-use/transportation planning procedure. The solution algorithm relies on Evans′ partial linearisation method, including a complex demand submodel. The integrated nature of this land-use/transportation model makes it attractive as one of the candidate tools for studies of network infrastructure and the urban environment.

18.3 A Prototype Combined Network Equilibrium Model for Comparative Urban Land-Use/Transportation Studies

The model of Kim (1989) mentioned in Section 18.2 is subject to an ongoing research collaboration. The long-term aim of our collaborative project is to carry out comparative studies of urban areas in various parts of the world. The initial focus is on our two "home" cities: Chicago and Stockholm.

First, the same model structure will be applied to both cities. A prototype model has already been formulated and applied to Chicago for the year 1980. The current status of the application to Stockholm 1980 and a comparison of the Chicago and Stockholm data sets are discussed in this paper. By applying the same model structure to different cities, we will investigate to what extent urban land-use/transportation patterns in very different environments can be explained by the same basic land-use/transportation interaction mechanisms. If the model exhibits quite different explanatory power in the two metropolitan areas, the nature and possible causes of these differences will be further analysed. Various ways of adapting the model structure to specific environments in order to improve its explanatory potential will also be investigated. Although our interest will initially focus on our home cities (which are very different in terms of size, economic structure, transport modes, etc.), the comparative analysis of urban growth patterns and land-use/transportation interaction mechanisms may later be extended to other cities, e.g. in other parts of Europe or in Asia. An application to Seoul is on the long-term agenda.

Within our collaborative effort we will also elucidate various formulations of system behaviour in terms of user optimisation, system optimisation and intermediary bilevel optimisation hypotheses, see also Section 18.5. Due to the complexity of the models it is not clear at the outset, whether implementations of bilevel urban formulations can be fully realised.

A compact formulation of the prototype model of our research collaboration was presented in an earlier paper, Rho, Kim and Lundqvist (1993), which also contained a brief outline of the solution algorithm and a summary of the model application to Chicago for the year 1980. We summarise the model formulation (to be called TRANSACT) before making some comments on the nature of the model.

18.3.1 Model Formulation

The model is formulated in the following way:

$$\min \quad \sum_k \sum_a \int_0^{f_a^k} c_a^k(x)dx + \sum_i \sum_r d_r^i E_r^i + $$

$$+\sum_i \sum_r (L \cdot \exp(\alpha_{1r} s_r^i + \beta_{1r}) + R \cdot \exp(\alpha_{2r} s_r^i + \beta_{2r})) x_r^i \tag{1}$$

where

$$f_a^k = \sum_r g_r \sum_i \sum_j \sum_p x_r^{ijkp} \delta_{ar}^{ijkp} \tag{2}$$

subject to

$$\sum_i E_r^i \geq E_r \tag{3}$$

$$\sum_{j \neq i} \sum_k x_r^{jik} + x_r^i \geq \sum_{j \neq i} \sum_k x_r^{ijk} + \sum_q a_{rq} x_q^i + E_r^i \tag{4}$$

$$-\sum_i \sum_j \sum_k x_r^{ijk} \ln x_r^{ijk} \geq S_r^1 \tag{5}$$

$$-\sum_i E_r^i \ln E_r^i \geq S_r^2 \tag{6}$$

$$\sum_r \exp(\alpha_{1r} s_r^i + \beta_{1r}) x_r^i \leq l^i \tag{7}$$

$$x_r^{ijk} = \sum_p x_r^{ijkp} \tag{8}$$

$$x_r^{ijkp}, x_r^i \geq 0; \;\; E_r^i, x_r^{ijk} > 0 \tag{9}$$

The exogenous variables are:

$E_r =$ the total export of commodity r from the urban area as a whole;
$a_{qr} =$ the amount of input q required per unit of output r, where r ranges from 1to r' and where 1 to r'-1 represents intermediary goods and r' represents the labour input;
$d_r^i =$ the unit composite cost of exporting commodity r from export zone i;
$g_r =$ the passenger car equivalent of road space required for a unit shipment of commodity r;
$\delta_{ar}^{ijkp} =$ incidence matrix with value 1 if route p of shipments of commodity r between zone i and zone j by mode k uses transportation link a, and with value 0 otherwise;
$l^i =$ the land available for production and residential use in zone i;
$S_r^1 =$ the level of spatial dispersion of the shipment pattern for commodity r;
$S_r^2 =$ the level of spatial dispersion of export for commodity r;
$L =$ the opportunity cost per unit amount of land at the urban periphery;
$R =$ the rental rate per unit amount of capital;
$c_a^k(x) =$ the generalised cost function denoting the cost per vehicle unit of shipments on transportation link a of mode k at the flow volume x;
$\alpha_{1r}, \beta_{1r} =$ the land input parameters of commodity r; and
$\alpha_{2r}, \beta_{2r} =$ the capital input parameters of commodity r.

The endogenous variables are:

$E_r^i =$ the amount of exports of commodity r from export zone i, where the designated export zones constitute a subset of all urban zones;
$x_r^i =$ the output of commodity r produced at zone i;
$s_r^i =$ the level of land-use intensity (or density) in production of commodity r at zone i;
$x_r^{ijk} =$ the amount of commodity r shipped from zone i to zone j by mode k
$x_r^{ijkp} =$ the amount of commodity r shipped from zone i to zone j by mode k usingroute p; and
$f_a^k =$ the traffic volume of mode k on transportation link a.

In the model formulation (1)-(9) above, equation (3) ensures that at least the required amount (E_r) of commodity r will be exported from all designated export zones ($\sum_i E_r^i$) in the urban area (see definition of E_r^i above). Equation (4) is a material balance constraint requiring that the sum of the total amount of commodity r shipped into each zone i ($\sum_{j\neq i}\sum_k x_r^{jik}$) and the production of commodity r in zone i (x_r^i) should at least cover the sum of the total amount of commodity r shipped to other zones ($\sum_{j\neq i}\sum_k x_r^{ijk}$), and the amount of commodity r consumed in zone i ($\sum_q a_{rq} x_q^i$), and to be exported (E_r^i) if the zone i is an export zone for commodity r.

Equation (5) ensures that the dispersion of commodity flows over zones and modes should at least attain a prescribed level. Equation (6) requires in a similar way that the distribution of export goods among export zones should exceed a certain level of dispersion. Equation (7) is the zonal land constraint. It ensures that the total land consumed for the production of commoditites r in zone i at the endogenously determined intensity of land-use (s_r^i), ($\sum_r \exp(\alpha_{1r} s_r^i + \beta_{1r}) x_r^i$), should not exceed the land available in that zone (l^i). The endogenous determination of land-use intensities is explained in (10)-(11) below. Since land consumption per activity amount is normally decreasing with density, α_{1r} will in most cases be negative. (On the contrary, capital consumption per activity amount is normally increasing with density, which is reflected by positive α_{2r} coefficients.)

The objective function (1) to be minimised by the model is the "user-optimal" total cost of the urban system. It contains the congested user travel costs (expressed as the sum of integrals of link costs over all links and modes in the usual user-optimal equivalent optimisation modelling tradition: $\sum_k \sum_a \int c_a^k(x)dx$)), the export handling costs ($\sum_i \sum_r d_r^i E_r^i$) and, finally, the total opportunity cost for land ($\sum_i \sum_r L \cdot \exp(\alpha_{1r} s_r^i + \beta_{1r}) x_r^i$) and capital ($\sum_i \sum_r R \cdot \exp(\alpha_{2r} s_r^i + \beta_2) x_r^i$). Thus the intensity of land-uses and the amount of commodities to be produced are determined as a result of interactions among producers seeking to locate near export zones and among all other mutually interdependent urban actors. These interactions include cost-minimising capital-land substitutions and transportation patterns with conventional gravity, logit and user equilibrium properties. The use of network infrastructure and the location of production and households are determined simultaneously based on an explicit modelling of density and congestion externalities. Hence, the structuring role of network infrastructure for urban form and location is central and the model focuses on the feed-back effects between land-uses and transportation. Activity densities and traffic congestion constitute key elements for assessing environmental impacts and sustainability of any urban system.

18.3.2 Model Characteristics

A number of major modelling traditions have been combined in the model. The internal production and consumption linkages of the urban economy are described by *input-output relationships* (4). The basic driving force of the urban economy is the level of exports, (3), in accordance with *economic base theory*. The spatial distribution of exports (among export zones), production and consumption is determined by *(entropy maximising) spatial interaction patterns of the gravity type* governed by (5)-(6). Equation (5) also gives rise to *modal choices of the logit type*. The first term of the objective together with (2) and (8) and the demand reflected by (3)–(4) make up a typical equivalent optimisation formulation of a *network equilibrium*. Including (5)–(6) extends the cost minimising network equilibrium into a *combined model* of trip distribution (gravity type), modal split (logit type) and route choice (user equilibrium). The zonal land constraints (7) determine land values as scarcity prices on the land market resulting from *competitive bidding for land* by the urban activities. Finally, given all behavioural aspects already described, the basic organising principle implied by the model is *cost minimisation*. The model produces an efficient urban structure with the size determined by export requirements and the transport flow patterns reflecting conventional behavioural assumptions.

From a mathematical point of view the entropy constraints (5)–(6) could equally well be included in the objective (premultiplied by its respective Lagrange multiplier). Instead of specifying "levels of dispersion" S_r^1 and S_r^2, the Lagrange multipliers of the entropy terms would then be estimated by e.g. the maximum likelihood technique.

The solution algorithm is deduced by stating the Lagrangian of the optimisation problem and setting the derivatives with respect to the production, flow and export variables equal to zero. These variables can then be expressed as functions of the land values λ^i, i.e. the Lagrange multipliers of (7). By assuming cost minimising land-capital substitution, the land-use intensity variables, s_r^i , can also be expressed as functions of the land value, λ^i , of each particular zone:

$$c_r^i = \min_{s_r^i} \{(L+\lambda^i)\exp(\alpha_{1r}s_r^i + \beta_{1r}) + R \cdot \exp(\alpha_{2r}s_r^i + \beta_{2r})\} \tag{10}$$

Setting derivatives with respect to s_r^i equal to zero gives the following optimal land-use intensities:

$$s_r^i(\lambda^i) = \frac{1}{\alpha_{2r} - \alpha_{1r}}\{\ln(-\frac{\alpha_{1r}(L+\lambda^i)}{\alpha_{2r}R}) - (\beta_{2r} - \beta_{1r})\} \tag{11}$$

To compute the land values, we are faced with solving a non-linear equation system (12)–(13) consisting of the land constraints with slack variables, y^i , representing vacant land, and a set of complementary slackness equations permitting positive land values only when there is no vacant land and vice versa:

$$\sum_r \exp(\alpha_{1r} s_r^i(\lambda^i) + \beta_{1r}) x_r^i(\bar{\lambda}) + y^i - l^i = 0 \qquad (12)$$

$$\lambda^i y^i = 0 \qquad (13)$$

When (12)–(13) are solved (for λ^i and y^i), the equilibrium interzonal commodity flows, the zonal export amounts, the zonal production amounts and the intensities of land-uses can be derived from the optimum conditions outlined above. Taken together, (12)–(13) and these optimum conditions constitute the transportation demand side which is incorporated into a generalised Evans′ algorithm for solving "combined" network equilibrium problems:

Step 0: Initialise traffic volumes on transportation links.

Step 1: Calculate congested interzonal shipment costs for the current link volume on transportation links.

Step 2: Solve the system of nonlinear equations, (12)-(13), to obtain an equilibrium solution for zonal land rents, zonal vacant land, interzonal commodity flows, zonal export amounts by sectors and zonal outputs by sectors.

Step 3: Calculate link volumes for interzonal shipments obtained in step 2. Note that the interzonal shipments in dollars obtained in step 2 must be converted into interzonal vehicle movements before calculating link volumes in terms of numbers of vehicles.

Step 4: Search a new main problem solution using a one-dimensional line search method.

Step 5: Test for convergence. If the solution has not converged, update traffic volumes and return to step 1.

See Rho (1988) for a detailed description of the solution procedure.

18.4 The Chicago and Stockholm Applications and Data Sets

In this section the Chicago data base is compared with the present Stockholm data base. The comparison both highlights the differences between the two metropolitan areas and provides information on the availability of various kinds of data for the Stockholm application. A summary is provided in Table 18.1. Many of the data items are easily available but there are some components that require special treatment in terms of data collection, estimations or "informed guesswork".

Table 18.1. Comparisons between the Chicago and Stockholm data bases (both referring to 1980)

Item		Chicago	Stockholm
Population		7.7 million	1.52 million
Employment		3.5 million	0.84 million
Number of zones		74	30
Public transit share (peak hour)		22%	56%
Number of nodes in road network		1060	397
Number of road links		2902	921
Number of volume/delay functions		1	12
Observed/estimated interzonal			
traffic flows:	– car	yes	yes
	– transit	yes	yes
	– truck	yes	yes[1]
	– rail	yes[2]	no
Sectors (employment share):		manufacturing (30%) trade (20%) services (50%) households	manufacturing (22%) private services (33%) public services (45%) households

18.4.1 Zones and Networks

The Chicago region with 7.7 million inhabitants and a working population of 3.5 million 1980 was represented by 74 zones. The Chicago Area Transportation Study's (CATS) highway sketch network was used in the model application. It includes 1060 nodes and 2902 links. The transit network was represented by a generalised cost matrix, since transit travel costs were assumed to be unaffected by the number of travellers. The generalised transportation cost is composed of in-vehicle travel time, in-vehicle travel cost (including parking fee or transit fee) and out-of vehicle travel time. Both in-vehicle travel time and in-vehicle travel cost on highway links were flow or speed related. Special weights were introduced to make transportation costs per person comparable to transportation costs per million dollars value of commodities. Commodity specific and mode specific constants were also included in the generalised cost function. These were calibrated to improve the goodness-of-fit with respect to average transportation distances and average modal shares.

The Stockholm metropolitan region had a population of 1.52 million 1980 and a work force of 0.84 million. In the present data set, 30 zones represent the Stockholm region (about 51000 persons and 28000 workers per zone as compared to about 104000 persons and 47000 workers per zone in Chicago). The total land area by zone is easily available, but land areas that are not used by the activities of TRANSACT need to be excluded (as was done in the Chicago data base).

Export zones for each sector in Chicago were selected by using location quotients: zones with overrepresentation of employment in a certain sector were

[1] Available between 25 municipalities for the year 1985.

[2] Estimated only.

designated as export zones for that sector. The export handling costs were also estimated using the calculated distribution of economic base employment over export zones, assuming that agglomeration of economic base activities is related to low export handling costs. Similar techniques have been used to specify export zones and export handling costs in the present Stockholm data base.

A highway network with 397 nodes and 921 links has been adapted to the system of 30 zones from a research network used earlier. The links constitute a selection of regionally important highways leaving out all local streets and minor roads. The link specification contains the type of volume/delay function that is applicable in each case. There are 12 types of volume/delay functions including access and egress links connecting the zones to the network. In previous applications with this network only travel times have been used. However, it is easy to include a distance related auto operating cost in the generalised cost of each link. This operating cost may later be related to velocity. Introducing an out-of-pocket cost component implies that the model is prepared for analysing the impacts of various highway toll schemes, which have been subject to intensive discussion. A transit generalised cost matrix is available covering in-vehicle time and out-of-vehicle time (including time for interchanges). The transit cost matrix can later be extended to cover transit fares. A large part of the travellers during peak hours use a monthly pass implying a low and flat average fare.

For calibration or estimation of parameters (e.g. the Lagrange multiplier of the entropy constraint (5) representing sensitivity to generalised costs) and coefficients (e. g. weights and coefficients in the generalised transportation cost measure) data on observed trip and transportation patterns are required. For Chicago observed auto, transit and truck OD matrices were available. For Stockholm observed peak hour auto and transit matrices 1980 are readily available but knowledge on freight transportation is more limited. Annual freight volumes between the 25 municipalities of the region have been estimated for 1985. The observed peak hour transit share in Chicago is 22% as compared to 56% in Stockholm.

18.4.2 Input, Output and Technical Coefficients

In the Chicago application the urban economy was represented by four sectors: manufacturing, trades, services and households. A non-survey technique was used to compute the regional input-output table. Total output by sector was calculated by using employment and wage statistics together with the labour input coefficient from the input-output table. Finally, "exports" was calculated from total outputs and the input-output table by subtracting intermediary deliveries from total output. This means that "exports" represents all final demand that is not included in the regional input-output table. For example investment deliveries need to be part of "exports". Private and public consumption is covered by the input-output table, provided that public services are included in the service sector. In Chicago manufacturing, trades and services account for about 30%, 20% and 50% of the total employment respectively.

A complete regional transaction table for the Stockholm region 1980 is available from earlier research, see Lundqvist (1981). This includes nine sectors with the public service sector being one of them. Initially, we distinguish the following four sectors: manufacturing (22% of total employment), private services (33%, incl. trades, transportation and housing services) and public services (45%, incl. national, regional and local public services). "Exports" (incl. interregional deliveries), total import shares and output levels could also be obtained from the estimated total transaction table.

18.4.3 Land and Capital Input Coefficients

The production technology is represented by fixed Leontief input-output coefficients as discussed above. Substitution is however permitted between land and capital resources. The land and capital input coefficients (see e.g. (1)) are estimated as functions of land-use intensity (building height). In the Chicago case the parameters of the land and capital input functions could be estimated using regression analysis. Land input per sectoral floor space unit and capital input per sectoral floor space unit were available from zoning ordinances and building cost manuals.

Comparable data for Stockholm are not readily available. Some regression results relating building cost per unit housing floor space to building height are available from earlier research, see Andersson and Samartin (1979). Studies of land requirements per unit floor space area in a number of housing areas has been conducted by the Regional Planning Office. Land and capital requirements by various industries and services as a function of the distance from the CBD were reported in a another location study. From these sources initial parameter values of land and capital input functions have been estimated. These preliminary data need to be validated and refined in future research.

18.4.4 Rents, Trip Generation Factors and Loading Factors

The remaining data items include agricultural land rent (L) and capital cost (R). In Chicago these were related to the average agricultural land value for the states of Illinois and Indiana (10% of land value) and to the federal discount rate respectively. A similar approach has been used in Stockholm.

The amounts of activities in the model are measured in monetary terms and peak hour trip generation factors have to be provided. In Chicago these trip generation factors (person trips and freight tons) by sector could be deduced from a CATS study of the Chicago central area and from CATS′ commercial vehicle survey data base. In Stockholm, the trip generation factors have been compiled to reflect the total number of observed peak hour personal trips and the estimated peak hour truck volumes. Finally, auto occupancy, the truck loading factor and the passenger car equivalence factor of trucks were easily available.

18.4.5 Computer Implementations of TRANSACT

The Chicago version of TRANSACT has been implemented on supercomputer for large scale applications. The Chicago application was run on a Cray XMP, requiring 50 minutes of computation time for five iterations. Earlier Rho and Kim (1989) reported results from a test problem with 4 sectors, 1 mode, 7 zones, 57 nodes and 172 links. On a CDC Cyber 175, 10 iterations were completed in about one minute.

The Stockholm version of TRANSACT has been implemented on a 486/66 PC. The travel time and auto operating cost functions have been changed and all freight volumes are allocated to trucks. The computation time for initial runs has been about one minute per iteration.

18.4.6 Results

Results from the full scale Chicago application have been reported elsewhere, see Kim (1989) or Rho, Kim and Lundqvist (1993). On an aggregated level (characterisation of zones in terms of concentrations of one or two activities) the model generated land-use pattern is very similar to the observed distribution of land-uses. Also the land values generated by the model are reasonably close to observed land rents, see Fig. 18.1 for the spatial variation of land values along two

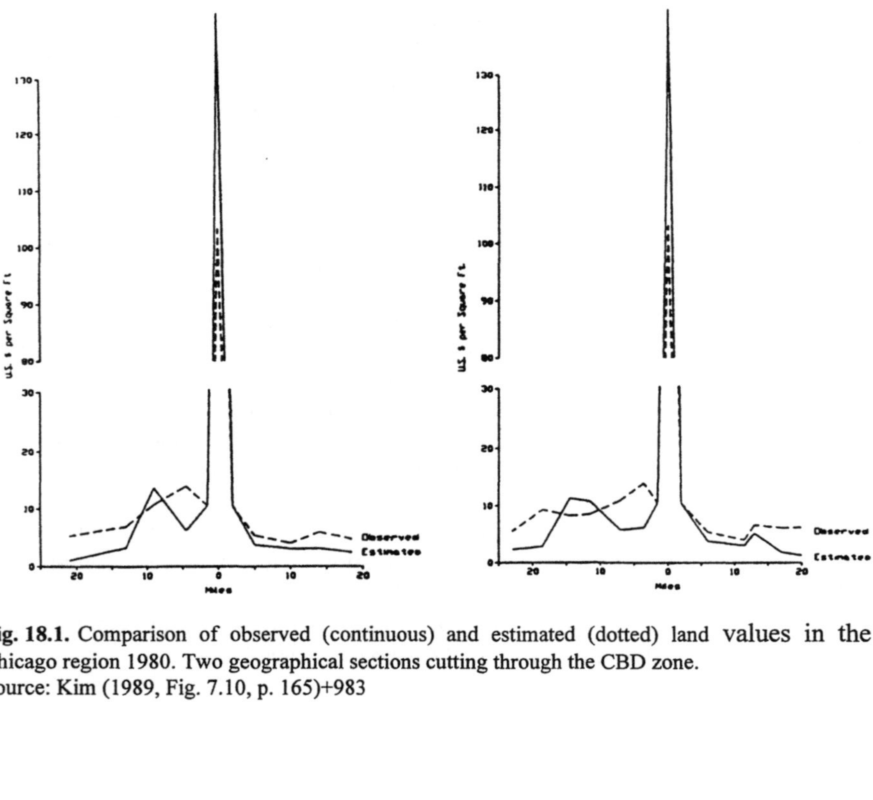

Fig. 18.1. Comparison of observed (continuous) and estimated (dotted) land values in the Chicago region 1980. Two geographical sections cutting through the CBD zone.
Source: Kim (1989, Fig. 7.10, p. 165)+983

geographical sections through the region and including the CBD zone. TRANSACT slightly overestimates the spatial variation of land values.

The goodness-of-fit is less exact for the estimated interzonal modal trip patterns with considerable over- and underestimations for many zonal pairs. This is to be expected for a "simultaneous" trip distribution and modal split model of the singly constrained type as represented in TRANSACT (see other demand structures in Lundqvist and Mattsson (1992) and Abrahamsson and Lundqvist (1998)).

The Stockholm version of TRANSACT has not yet been fully implemented and only the first model runs have been carried out. The first results indicate that some of the data need to reviewed (land and capital input coefficients and the specification of transportation cost per unit of shipment). The level of interzonal transportation is too low and the spatial distribution of activities is too concentrated. Quite naturally, there is a clear need for a comprehensive estimation and calibration process and a reconsideration of some items of the initially compiled data set.

18.5 TRANSACT and Sustainable Development

There is a renewed interest in comprehensive urban modelling, partly motivated by increasing environmental concerns and linkages to geographical information systems, see Batty (1994) and Wegener (1994; 1998). Lee, the author of the seminal article "Requiem for large scale models" in 1973, also sees a role for comprehensive urban models in strategic planning and in environmental impact assessments (EIA), see Lee (1994).

As touched upon in Section 18.1, the urban system interacts with the local and global environment in many ways: flows of materials, energy, pollution etc. Environmental sustainability of urban systems has to be judged on the basis of at least the following aspects:

- the regenerative capacity of natural systems;
- environmental qualities of land, water and air;
- resource depletion;
- import and export of pollution; and
- global climate effects.

The first three aspects are relevant for both the intraurban environment and its neighbouring regions. The last three aspects also cover impacts on a wider geographical scale. Here, the open systems character of the urban economy becomes very explicit.

The strength of TRANSACT in environmental analysis is that it covers both localised activities (point sources) and transport interactions (diffuse sources). The density of land-uses is determined in TRANSACT, which has important bearings on the concentrations of emissions and waste as well as on the potential for certain infrastructure systems (e.g. district heating). The model covers both personal transportation and freight deliveries between all urban activities. The flows of exports and imports are also represented in the model. The comprehensive treatment of localised activities and transportation flows provides a firm basis for evaluation of resource consumption and global climate effects. By linking TRANSACT to environmental systems models (i.e.models of emissions, dispersion, immissions of pollutants used in Environmental Impact Assessment (EIA), see Wegener (1998)), the intraurban and interregional impacts on natural systems and environmental qualities may be judged taking imports and exports of materials, energy and pollution into account. Measures to preserve the regenerative capacity of natural systems and environmental qualities by reserving green spaces can be accounted for in the land-use constraints. The user optimal flow pattern in TRANSACT can easily be changed to a system optimum by substituting social marginal costs for the average transportation cost functions. Charges for environmental externalities of transportation can be included in both modes of operation. Land-use constraints and pricing schemes in transportation are two examples of how policies based on environmental impact assessments, linked to TRANSACT, might affect location behaviour.

The weaknesses of TRANSACT are mainly of at least three kinds. The model is fairly complex and has only been applied in aggregate strategic planning contexts. Hence, the sectoral detail is not enough for distinguishing localised activities according to their differential environmental impacts. The number of sectors must be increased and defined according to the needs of EIA and the spatial resolution has to provide enough detail. A more complete treatment of external trade (interregional and international) patterns may also be desired from an environmental point of view. Secondly, TRANSACT is a long-term equilibrium type of model, providing a "snapshot" picture of the future urban structure irrespective of its history. In more short-term perspectives, the model needs to take constraints related to the base year land-use pattern into account and the necessary investment costs for transforming historical land-uses to future "market oriented" equilibria. Thirdly, the present version of TRANSACT is only capable of treating environmental externalities related to transportation flows or settlement densities. Accounting for other agglomeration economies and diseconomies of localised activities requires further developments of the model.

Although TRANSACT can be used in a normative mode by inserting social marginal transportation costs (see above), it needs to be extended to a bilevel formulation in order to be capable of treating normative planning goals and user optimal behavioural patterns at the same time. A less ambitious approach would be to iterate between a normative model of urban settlement structure and network design on the one hand and a disaggregate version of TRANSACT (user optimal mode) on the other hand, see Lundqvist (1997). Such more complex model systems are needed in order to explicitly account for environmental goals and the trade-offs between these goals and other urban development objectives (i.e. interpretations of sustainable development), while retaining the market oriented character of behavioural patterns in the present version of TRANSACT.

Acknowledgement

Permission to reproduce Fig. 18.1 is gratefully acknowledged from Kluwer Academic Publishers. This research has been supported by The Swedish Transport & Communications Research Board and The Swedish Council for Building Research.

References

Abrahamsson, T. and Lundqvist, L. (1998), 'Formulation and estimation of combined network equilibrium models with applications to Stockholm', forthcoming in *Transportation Science.*

Andersson, R. and Samartin, A. (1979), *Interdependence among Housing, Heating and Transportation in Cities*, Document D9:1979, Swedish Council for Building Research, Stockholm.

Batty, M. (1994), 'A chronicle of scientific planning: The Anglo-American experience', *Journal of the American Planning Association*, vol. 60, pp. 7–16.

Beckmann, M.J., McGuire, G.B. and Winsten, C.B. (1956), *Studies in the Economics of Transportation*, Yale University Press, New Haven, CT.

Boyce, D.E. and Zhang, Y.-F. (1998), 'Parameter estimation for combined travel choice models', Chapter 10, this volume.

Evans, S.P. (1976), 'Derivation and analysis of some models for combining trip distribution and assignment', *Transportation Research*, vol. 10, pp. 37–57.

Kim, T.J. (1989), *Integrated Urban Systems Modeling: Theory and Applications*, Kluwer, Dordrecht.

Leblanc, L.J., Morlok, E. and Pierskalla, W. (1975), 'An efficient approach to solving the road network equilibrium traffic assignment problem', *Transportation Science*, vol. 9, pp. 309–318.

Lee, D.B. (1994), 'Retrospective on large-scale urban models', *Journal of the American Planning Association*, vol. 60, pp. 35–40.

Lundqvist, L. (1981), 'Applications of a dynamic multiregional input-output model of the Swedish economy', *Papers of the Regional Science Association*, vol. 47, pp. 77–95.

Lundqvist, L. and Mattsson L.-G. (1992), 'Modelling travel demand in an extended metropolitan region', Paper presented at the 32nd European Congress of the Regional Science Association, Brussels.

Lundqvist, L. (1997), 'Strategic urban design and land-use/transportation markets: Model formulations and applications to Stockholm', Paper presented at the 37th European Congress of the Regional Science Association International, Rome.

Nagurney, A. and Zhang, D. (1998), 'Introduction to projected dynamical systems for traffic network equilibrium problems', Chapter 8, this volume.

Rho, J.H. (1988), *Implementation and Evaluation of a Nonlinear Three Dimensional Urban Activity Model*, PhD thesis, University of Illinois at Urbana-Champaign, Urbana, IL.

Rho, J.H. and Kim, T.J. (1989), 'Solving a three-dimensional urban activity model of land use intensity and transport congestion', *Journal of Regional Science*, vol. 29, pp. 595–613.

Rho, J.H., Kim, T.J. and Lundqvist, L. (1993), 'Integrated land-use transportation model: Application to Chicago and outline for Stockholm', in Lakshmanan, T.R. and Nijkamp, P. (eds), *Structure and Change in the Space Economy*, Springer-Verlag, Berlin, pp. 259–279.

Safwat, K.N.A. and Magnanti, T.L. (1988), 'A combined trip generation, trip distribution, modal split, and trip assignment model', *Transportation Science*, vol. 18, pp. 14–30.

WCED, World Commission on Environment and Development (1987), *Our Common Future*, Oxford University Press, Oxford.

Wegener, M. (1986), 'Transport network equilibrium and regional deconcentration', *Environment and Planning A*, vol. 18, pp. 437–456.

Wegener, M. (1994), 'Operational urban models: State of the art', *Journal of the American Planning Association*, vol. 60, pp. 17–29.

Wegener, M. (1998), 'Applied models of urban land use, transport and environment: State of the art and future developments', Chapter 14, this volume.

19 Development of a Compact Urban Simulation Model

John R. Roy
Leorey O. Marquez
Commonwealth Scientific and Industrial Research Organisation
Division of Building, Construction and Engineering
P.O. Box 56, Highett
Victoria, 3190, Australia

Michael A.P. Taylor
Transport Systems Centre
University of South Australia
The Levels, South Australia, 5095
Australia

Takayuki Ueda
Department of Civil Engineering
Gifu University
Yanagido 1-1
Gifu, 501-11, Japan

19.1 Introduction

Recent research, such as in Wegener (1996), is revealing that quite radical policies may be required if the transport sector in most Western cities is to do its share in meeting the 2005 Greenhouse targets, to which most national governments have committed themselves. On the other hand, during recent years, planners have acquired their skills in implementing integrated land-use/transport models in an atmosphere of quite gradual change. Thus, with reliable guidelines not available for handling this new situation, and with the large data requirements and computational effort needed for each test run of the integrated models, a fresh approach is called for. In response, an urban simulation model SUSTAIN (Sustainable Urban STructure And Interaction Networks) is being developed to act as a preliminary design/sensitivity analysis tool to streamline subsequent application of one of the integrated models. The package evolved from some preliminary analysis in Roy (1992), itself stimulated by early work in Sweden by Marksjö (1970) and investigations by Anderson et al. (1986).

In SUSTAIN, the geometry of the urban form and its transport network is appropriately simplified, without being simplistic, to permit efficient analysis of a wide set of land-use/transport development and policy scenarios. At the same time, the planner gains insight interactively on the sensitivity of system performance to alternative strategies. Also, due to the economy in definition of the city and its networks, the model can transcend the usual limitations of the integrated models, by not only allowing the usual adjustments of land-uses to alternative transport network conditions, but also by simultaneously performing some optimisation on the locations and capacities of major links of the transport network. Of course, such coordinated land-use and network design is also possible for the large integrated models, but is much more tedious. The expectation is that such simultaneous adjustments may be required to meet effectively the stringent 2005 guidelines. This paper describes the current development of SUSTAIN, which is now being accomplished as a joint effort between Australia and Japan (Roy et al., 1996).

The first version of the model, SUSTAIN Mark I, analysing geometrically-idealised cities with a radially symmetric distribution of housing and jobs and radially symmetric transport networks, is meant to be used as an educational tool, whereby users efficiently create widely different urban configurations and associated transport networks and monitor their environmental/energy performance. On the other hand, SUSTAIN Mark II is aimed at preliminary design of an actual city and its evolution over time. Although it maintains a relatively simplified geometrical representation of the city and its networks to simplify simultaneous land-use and transport network design, it lifts the radial symmetry restriction, both for land-use and transportation. The first part of the paper defines the city in terms of its inhabitants, housing, jobs and transport networks. Then, a summary of the modelling of the key urban land-use and transport processes is described, including the structure of the C++ object-oriented program and its use of the GIS package MAPINFO. Finally, some test results are summarised and some future enhancements are sketched.

19.2 The Urban Form and its Networks

The urban form and networks are described in terms of population, housing, employment, private transport and public transport, as follows.

19.2.1 Population

The city has a given total population, which consists of households of varying size and varying job participation rates. In all comparison cities created, the urban area adapts such that it precisely absorbs this given population. Then, in the forecast periods, the city is allowed to expand (or contract) to absorb designated new populations.

19.2.2 Housing

As seen in Fig. 19.1, three rings of housing are defined for Mark I, including (i) an inner core, (ii) a middle ring and (iii) an outer ring, with discretely decreasing densities as one moves outward from the Central Business District (CBD). In the experimentation phase, Mark I, where different cities are 'created' and their sustainability properties compared, the housing density is constant within each ring, only varying between rings. However, in the projection phase, Mark II, where one gains insight on the sequential effectiveness of alternative policies, each discrete analysis zone has a potentially different housing density and land price, depending on the results of the utility/bidding model for housing land, based on the combined relative accessibility of each housing zone to both CBD and suburban jobs. Also, housing residents are divided into three income groups, each of which defines a budget available for housing plus a composite good. These income groups correspond with the three occupational groups defined for employment. Finally, in the experimentation phase of SUSTAIN (see next section), an annulus of locally higher density housing iscreated around each ring

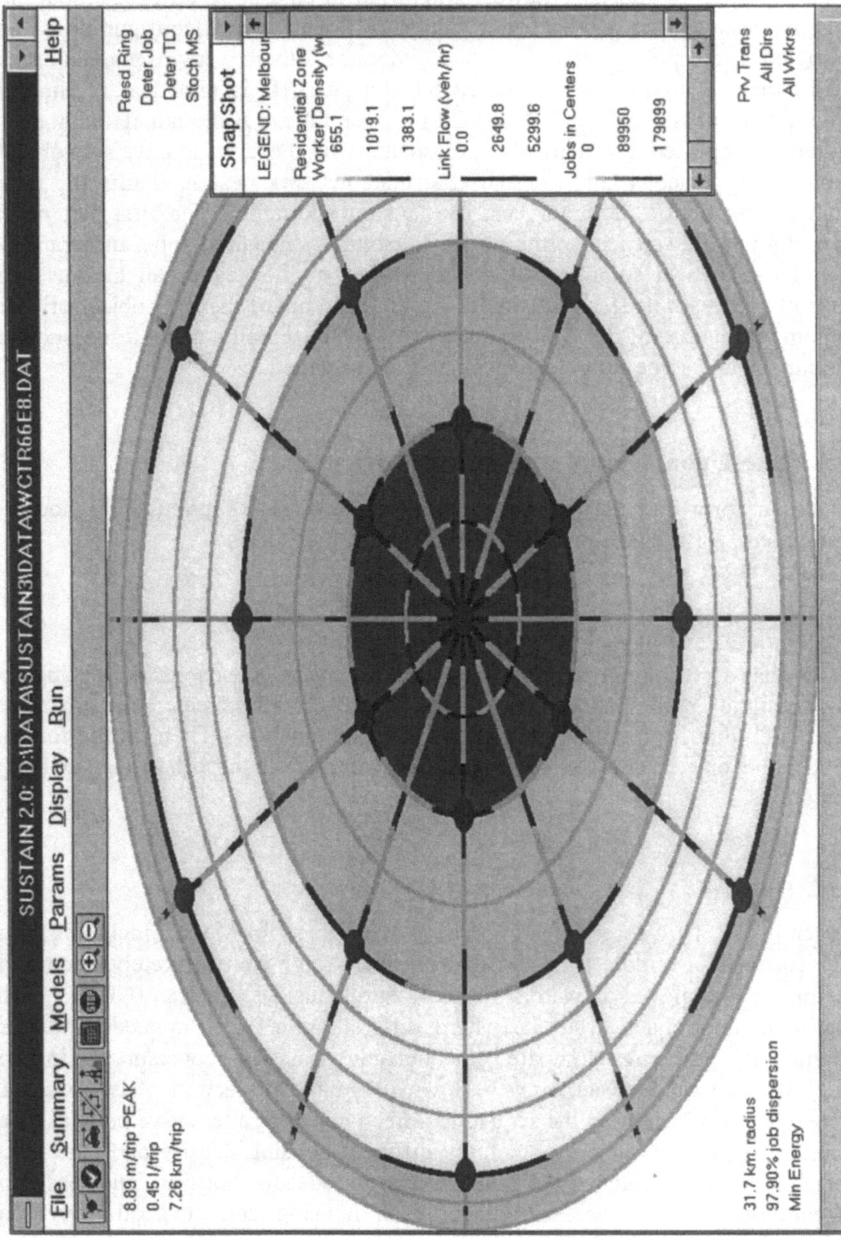

Fig. 19.1. Typical city for SUSTAIN Mark I

road to reflect its enhanced accessibility. In the projection phase, this density is evaluated endogenously using the housing location submodel.

19.2.3 Employment

In defining three categories of worker in terms of their jobs, we have considered two principles: (i) to segment according to different income levels and associated housing budgets and (ii) to segment according to job selectivity and corresponding sensitivities to commuting travel times and costs. This resulted in three groups of worker as follows:

(i) industrial workers – low to middle income;
(ii) service workers – middle income; and
(iii) knowledge-based workers – middle to high income.

Groups (i) and (ii) are expected to be more sensitive to travel costs than group (iii), have more homogeneous job opportunities and thus a higher value of the travel impedance parameter β. On the other hand, group (iii) is expected to be much more selective about job options, with less concern for travel cost, but higher concern for travel time, usually yielding a lower value of β.

Clearly, with the very rapid growth of service jobs in most cities and their potential for suburbanisation, the model needs to handle a combination of CBD and suburban employment. To allow job dispersion in a simple manner, employment subcentres are located in one or more rings about the CBD. Whilst the number of subcentres in each ring is optional, they remain the same, or double in quantity, as one moves outwards. For instance, one may define three rings of subcentres in which the inner ring contains 4, the middle ring also contains 4 and the outer ring has 8.

Note that, the urban activity system is restricted to the location of housing and employment, with the above three groups choosing jobs in their category. Interaction between these two sets of activities then represents the journey to work. This is not so restrictive as it may appear at first sight as (i) both the subcentres and the CBD, in relation to the current strong growth in the service sector, are natural locations for retailing and (ii) with the increasing long-term involvement of women in the work force, many shopping trips are work-based or performed on the way home from work, rather than being so predominantly home-based, and are thus included implicity in the journey-to-work.

19.2.4 Road Network

The location of the employment subcentres automatically sets priorities for specification of the road network. Thus, circumferential (orbital) ring roads (freeways or arterials) link the subcentres with each other. Similarly, radial freeways and arterials pass through the subcentres and link the different rings with each other and with the CBD. Of course, other radial roads can be defined which do not pass through subcentres. In addition, intermediate circumferential routes can be specified. These latter roads are usually at the urban arterial level.

Radial freeways are not constrained to continue from the outskirts of the city to the CBD at constant width. Land acquisition costs may require them to reduce from say eight to finally four lanes as the inner area is approached. In other cases, they may stop at an inner ring road, discouraging entry into the inner core. Because of the economies of network definition implicit in the radially symmetric structure of SUSTAIN Mark I, such capacity variations are easy to introduce and to change interactively.

Whilst the above economy in network definition of SUSTAIN Mark I encourages the user to investigate and obtain insights on the complex interactions between network capacities, congestion and land use adjustments, it is not directly applicable to many actual cities, where the radial symmetry assumption for land uses and the associated transport networks may be very restrictive. For instance, Port Phillip Bay in Melbourne subtracts about 90 degrees from the available urban area, and development is skewed towards the South and East. SUSTAIN Mark II removes the radial symmetry restriction, yielding a convenient preliminary design tool for actual cities. It allows a radial/circumferential network centred on the CBD, but where the circumferential links do not need to go through 360 degrees at constant width, but can be partial, with different capacities as one proceeds circumferentially.

In the use of SUSTAIN Marks I and II for strategic level planning, the network just encompasses freeways and arterials, with roads of a lower hierarchy only being represented implicity in obtaining shortest path routes. For these lesser roads, average speeds are provided exogenously and are not subject to congestion feedback.

19.2.5 Public Transport

As for the road network, the locations of the main subcentres themselves define the routes of the major public transport lines. These routes can be serviced selectively either by (i) transit, (ii) light rail or (iii) buses, where the latter also contribute to the loading of the road network. Access to terminals for these main routes can be either by (i) car, (ii) bicycle/walk or (iii) para transit. The latter mode deserves special attention in the low density suburban areas of U.S., Canadian and Australian cities, where computer scheduling technology can co-ordinate vehicles, such as mini-buses, for trips with both dispersed trip origins and dispersed trip ends. Note that, for both public and private transport, primary energy consumption (e.g. including petrol refining losses, as well as electricity generating and distribution losses) is the basis of the energy calculations.

19.3 The SUSTAIN Analysis System

The discussion is subdivided into the theoretical foundations, program structure, elements of model operation, location of employment, location of housing, the interaction between housing and employment and associated traffic assignment for the journey to work.

19.3.1 Theoretical Foundations

Whereas urban economists may criticise Lowry-type models because of absent or exogenous prices and a demand-side orientation, practising urban modellers are also critical of urban economics models, which often use idealised geometries, ubiquitous networks and non-overlapping employment catchments. SUSTAIN endeavours to harness the strengths of both traditions, in a manner consistent with its preliminary design role.

The first main requirement is to nest short run (price adjustment) processes within long run processes of locational change. Thus, the housing location model, developed from the urban economics analysis of Mohring (1993), satisfies short run utility maximisation within a longer run framework of locational change, where a uniform level of realised utility for each household within an income group is eventually attained. A similar Nash equilibrium long run adjustment process, derived from Harris and Wilson (1978), is used to locate the retail and service activity. As the classical retail choice models do not contain prices explicitly (that is, they *implicitly* assume that goods prices are constant between retail centres), the short run adjustments just represent the influence on demand of floorspace changes. Nevertheless, the locational equilibrium of the Harris and Wilson retail model can be interpreted as a situation where, assuming free entry, constant rents and an exogenous total floorspace, a uniform profit per unit floorspace is achieved at all retail centres, analogous to the uniform utility condition in the housing demand model.

Although the urban economics analysis of Yinger (1992) deals innovatively with a multi-centred city, there is no overlapping of job catchments for competing employment centres, with all workers choosing the closest job. In SUSTAIN, this limitation is overcome by using the *expected* or log-sum travel cost from the trip distribution model to represent dispersion of job choice to alternative subcentres from each residential zone. Despite the traditional use of such log-sum costs to ascend to the next level of a nested logit model, their additional interpretation as expected costs justifies their use in a utility maximisation model as a combined access measure to all job centres from each housing zone.

In specifying idealised cities, an important challenge is to nevertheless generate realistic depictions of behaviour. At the same time, for both idealised and actual cities, plausible bounds on city performance are necessary to determine the scope of adjustment of travel and locational behaviour to alternative policies, such as the pricing of fuel. SUSTAIN establishes these bounds for all behavioural processes independently, as well as yielding reasonable intermediate results, in the following ways:

Lower Bound–Deterministic. This is the case usually considered in urban economics, where workers choose the nearest job and cheapest travel mode and households choose their maximum bid lot.

Upper Bound–Random. At the other extreme, workers are taken to be indifferent to travel times and costs, as well as to bid rent differentials, choosing from each spatial option just in proportion to the relative supply.

Intermediate Cases – Stochastic. Upon solution for the simpler upper and lower bounds (e.g. of average trip time), the model is calibrated to a chosen intermediate value (e.g. 0.6 (Lower) + 0.4 (Upper)), reflecting typical trade-offs between quality of choice and travel costs.

Whereas in the experimentation phase Mark I, in which the user seeks to identify well-balanced patterns of land use and the supply of networks, the model is iterated out to a long run locational equilibrium together with an equilibrated transport network, the sequential use of the model over time in Mark II includes a one period lag between transport network changes and the long run locational adjustments. This also means that network equilibrium does not occur simultaneously with locational equilibrium, but is just achieved once at the end of each time period, yielding given trip times for the locational processes of the next time period.

19.3.2 Program Structure

The SUSTAIN program is an object-oriented program using the C++ language with Windows 3.1. This structure allows portability of program 'objects', vital in the current shared development between Australia and Japan. The Windows 3.1 format makes it attractive for potential interactive use by students and practicing planners. An interface with the GIS package MAPINFO enhances the preparation of the data input files and the mapping of the idealised actual city and its networks. For more detail about aspects of the object-oriented programming philosophy used in SUSTAIN, the reader is referred to Roy et al. (1996).

19.3.3 Elements of Model Operation

The model is operated in two phases (i) the experimentation phase and (ii) the projection phase, now described in order.

Experimentation Phase (Mark I). The aim of this phase is to allow the user to 'create' and arbitrarily change cities of his choice, without consideration of how that city's land use configuration may have evolved. This phase is primarily educational, aiming to identify typologies of energy-efficient urban forms and their associated transport networks. The locations of job centres and density categories of housing are specified by the user and can be changed interactively, with network equilibrium being achieved with respect to each pattern. Alternatively, based on experience, the user may assign desired maximum speeds and numbers of lanes to the various major links of the network, obtain the corresponding shortest path link volumes and associated speeds, correcting iteratively on lane capacities until the desired speeds are closely achieved. Also, the user can activate an automatic optimisation on the location and capacity of the major ring roads, a feature not available in the comprehensive models. SUSTAIN moves each ring road incrementally, plotting a curve of associated energy use or CO_2 generation against ring road radii, allowing optimal radii to be identified at the minimum CO_2 or energy consumption.

Projection Phase (Mark II). If, in the experimentation phase, the user has created cities with broad similarities to his own city, but with better environmental performance, he is well-placed to proceed further. In the projection phase, guided by his experimentation, he examines the sets of alternative policy scenarios which may lead the actual city from its current state toward one of the favored states identified earlier. The model runs sequentially over time, enabling some policies to be introduced incrementally. From the results of Wegener (1996), it is clear that much 'trial-and-error' experimentation is required on the intensity and sequence of introduction of the various policies within a policy scenario, if there is to be any hope of meeting the 2005 targets. With practically no examples in the world of large cities which might be classified as both environmentally and socially sustainable, even if these criteria are defined quite narrowly, models need to be flexible enough to encourage development of imaginative alternatives. The simplified SUSTAIN geometry allows such imaginative changes to be introduced very easily.

19.3.4 Location of Employment

Whilst the CBD and the subcentres denote points where employment can potentially locate, the projection phase allocates changing amounts of employment of the three categories to subcentres according to various guidelines, as follows.

Industry Jobs. As jobs in secondary industry are relatively immobile and a decreasing fraction of total employment, and in some cities are decreasing in absolute terms, the model is unsophisticated here. Such jobs are either automatically located in proportion to existing industry jobs or the user can intervene to make his own allocation.

Service and Knowledge-based Jobs. The first step is calibration of a retail trip model on the base period housing and floorspace distribution. Letting N_{ik} equal the number of households of income group k in zone i and s_k the unit trip generation rate for such households, the number of trip origins O_i is $(\sum_k s_k N_{ik})$. If

W_j is the floorspace in subcentre j and $\bar{g}_{ij}$ the generalised (cost + time) travel cost between zone i and subcentre j (at this stage, taken as exogenous off-peak private transport costs) it is possible to model the number of trips S_{ij} between i and j and calibrate the gravity parameter $\bar{\beta}$ in the origin-constrained model

$$S_{ij} = O_i\, W_j \exp - \bar{\beta}\ \bar{g}_{ij} / \sum_j W_j \exp - \bar{\beta}\ \bar{g}_{ij} \tag{1}$$

In the model operation through sequential time periods τ, the travel costs $g_{ij}^{\tau-1}$ of the previous time period are always used. Within each such time step, the model iterates in 'Lowry' fashion between housing and job location, until the service job locations entering the housing model produce virtually no change in housing location compared with the previous iteration. Each such housing/job iteration is designated as m. Assuming, as discussed earlier, that a long run Nash locational equilibrium occurs when demand ($\sum_i S_{ij}$) is equi-proportional to supply W_j at each subcentre j (that is, an equi-profit criterion per unit floorspace if prices and rents are taken as uniform), we may write the equilibrium flows as

$$S_{ij} = (\sum_k s_k\, N_{ik}^m)(\sum_i S_{ij}) \exp - \bar{\beta}\, g_{ij}^{\tau-1} / [\sum_j (\sum_i S_{ij}) \exp - \bar{\beta}\, g_{ij}^{\tau-1}] \tag{2}$$

where N_{ik}^m is the housing distribution from the housing model at iteration m of the Lowry housing/job balance process. As the unknowns S_{ij} occur both on the left-hand and right-hand sides of (2), the solution is obtainable via successive substitution. Finally, jobs D_{j2}^m and D_{j3}^m are allocated to centres j from the totals D_2 and D_3 in proportion to 'floorspace' $(\sum_i S_{ij})$.

Although in Roy (1996), a retail demand model consistent with microeconomic theory has been formulated, it is considered that the simpler procedure (1) and (2) is appropriate for the SUSTAIN preliminary design task. A future challenge is to locate the knowledge-based jobs in relation to *both* journey-to-work access and access to face-to-face contacts.

19.3.5 Location of Housing

The Lowry model framework is inadequate here, neither considering housing prices endogenously nor letting them adjust over time. The deterministic version of our model is based on the Alonso/Mills tradition, generalising some recent work by Mohring (1993). We make four main enhancements of Mohring's model (i) jobs are located not just at the CBD, but at suburban subcentres, using a log-sum definition of expected access to employment (ii) three types of household k compete for housing, (iii) as specified by Ueda, our discrete zoning system causes the incremental equilibrium condition of Mohring's continuous model to be replaced by a condition maintaining constant levels spatially of the maximum realised utility $\bar{U}_k$ of each household k across the urban area and (iv) the maximum bidder is always successful at any preferred location. This implies that the poorer households (k=1) either obtain smaller lots, less accessible lots or both.

Proceeding to the notation, let: I_k = Average budget for land, commuting plus composite good of group k household, L_{ik} = Unit lot size for household of group k in zone i, X_{ik} = Consumption of composite good by household of group k in zone i, r_{ik}=Unit rent of household of group k in zone i, $\bar{g}_{ik}$ = Expected generalised commuting cost of household of group k from zone i, D_{jk}^{m} = Jobs of group k in zone j for Lowry loop m, $g_{ijk}^{\tau-1}$ = Generalised commuting cost of household of group k between zones i and centre j in previous time period, R_a = Rent of peripheral developed agricultural land, N_k = Number of locating households of type k (in time period τ), N_{ik}^{m} = Unknown number of households of type k in zone i at Lowry loop m, A_i = Available land in zone i, $\Gamma_{ik}^{L}, \Gamma_{ik}^{X}$ = Utility function weighting factors $[\Gamma_{ik}^{L} + \Gamma_{ik}^{X} = 1]$

In order for Γ_{ik}^{L} and Γ_{ik}^{X} to be assigned reasonable experimental values, key variables are made dimensionless, as follows:

$$X'_{ik} = X_{ik} / I_k \ ; \qquad L'_{ik} = L_{ik} / \bar{L}_k$$

$$\bar{g}'_{ik} = \bar{g}_{ik} / I_k \ ; \qquad r'_{ik} = r_{ik} \bar{L}_k / I_k$$

where $\bar{L}_k$ is the observed average lot size for group k.

Assuming a Cobb-Douglas utility function in terms of land and the composite good, the household allocation problem for a given rent under the budget constraint for a group k household in zone i is

$$Z = \underset{L'_{ik}, X'_{ik}}{\text{Max}} \quad \Gamma_{ik}^{L} \log L'_{ik} + \Gamma_{ik}^{X} \log X'_{ik} + \lambda_{ik}[1 - (X'_{ik} + r'_{ik} L'_{ik} + \bar{g}'_{ik})] \tag{3}$$

where λ_{ik} is a Lagrange multiplier. This yields

$$X'_{ik} = \Gamma_{ik}^{X}[1 - \bar{g}'_{ik}] \ ; \quad L'_{ik} = \Gamma_{ik}^{L}[1 - \bar{g}'_{ik}] / r'_{ik} \tag{4}$$

From the following trip distribution model (9), the quotient $(\sum_j T_{ijk}) / (\sum_{ij} T_{ijk})$ represents the probability that a commuting trip emanates from i for group k. If, in the context of a hierarchical logit model, the numerator is expressed as exp (– $\beta_k \bar{g}_{ik}$), we obtain

$$\bar{g}_{ik} = -(1/\beta_k)\log\{O_{ik}A_{ik}(\sum_j D_{jk}B_{jk}\exp-\beta_k g_{ijk}^{\tau-1})/E_k\} \tag{5}$$

where $E_k = \bar{O}_k\,\bar{A}_k\,\bar{D}_k\,\bar{B}_k\,n$, in which $\bar{O}_k$ and $\bar{A}_k$ are the averages of O_{ik} and A_{ik} over i, $\bar{D}_k$ and $\bar{B}_k$ are the averages of D_{jk} and B_{jk} over j and n is the number of employment subcentres. This generalises the normalisation criterion suggested by Fisk and Boyce (1984), ensuring that if all centres were equally accessible and equally attractive, the expected cost $\bar{g}_{ik}$ would equal the actual uniform access cost $g_{ik}^{\tau-1}$. Note also, that in the similar use of a binary logit model for modal split with parameter λ_k, $g_{ijk}^{\tau-1}$ is itself determined (after the first iteration) as the expected cost over the two modes via

$$g_{ijk}^{\tau-1} = -(1/\lambda_k)\log\{(\sum_{m=1,2}\exp-\lambda_k g_{ijkm}^{\tau-1})/2\} \tag{6}$$

where $g_{ijkm}^{\tau-1}$ is the generalised travel cost on mode m.

Substituting (4) into (3) to obtain the indirect utility function, and specifying a constant (unknown) maximum utility level $\bar{U}_k$ which is to be realised by each member of group k, the corresponding rents r_{ik}' come out as

$$r_{ik}' = \Gamma_{ik}^{L}(1-\bar{g}_{ik}')\exp-[(\bar{U}_k - E_{ik})/\Gamma_{ik}^{L}] \tag{7}$$

where $E_{ik} = \Gamma_{ik}^{X}\log[\Gamma_{ik}^{X}(1-\bar{g}_{ik}')]$. Using the rent from (7), the lot size L_{ik}' is readily obtained from (4).

The utility level $\bar{U}_k$ is adjusted iteratively to accomplish market clearing with respect to group k, whilst the urban periphery adjusts such that the rent paid at the outermost zones is just greater than the agricultural rent R_a. In this deterministic case, the market rent r_i is the highest bid rent r_{ik} offered by any group k, that is, $r_i = \max_k(r_{ik})$, where group k obtains all the land in that small zone. If $(i)\bar{k}$ is the set of zones where k is the highest bidder, the current number $\tilde{N}_k$ of households accommodated is

$$\tilde{N}_k = \sum_{(i)k} A_i / L_{ik} \tag{8}$$

which is brought to the market clearing result $\tilde{N}_k = N_k$ by adjusting the utilities $\bar{U}_k$ simultaneously for all groups k. For the successful household class k in each zone i, the number N_{ik} of households accommodated is A_i/L_{ik}.

In practice, all households in group k are not identical, and the deterministic bidding rule is too rigid. Thus, as a further option, Ueda has introduced a random bidding capability, formulated in such a way that at zero dispersion, the above deterministic result is obtained. The reader is referred to the analogous formulation of Martinez (1992).

19.3.6 Interaction between Job Choice and Housing Choice

Given the jobs D_{jk} of class k in zones j and the workers O_{ik} generated by the households N_{ik} of class k in zones i, one may calibrate and apply the Wilson-type doubly-constrained gravity model as

$$T_{ijk} = O_{ik} A_{ik} D_{jk} B_{jk} \exp - \beta_k g_{ijk} \tag{9}$$

where the balancing factors A_{ik} and B_{jk} are given in the usual recursive form. Then, the modal split and trip assignment steps can be implemented. Whilst the former is a quite routine binary logit model, the latter is decribed in detail in the next section.

Whereas the above procedure is in the Lowry tradition, it blurs any distinction between those who, in any one period, change house, change job, change both house and job or stay put. Both Mackett and Wegener have criticised this approach, as with O_{ik} including existing plus new minus departing households and with D_{jk} representing the same net total for jobs, there is no enforced consistency with observed levels of housing mobility and job mobility. At this stage, not being sure of the implied degree of approximation, we have retained the above simple approach. However, if this can be shown to produce errors of a greater order than those already present in this preliminary design model, we will make the required improvements.

19.3.7 Equilibrium Traffic Assignment in SUSTAIN

Traffic loads on the SUSTAIN road network are modelled using a deterministic equilibrium traffic assignment procedure. The required input for this procedure is a node-link network description and one (or more) origin-destination trip matrices. Note that, with the use of travel conditions of the previous time period in the location models, the assignment model is not activated at each Lowry loop m, but just once at the end of each time period τ.

The equilibrium assignment model for fixed (inelastic) travel demand is an expression of Wardrop's first principle, whereby each driver attempts to use the minimum travel time path for her/his journey, and the solution to the assignment problem is theoretically stable. The equilibrium assignment model formulation provides a useful macroscopic simulation of travel on a metropolitan network, and may be written as the following non-linear optimisation problem, for which a convergent solution may be found (as indicated, for example, in Taylor (1984)):

$$Z = \min\{\sum_e \int_o^{q(e)} c_e(x)dx\} \tag{10}$$

subject to the continuity of flow constraints

$$T_{ij} = \sum_{r} X_{rij} \qquad \forall i, j \tag{11}$$

and

$$q(e) = \sum_{ijr} \delta_{eijr} X_{rij} \qquad \forall e \tag{12}$$

where

δ_{eijr} = 1 if and only if e is in path r from i to j

= 0 otherwise

X_{rij} is the number of trips using path r between i and j, q(e) is the total flow on link e and the function $c_e(q)$ is the congestion function for link e.

A number of functional forms relating travel conditions to traffic flows at the link level are available (e.g. see Rose et al. (1989) for some reviews of such functions). One suitable function is the Davidson congestion function, which in its most practical form is

$$c = c_0 \left\{ 1 + J \frac{\mu}{1-\mu} \right\} \qquad \mu < \rho$$

$$= c_0 \left\{ 1 + J \frac{\rho}{1-\rho} + \frac{J}{(1-\rho)^2} (\mu - \rho) \right\} \qquad \mu \geq \rho \tag{13}$$

where c is the link travel time, c_0 is the free-flow link travel time, μ is the volume-capacity ratio and J is an environmental parameter that reflects the road type and abutting land use development (and hence the level of friction within the traffic stream). The volume-capacity ratio is defined as the ratio of traffic volume (q) to link capacity (S). The linear extension of the curve for $\mu \geq \rho$ (where $\rho < 1$ is a pre-determined constant, usually in the range (0.85, 0.95)) provides a finite definition of the function for all finite volume-capacity ratios. It also allows for over-saturation of the link (see Taylor, 1984), and has been implemented in SUSTAIN. The parameters are based on observed data for different classes of urban roads, with the following exceptions:

(a) 'local street' links exist in a SUSTAIN network only implicitly to allow for traffic movement from the (nominal) zone centroids to the surrounding arterial road network, and thus the Davidson function parameters for this road type are set not only to allow for such movement, but allow for it completely, yielding local access movement to and from the arterial road network at a constant speed of 20 km/h, and

(b) a 'guideway' road type has been included as an option in the model, to allow for consideration of future automated or 'intelligent highway' links in the network. The parameters representing the guideway road type are nominal only, but reflect (i) an increase in lane capacity of such a facility, due to the reduced inter-vehicle headways possible, and (ii) the reduced build-up of delay on the guideway, due to its automatic control mechanism, which is reflected in a small value of the J parameter. On the other hand, free speed on the guideway is taken as equivalent to that on a conventional freeway, reflecting the vehicle technology.

SUSTAIN Mark I models the travel demand pattern for the circular city by making full use of the inherent radial symmetry of that city structure. Considerations of symmetry reduce the circular city to a 'wedge' between adjacent radial routes, with segments of the orbital routes between them. The traffic assignment submodel takes advantage of this computationally, for it needs to find alternative routes across the entire city. For instance, for cross-town trips, the choice between the use of a pair of arterial roads (one to approach the city centre and one to move from that centre to the destination) or a set of orbital and radial routes (possibly using radial links to move firstly to an outer orbital and then to come back towards the destination) cannot be readily assessed from the wedge alone. At the same time, the radial/circular geometry permits a *compact* network and zonal description, which, in particular, facilitates *network design coordinated with land use changes.*

Thus, the model needs to generate a node-link-centroid description of the city structure, to which it can assign the travel demand revealed by the O-D trip matrices. An automatic network coding procedure has been developed to this end, in which the model is fed the following information: (a) the outer radius of the city; (b) the number of radial routes in the city. These are assumed to be distributed symmetrically; (c) the number of orbital routes in the city and the spacing between orbitals along the radial routes; (d) the number of zone centroids between the radials, and (e) the number of zone centroids between the orbital routes.

The SUSTAIN city wedge is then made up of the area between adjacent radials, including segments of each orbital route and zone centroids. Once this basic geometric structure of the radial-orbital network is defined, the user then specifies the road type and capacity (number of lanes) of each link (road segment) in the wedge. This information is then used to code the full city network. Note that any road segment can be omitted, by specifying it as having zero lanes. Although the link (road segment) remains in the network configuration, it has no capacity and thus attracts no traffic. In this way the effect of the removal of particular link, or the need for inclusion of a given road segment, can be examined by the model.

19.4 Results and Conclusions

Because of the concentration on model development, the number of applications is still not large. Nevertheless, it is possible to summarise results from a study by Roy et al. (1995), where the influences on urban energy efficiency of both changes in urban residential density and job decentralisation were simulated. Whereas SUSTAIN was used to define the city, handle housing location and model travel behaviour, employment was normatively allocated about housing using the TOPAZ optimisation module developed by Brotchie and others. The performance of low density 'sprawl' cities with dispersed employment and a strong freeway network was compared with that of medium density 'consolidated' cities with a high standard of transit. As expected, so long as attractive public transport was available in the denser cities, any incipient congestion caused by loading an increased number of commuters onto the same configuration of roads was mitigated by a strong modal shift to transit. In fact, in the rather extreme levels of consolidation which were examined (e.g. complete infill for an addition of one million to a city of four million population), per capita transport energy consumption reduced by 17% in comparison with the low density city. However, such policies work very slowly. If faster-working TDM (Transport Demand Management) policies, such as road pricing or increased fuel prices, were also to be introduced, the study illustrated that the resulting long run tendencies for workers to either change house or change job dramatically reduces the energy benefits of consolidation. At the same time, the best sprawl scenario with jobs located via the TOPAZ optimisation module was very interesting. The city was characterised by a large CBD with high modal split to radial transit, together with a strong concentration of outer suburban jobs located at such radii that efficient *reverse commuting* by car occurred on outer radial links, together with efficient ring road access to subcentres. If para transit modes (e.g. mini buses) were readily available in outer areas to encourage use of express buses to the outer circumferential subcentres, the primary energy consumption of such a low density city may be less than that for a transit-oriented medium density city, unless electricity generating energy losses could be reduced considerably.

As SUSTAIN moves more into its Mark II role of the preliminary analysis of actual cities, one may have to increase the sophistication of the housing and employment location models. The housing model may have to introduce bias terms to relate rents to factors other than job accessibility (e.g. access to schools, parks, beaches, etc.). Alternatively, a hedonic approach could be considered. For the retail model, prices and agglomeration economies could be added, as well as endogenising trip generation (Roy, 1996). The latter feature would become increasingly important for urban energy evaluations, where the impacts of more frequent shorter trips (caused by more service decentralisation) would need to be compared with those of less frequent longer trips and associated modal split changes.

The increasing availability of cheaper telecommunications and the Internet is expected to have large effects on travel and locational behaviour. For instance, despite the relative increase of work-based shopping trips enhancing the role of the work trip, an increase in telecommuting will work in the opposite direction. However, whilst it is likely that telecommuting will continue to increase, work activity remains the most demanding task confronting individuals, and a certain frequency of face-to-face contacts will probably be indispensable. If both tele-shopping and telecommuting nevertheless increase rapidly, and the ageing population and increasing automation in industry provides more and more leisure time to the community, the share in urban energy of social and recreational travel may increase dramatically. Also, in cities such as Tokyo, car traffic remains heavy throughout the day, mainly because of the car's flexibility in making dispersed business contacts. The modelling of social, recreational and business trips has not been high on the agenda of transport planners, and is not included in the current version of SUSTAIN. However, these trips do seem likely to increase in relative importance, and model formulation and calibration should be encouraged. The modelling of urban freight deliveries is another field which needs more research, especially in regard to alternative scheduling arrangements.

Acknowledgments

The authors would like to thank Anton Weller, CSIRO, for valuable assistance in the C++ programming of SUSTAIN. They also appreciated constructive advice from the editors and reviewers.

References

Anderson, M., Roy, J.R. and Brotchie, J.F. (1986), 'Some aspects of adaptability of alternative urban configurations', *Environment and Planning B*, vol. 13, pp. 305–318.

Fisk, C.S. and Boyce, D.E. (1984), 'A modified composite cost measure for probabilistic choice modelling', *Environment and Planning A*, vol. 16, pp. 241–248.

Harris, B. and Wilson, A.G (1978), 'Equilibrium values and dynamics of attractiveness terms in production-constrained spatial interaction models', *Environment and Planning A*, vol. 10, pp. 371–388.

Marksjö B. (1970), 'Gravitationsmodellen' (The gravity model), in *Matematiska modeller för fysisk planering*, Meddelande 3, Regionplanekontoret, Stockholm.

Martinez, F. (1992), 'The bid-choice land use model: An integrated economic framework', *Environment and Planning A*, vol. 24, pp. 871–885.

Mohring, H. (1993), 'Land rents and transport improvements: Some urban parables', *Transportation*, vol. 20, no. 3, pp. 267–283.

Rose, G., Taylor, M.A.P. and Tisato, P. (1989), 'Estimating travel time functions for urban roads: Options and issues', *Transportation Planning and Technology,* vol. 14, no. 1, pp. 63–82.

Roy, J.R. (1992), 'Transport efficiency in cities with subcentres', *Selected Proceedings of 6th World Conference on Transport Research,* vol. 1, pp. 291–302.

Roy, J.R. (1996), 'A more consistent retail demand model: Integration of discrete and continuous choice', submitted to *Geographical and Environmental Modelling*.

Roy, J.R., Marquez, L.M. and Brotchie, J.F. (1995), 'A study on urban residential density versus transport energy consumption', *Proceedings of Eastern Asia Society for Transportation Studies (EASTS),* vol. 1, no. 1, pp. 11–24.

Roy, J.R., Marquez, L.M., Taylor, M.A.P., Ueda, T. (1996), 'SUSTAIN – A model investigating Sustainable Urban STructure And Interaction Networks', in Hayashi, Y. and Roy, J.R. (eds), *Transport, Land-Use and the Environment*, Kluwer, Dordrecht, pp. 125–145.

Taylor, M.A.P. (1984), 'A note on using Davidson's function in equilibrium assignment', *Transportation Research B*, vol. 18, no. 30, pp. 181–199.

Wegener, M. (1996), 'Reduction of CO_2 emissions of transport by reorganisation of urban activities', in Hayashi, Y. and Roy, J.R. (eds), *Transport, Land-Use and the Environment,* Kluwer, Dordrecht, pp. 103–124.

Yinger, J. (1992), 'City and suburb: Urban models with more than one employment center', *Journal of Urban Economics*, vol. 31, pp. 181–205.

20 An Interactive Computer System for Land-Use Transport Analysis

Geoffrey G. Roy
Department of Engineering
Murdoch University
Murdoch, WA 6150, Australia

Folke Snickars
Department of Infrastructure and Planning
Royal Institute of Technology
S-100 44 Stockholm, Sweden

20.1 Introduction

Computer-based modelling provides a range of tools for the analysis of urban and regional systems. The study of these systems has been central to the evolution of modern regional science and the availability of computer technology has allowed many models to be built and computations performed that would have otherwise been impossible. The computer has therefore opened up many opportunities for regional scientists. The question that is posed therefore concerns the usefulness of the classical modelling methods and the expectation placed on the emergent modelling generation.

Many computer systems are aimed at being comprehensive, i.e. they attempt to realise complex models which embody a wide range of attributes and inter-relationships. To be meaningful and well founded they need to include a wide range of attributes including representative supplies and demands, and constraints on development. In essence this means that we must not only model land-use and the relationships that exist between land-use elements, but also the means of communication amongst these elements. Modelling methods should thus involve both land-use and transportation systems and their interrelationships.

While regional scientists may be able to maintain a grasp of their complex models, practicing planners are often less experienced and the more abstract concepts are not always fully understood and appreciated. Modelling systems often imply theoretical concepts and data requirements which require specialist staff to implement and maintain the tools. In view of the substantial development that has occurred in the modelling field surprisingly little has become best application practice.

Our aim in performing the current research is to improve the accessibility of computer technology to users who may not have the time, resources or technical skill to implement a fully fledged modelling system. In the paper, we present the conceptual framework and computer implementation of a highly interactive modelling system for combined land-use and transport analysis called MacNetwork. We intend that the MacNetwork modelling system will provide a toolbox for performing a range of simple analyses on urban and regional systems, and allows models to be swiftly developed, validated and applied without the normal overheads.

Naturally some theoretical compromises are necessary to achieve acceptance and ease of use in such an applied context. The range of capabilities and limitations of MacNetwork in this context are the focus of this paper.

Traditionally regional modelling systems have been developed by and for use by back-room technocrats. While many sophisticated modelling systems have been developed, most have not survived outside the direct control of the original developers. When we developed the ISP (Interactive Spatial Planning) system, Roy and Snickars (1993), a modelling system for urban and regional planning integrating land-use and transportation using an entropy-based forecasting method, the ambition was to streamline the system for practising planners. While we were partly successful, from our experience with the ISP system it is clear that complex models (even if the system itself is highly interactive) are not easily accepted by many planners.

The underlying problems of long term acceptance concern the complexity of the models and the computer systems in which they are implemented. While many such models may have well founded theoretical bases, they often do not instil a high level of confidence in the planners and decision-makers who make the ultimate decisions for urban and regional policy. Proper packaging of scientific knowledge is thus important in promoting the knowledge base of policy decisions.

The concept of small scale (desk-top) modelling strategies has been explored by a number of authors, see for instance Newton et al. (1990). Here we saw for the first time the view that not all modelling for urban and regional planning need involve complex and highly coupled modelling methods. The trade-off between complexity and accessibility was clearly identified and a number of useful, though relatively simple, micro-computer based systems were described.

In more recent years we have seen the rather universal penetration of Geographic Information Systems (GIS) into planning practice. For instance, Scholten (1988), Batty (1993), Openshaw (1993) and Arentze et al. (1996) describe the range of capabilities and potential applications of GIS in urban and regional planning.

The basis for the work presented in this paper is predicated on the assumption that techniques from interactive computer graphics (as well typified in comprehensive GIS systems like ARC/INFO) provide important tools for planners that are not generally present in classical modelling systems. We are also aware from experience that simple spread-sheet packages are in wide-spread use in planning agencies. This is not because of their sophisticated computational capabilities, but rather because of their immediate accessibility. The planners can do it themselves, relatively quickly and without the need to ask the expert. It may be that this type of computer technology offers an unfounded degree of confidence, but nevertheless it indicates the impact of inexpensive and highly accessible technologies on planning practice.

One of our tasks as researchers is to identify how some of the more well-founded regional science techniques and theories can be made available to planners without large-scale overheads. This goal is the motivation for the development of MacNetwork. This micro-computer based system is aimed at offering a range of computational tools in an environment that can be fully managed by planners and decision-makers themselves. We also feel that the tool will have a constructive role to play in educational programmes.

MacNetwork is not really a GIS, or a spread-sheet package, or a comprehensive land-use or transportation modelling system, but it embodies elements of each. In some ways it might be considered a compromise, in other ways it does not have a correspondence at all. We think that it provides an interesting set of capabilities that could well find application in many urban and regional planning tasks.

20.2 Overview of MacNetwork

MacNetwork is a desktop computer (Apple Macintosh) package intended for the modelling and analysis of urban and regional systems. The modelling functions include two principle elements.

20.2.1 Communication Networks

The communication networks represent the transportation system within a region and are defined by a set of nodes and links between these nodes. The links define the paths of communication (e.g. road, rail, or even data links) and the nodes the places where the links meet. Nodes are thus usually associated with regional centres which approximate those places where commuters (vehicles, people, information) can enter or leave the network. Nodes can also simply be points of intersection between links in the different transportation systems.

20.2.2 Development Zones

Urban activities can take place on land that is located within a development zone. Development zones are attached to nodes in the network as they describe the potential for nodes to service the demand for land-use activities and their relationships to the communications networks. Not all nodes need to be designated as connection points between land-use and transport.

MacNetwork is highly interactive and it is intended to be used in this mode. The user sketches the components of the system and annotates the elements with appropriate data values where these cannot be inferred directly from the drawn geometry. Analysis can be done immediately, and progressively, as the model is developed. There are no overheads of having to prepare large tables of data.

The model can be easily edited to change any item, both geometric and other properties. Further analysis can be done immediately with the results available directly on the screen or exported in tabular form to most spread-sheet packages for more detailed analysis and reporting if required. It is also possible to import data from a spread-sheet package into MacNetwork, for instance, as a result of statistical analyses or forecasting performed outside the system, using data partly or completely generated within the system.

The main objective in the design of MacNetwork was to achieve an interactive online modelling system. We suggest that this will contribute to improving the overall value of regional science in practical planning applications. We also suggest that MacNetwork will have a value in allowing a user to gain qualitative insights into the region being studied that are not possible with conventional tools. While some of the analysis options are simple, they are tedious to perform by hand or only available in the commercial GIS systems or special purpose transport modelling systems. In such cases there are often significant overheads in setting up the data and learning how to use the systems before any results can be seen. MacNetwork is designed to work in parallel with spread-sheet or statistical analysis packages. The system is intended to be applicable at a range of different geographical scales of urban systems design and regional analysis.

20.3 Using MacNetwork for Transport Analysis

The most straightforward way of portraying the capabilities of MacNetwork is to describe the steps in the development of a simple model. We will use a hypothetical model and keep the explanations short ensuring that the reader can readily observe the steps and encompass the results obtained. As we proceed the various capabilities of MacNetwork will be described.

To begin, a region exists in a geographical context. The land-use and transportation network must accommodate the existing geographic features. Our hypothetical region, Alphaville, is shown in Fig. 20.1.

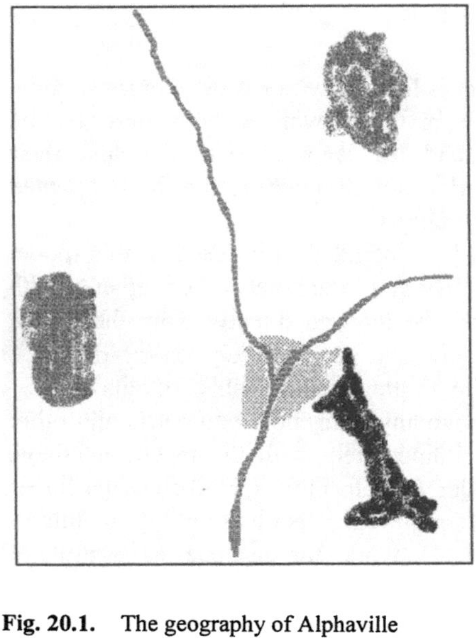

Fig. 20.1. The geography of Alphaville

Here we see the central river system, the protected forest areas (west and north east), the ridge (south east) and the wetlands area near the junction of the rivers. Fig. 20.1 is the map that is prepared for MacNetwork, either from scanning an actual map, or from a drawing made for the purpose, or some combination of both.

The map displayed here is quite simple, but for substantive applications a more complex and informative map using colour is available. The usual starting point might typically be a scanned map of the region showing the major transportation networks, the significant land-forms and the existing land-use activity. From such a map the planner gains not only considerable insight into the system to be modelled, but also an acceptably accurate set of spatial references upon which the MacNetwork model can be defined. Once the details of the land-use patterns and transport networks have been determined, the user may find it more useful to have a background map with less detail, perhaps just including the key topographical features.

The next task is to locate the nodes which are intended to describe the centres of urban activity and the places where the major communication links connect, as shown in Fig. 20.2. The grid lines are displayed to provide a guide to correctly locate the nodes relative to the background geography and to each other. Here the grid is 2 km north-south and 1,5 km east-west. The nodes are located by choosing the appropriate menu option then positioning the cursor at the required place and clicking the mouse button. Nodes can be edited (moved and deleted) as required and given a variety of internal properties. The scale of the grid can be interactively altered and adapted to the spatial resolution of the application.

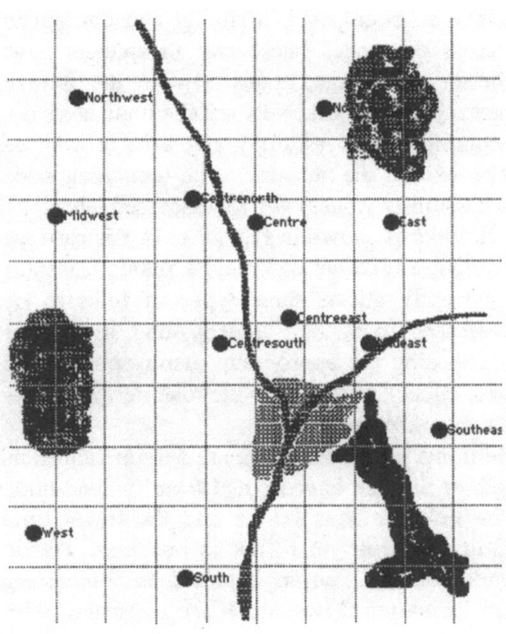

Fig. 20.2. Adding the nodes

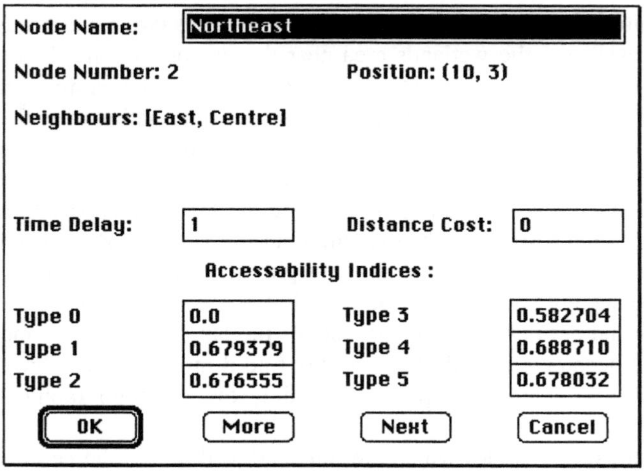

Fig. 20.3. Node property information

The properties of nodes which can be set by the user include their name, and the time and distance costs of transiting the node, as shown in Fig. 20.3. In this case the user has given the nodes names appropriate to Alphaville. Nodes also have time and distance properties to represent the fact that there are often costs in terms of time and/or distance incurred in crossing a node. The time cost for the node is an estimate of how long it takes (in minutes or some other time unit) for a commuter to pass through the node. The distance cost is a similar concept, but in terms of the distance travelled to cross the node. These two parameters have default values of zero and they can be set and changed at any stage by the user. In this case we have given a value to the time cost. In MacNetwork the node costs are incurred when moving from an in-coming link to an out-going link, as well as when entering the network (but not on exiting the network). The remaining node property items shown in Fig. 20.3 (accessibility values) will be discussed later.

The next step is to add the network links as shown in Fig. 20.4. In this case we have defined two transportation networks: a road network, and a public transport (e.g. train) network. MacNetwork currently allows three types of links to be defined, these may be used to represent road, rail, underground or similar networks. The links are defined by choosing the appropriate menu option, then marking the two end nodes (two mouse clicks). In Fig. 20.4 the road network links are shown by black lines and the rail network by grey lines.

Links may also be given some additional properties (there are default values) as shown in Fig. 20.5. The default length of the link is computed from the end-node positions. The average speed for the link can then be set and the travel time computed (or vice versa). The default flow rate on a link is one unit. Traffic forecasts are external to MacNetwork but data can be exported to forecasting models and imported again to obtain forecasted flows which are consistent with the land-use and transport information.

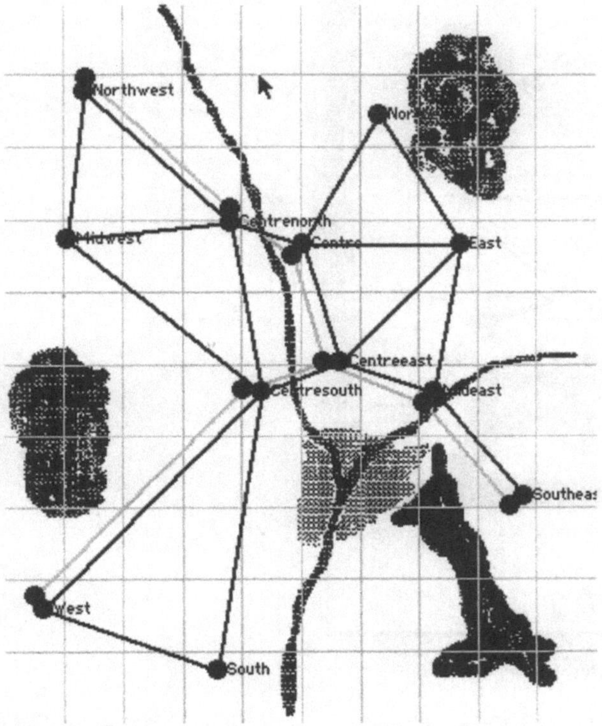

Fig. 20.4. Adding network links

The type of transport link can also be set or changed. Network links can be edited (deleted, modified) as required using menu options. Sub-networks can be attacted to one another by connecting several link types to the same node.

The two transportation networks in Fig. 20.4 are interconnected, where twin nodes are defined, by a short link which is given a travel time value and a zero distance. These links are used to represent the time-cost to change travel modes.

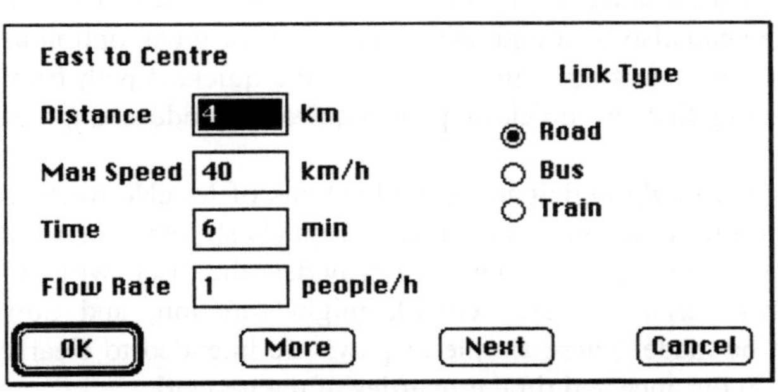

Fig. 20.5. Setting and examining the link properties

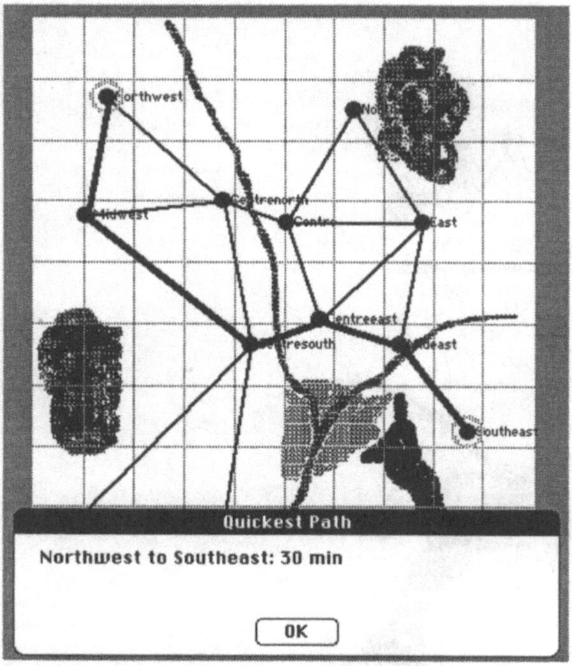

Fig. 20.6. Displaying the shortest path

The various communication networks may, or may not, be interconnected, depending on the types of analyses required.

Both nodes and links are represented by dynamically allocted data structures which faciltiate efficient access both from screen-based selection (with the mouse) as well as the needs to access adjacent nodes for the shortest-path computations.

We now have the communication network defined and a range of analyses can be undertaken. For example, Fig. 20.6 shows the shortest paths from node Northwest to node Southeast on the road network. The path length of 16 km is displayed and the path is highlighted on the network.

The shortest path between any two nodes is computed by selecting the appropriate menu option then marking the two nodes with the mouse. The path with the shortest travel time can also be found using an alternative menu option as shown in Fig. 20.7. Using both networks, Fig. 20.8 shows the quickest path from Northeast to West, indicating that the quickest path requires a mode change at Centre.

We see here the result we might anticipate in applications of MacNetwork to policy analysis. Due to the difference in travel times the quickest path is not the same as the shortest path. This type of analysis provides the user with an understanding of how this variant of the network might function, and how commuting choices might be made. These simple displays are intended to clearly show the difference between distance and time properties of the network.

The analyses in Fig. 20.6, 20.7 and 20.8 are of interest when we have realistic link properties. The link lengths are set by default to be the actual length as computed from the co-ordinates of the end-nodes. The actual length can be interactively changed to suit the users need. For instance, a simple straight line may be sufficient to represent the link on the screen, but it may be a tortuous road with many turns and thus the actual length must be specified explicitly. Once defined the length of a link can be fixed so that it will not change even if the end nodes are moved about.

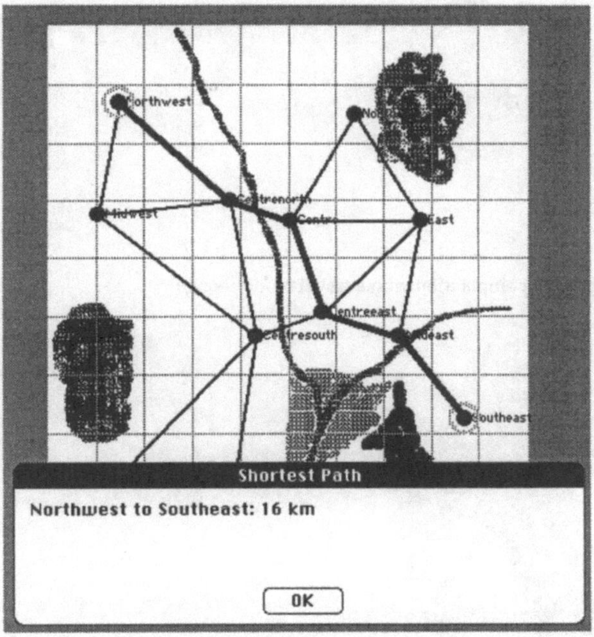

Fig. 20.7. Displaying the quickest path

To help the user observe the state of the network, the width of the drawn link can be used to denote a link property (e.g. length, speed, travel time, traffic flow). An example display is shown in Fig. 20.9, for the road network alone.

Another analysis concerns how far one can travel on the network in a given time (or distance). Such information is important in understanding the potential for commuters entering certain nodes to reach other nodes, or vice versa. This analysis is carried out in MacNetwork with the aid of a contouring analysis, based on time or distance. Fig. 20.10 shows a display of the 8 km contours from the node Centre on the road network. We can clearly see which nodes are within range, and which are not. The contours are computed along the network itself.

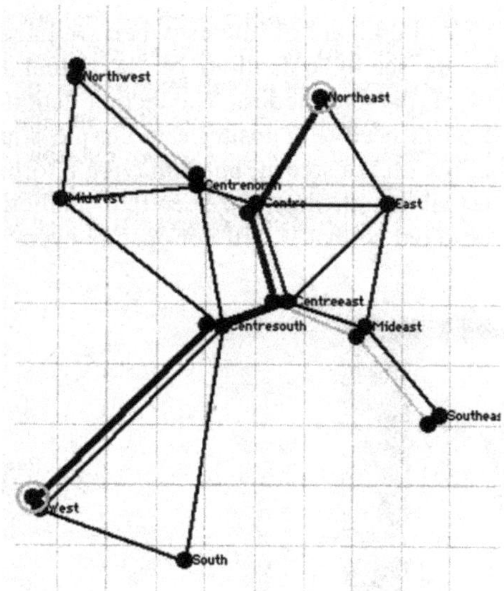

Fig. 20.8. A quickest path computation with a travel mode change

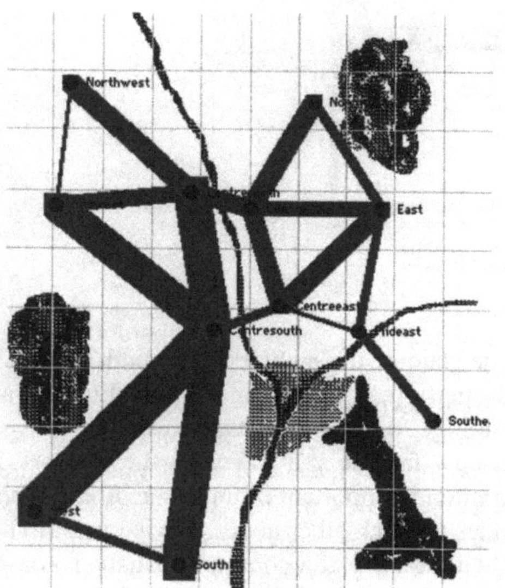

Fig. 20.9. The network drawn with link widths proportional to traffic flow

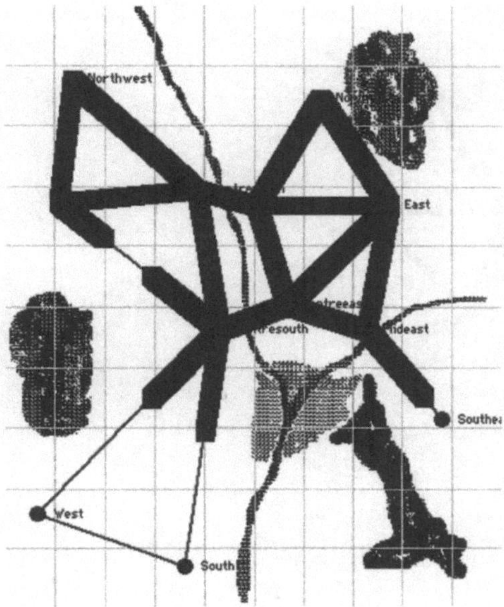

Fig. 20.10. The 8 km contour from centre

Fig. 20.11 shows a time-based plot of the 15 minute time contour from the node Centre, when using both rail and road networks and allowing mode interchange. In this case we can readily see the reachability of various parts of the region and how the two networks interact. Note that the cost of changing modes is accounted for in this case by placing appropriate time costs on the short network interconnecting links.

Contours for distance or time can be displayed for any network node, and can be overlayed as shown in Fig. 20.12. Here we can clearly see how an observer (commuter) located at a node will see the network in terms of the time or distance to reach other parts of the network. The figure shows a display with four overlayed time contours (at 5 minute intervals with varying thickness lines) from the node West. Here we can clearly see the travel time implications for a commuter originating a trip from West using the road network alone.

While these network analysis options involve quite straightforward computations they are none-the-less tedious to perform by hand even on small networks like Alphaville. For more complex examples computer methods are generally required. There are complex software products (GIS systems and transport analysis systems) which will undertake these types of analyses. MacNetwork enables the analyses to be done on the planners desk-top computer with minimal overheads and data preparation time. While the results may be approximate we suggest that they are still useful in describing the nature of the network for quick analysis and appreciation, and could prove useful to planners attempting to understand the properties of actual or planned regional systems.

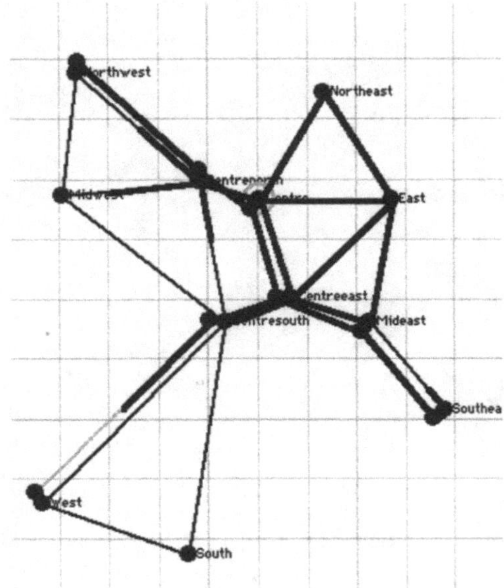

Fig. 20.11. The 15 minute time contour from centre on both road and rail networks with mode changes

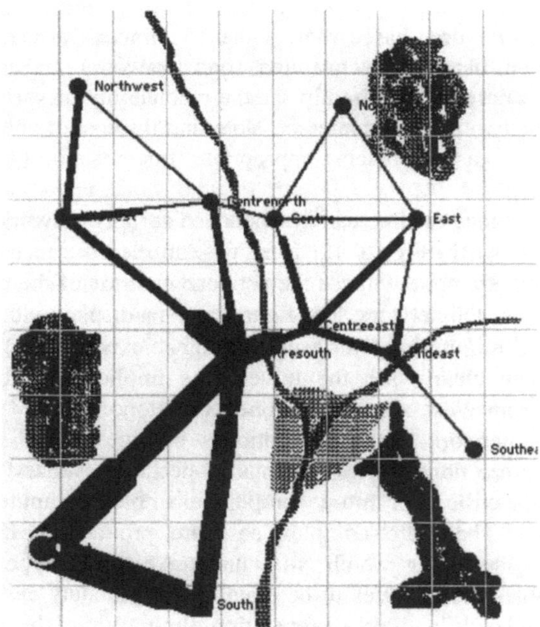

Fig. 20.12. The 5, 10, 15 and 20 minute contours from West

20.4 Using MacNetwork for Land-Use Analysis

The second major feature of MacNetwork concerns the representation of urban development through the location of various land-use activities. Urban activity is attached to nodes which provide the means of connecting it to the communication networks. A group of land-use cells attached to a common node is referred to as a development zone. Land-use is designated in one of six classes. Type 0 is generally used to designate empty but available land, and types 1 to 5 to designate land uses such as low-density residential, high-density residential, commercial, industrial, and green space.

Land-use activities are allocated to square cells of some nominal size, depending on the scale of the region. MacNetwork defines cells on a 100 by 100 zonal grid, so for the Alphaville application the cell size is approximately 150 metres square. Each cell is attached to the node which represents the location where commuters will enter or leave the communication networks from the development zone.

If the user is intending to model an existing regional system like Alphaville then the first task is to mark on the map the approximate location of the urban development cells and designate the dominant land-use they represent. The representation is intended to be schematic as indicated by the above-mentioned designation of activity classes. By this we mean that the numbers of cells allocated to each land use should approximate the intensity of actual land-use activities in the chosen units (production, persons, employment, hectares, square metres). It should also approximate the location of the selected activities so that the broad scale locational properties are observable. Thus, residential, industrial and commercial land-uses should be represented where they appear on the ground.

While all land cells are the same size on the screen, each land-use type in each development zone will have a unique scaling coefficient which converts a unit cell into the required unit land-use activity value. In this way it is possible to provide a realistic approximation to a range of urban activities which may have quite different land area demands. For the Alphaville model, the scale is such that all cells will have a meaningful association to actual land-use activities on the ground.

For larger scale systems, this approach will generally be too detailed. We can then allocate a single cell to each land-use type in a development zone and apply the appropriate average conversion coefficient to provide the correct quantity of land-use activity. The conversion factor represents a simple default method of linking land-use to urban activity levels. When making forecasts external to MacNetwork using information generated within the system it is of course possible to use any type of statistical analysis, giving rise to activity level data which can then be imported into MacNetwork.

To add a development zone, the user first selects the required node, then using menu options, cells with the required land uses can be added as shown in Fig. 20.13 for the node Northeast.

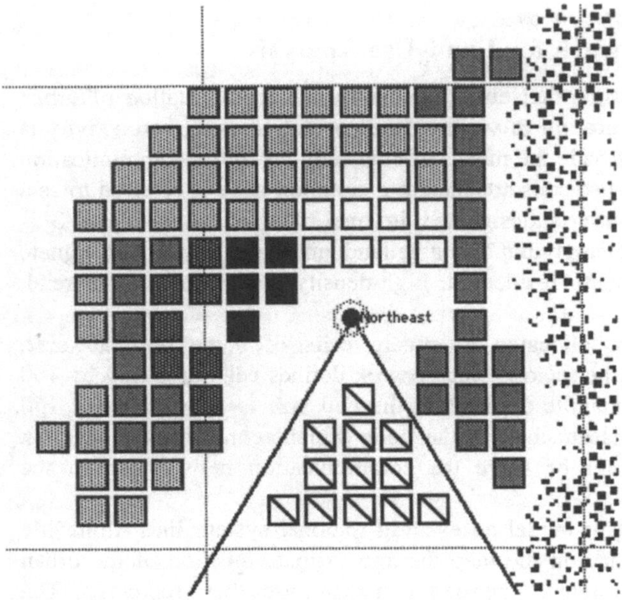

Fig. 20.13. Creating a development zone and adding land-use cells

Cell Type Name	# cells	z coeff	Value
Type 0:	12	0.0	0.0
Type 1:	10	1.0	10.0
Type 2:	38	2.0	76.0
Type 3:	29	3.0	87.0
Type 4:	7	1.0	7.0
Type 5:	5	1.0	5.0
Total:	101		185.0

Internal Time Delay: 2

Internal Distance: 1

Ok Cancel

Fig. 20.14. Summary statistics for the development zone at Northeast

The type 0 (empty) cells are designated by the cell with the single diagonal. Various levels of shading are used to designate the types 1 (white) to 5 (black). The user will just add cells to the required number and location to give a representation of the urban development at and around this node. The empty cells may be used to predefine the limits to the available land for development. Empty cells can be interactively converted into other types. A summary of the data for the node Northwest can also be displayed in a dialogue box as shown in Fig. 20.14 (the conversion factor is denoted by the letter z).

In the figure we see the computations which convert the number of cells into the appropriate land-use units via the conversion coefficient, which can be set individually for each development zone. We also see two additional properties which the user can set, the internal travel times and distances. These are intended to represent the internal commuting time and distance for the development zone itself. These properties are used in accessibility computations to reflect the fact that movement within the development zone itself will also impose a cost on the accessibility to activities within its own borders. These costs are approximated by the internal travel times and distances.

The user can proceed to define development zones at any, or all, nodes. For the case of Alphaville we see the overall result in Fig. 20.15. At this scale we cannot visually distinguish the different land-uses, but we can see the overall urban development pattern. The MacNetwork system functions zooming and panning can be used to examine each development zone in detail as illustrated in Fig. 20.13.

Once the development zones are defined we can undertake further analyses of the urban system. Some of these can be done interactively within MacNetwork whereas others demand the export of data generated within the system to an external spread-sheet package. In particular, a default method of computing accessibilities to each of the urban activities from each of the development zones has been implemented into the MacNetwork system prototype. This information may be used to determine, or influence, any further extension or changes to the allocated land-use activities within the development zones.

Accessibilities are meant to represent the relative level of access that that node has to each of the land-use activities. The information we have available in the current version of MacNetwork to interactively perform the accessibility calculations is a combination of the node and link properties from which shortest/quickest paths can be computed for each pair of nodes, and activity values for each land-use type at each node. Employing this information we could implement a range of accessibility measures, but for the moment only a limited number of measures have been included.

The accessibilities can be computed either on a distance, or on a travel time, basis using the following expression:

$$a_{ik} = \frac{\sum_j n_{jk} z_{jk} \exp(-c_k d_{ij})}{\sum_j n_{jk} z_{jk}}$$

In the formula, a_{ik} is the accessibility from development zone i to land-use activity k, n_{jk} is the number of cells of land-use type k in development zone j, and z_{jk} is the density coefficient to convert a cell of type k to the land-use activity level in

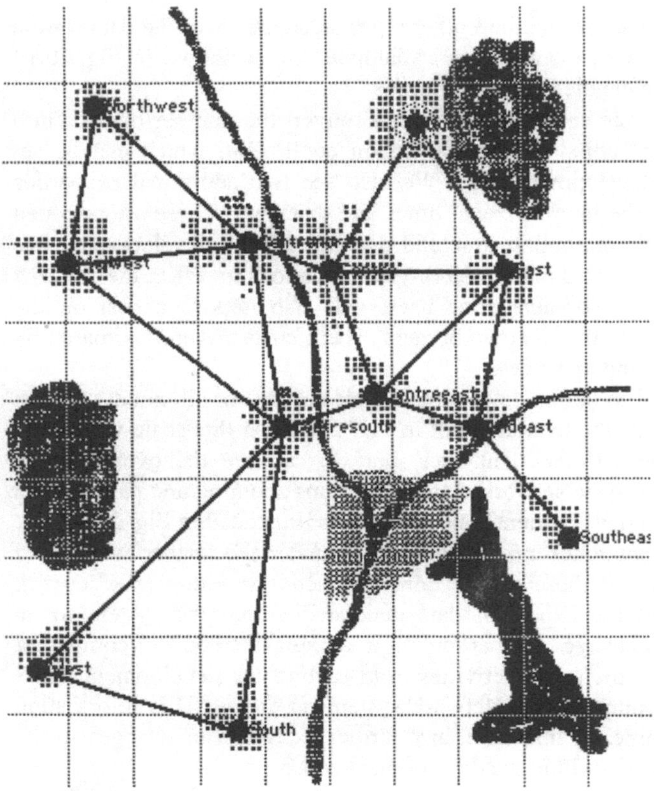

Fig. 20.15. The development zones for Alphaville

zone j. Also, c_k is the distance-deterrent coefficient for land-use type k and d_{ij} is the shortest distance, or travel time, from node i to j.

These simple accessibility measures will allow a user to compare each development zone as regards their potential to attract new urban activity or to generate demands for commuting on the network. The user can choose to use either distance- or time-based assessments. The distance-deterrent coefficients determine for each land use type the decline in perceived accessibility as a function of commuting distance, or time, and can be set by the user for each land-use type.

The accessibility calculations can be viewed as providing the user of MacNetwork with information which helps characterise the properties of the regional system. To see the results of the computations, each node can be interrogated and the results displayed, as shown earlier in Fig. 20.3. The user can also display the results graphically by drawing the network with the size of the nodes representing the relative values of the accessibility indices as shown in Fig. 20.16.

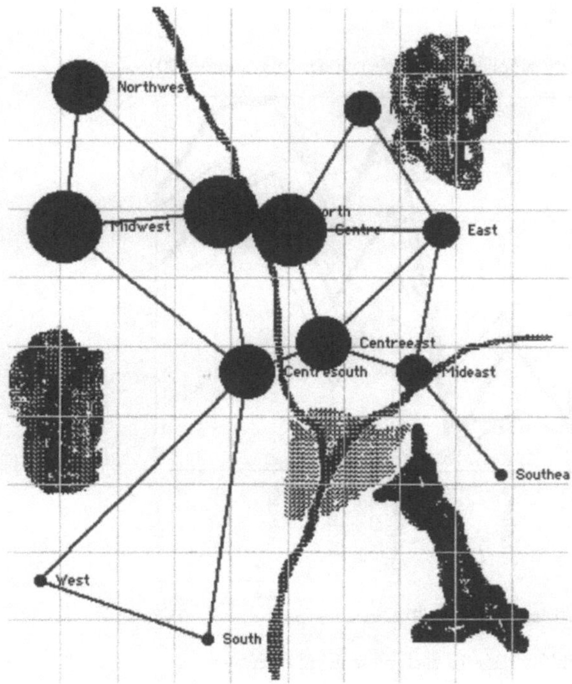

Fig. 20.16. Nodes drawn to show relative accessibility to land-use type 5

While MacNetwork provides a range of graphic displays which enable the user to visualise most of the data and properties of the model, there will always be cases where further reporting and analysis is required. To facilitate this process, MacNetwork provides options to export data and results into external files containing tables of data which can be imported directly into standard spread-sheet packages and any other suitable package which can import a tab-delimited text file.

In this fashion the interactive properties of MacNetwork may be extended to established spread-sheet environments where business graphics packages can be employed to provide further planning analyses. Since calculations made external to MacNetwork based on exported data can be imported back into the system after alteration further accessibility or other analysis may be performed by reserving accessibility indicator slots for such externally manipulated data.

From a table of node data, Fig. 20.17 shows a graph plotted using Excel, depicting the accessibility indices of each development zone for each of the land-use types 1 to 5. Here we can use the more flexible graphing capabilities of Excel to show the variation of accessibilitiy indices for all land-uses at each development zone. We can now clearly see which zones are well, or poorly, endowed with opportunities. This information can be used for some further formal, or informal, analysis which might indicate changes to the land-use activities.

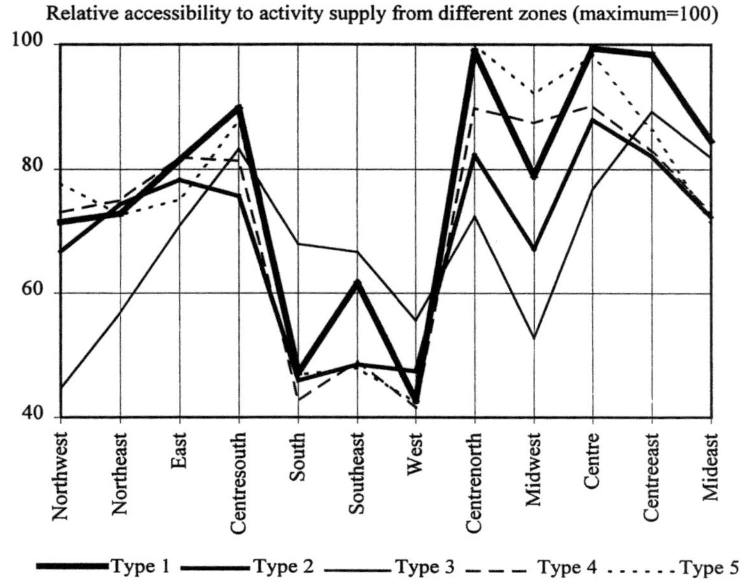

Fig. 20.17. Graph showing accessibility data for each node using Excel

Accessibility measures of this type are common ways of assessing the relative attractiveness of various parts of the regional system to activity development. In the context of MacNetwork, we have described these attractivities in simple terms. Indirectly, attractivity determines growth and change in the regional system. It would then seem natural to extend the capabilities of Mac Network to encompass the dynamic allocation of land-uses to zones, and the subsequent allocation of traffic to links. It would also be warranted to consider the process of growth and decline of urban activities in general.

At this stage, we have not taken the step to include these dynamic mechanisms into MacNetwork. Instead, we have dealt with the dynamics of change in a separate computer model of cellular automata type under the name of CityLife, Roy and Snickars (1996), see also Câmara et al. (1996), Sanders (1996), Xie and Batty (1996) and Roy et al. (1997). Our goal of producing a simple desk-top tool would be counteracted by the increase in system complexity associated with a full-fledged dynamic urban model. We would prefer to see the user apply a range of external sources of information to undertake dynamic analysis with the help of MacNetwork.

20.5 Conclusions

MacNetwork has been described via a guided tour through the capabilities of the computer system. The system has been designed through direct programming input. This gives the possibility to redesign the system, add new features, and include new indicators to evaluate the performance of regional systems without relying on the predetermined capabilities of commercial software systems. We would suggest that a package like MacNetwork has a number of potential uses:

- Assist planners in performing analyses of land-use/transportation systems with the view towards improving their knowledge of the region under examination;
- Enable regional analysts to develop models of regional systems which may be used to demonstrate the need for thorough study or to identify important system properties,
- Contribute to teaching courses in regional modelling and quantitative regional analysis relating to the basic principles of regional systems and their behaviour.

In a more general sense MacNetwork provides a set of tools for decision support, where the users can be the planners themselves. It is not necessary to be a computer expert to use the system. It has a level of complexity comparable to the commonly used spread-sheet packages. It is our belief that simple packages, like this one, are likely to find greater acceptance in planning practice than some of the more comprehensive black-box alternatives.

MacNetwork is not intended to replace the comprehensive land-use and transportation modelling systems. It does, however, provide a level of integration between these two principle components of urban and regional modelling, but with some level of compromise in sophistication. It does offer a set of tools to quickly and easily develop models of a range of regional systems using simple interactive drawing techniques. A user is then able to undertake analyses of activities and networks which will provide insights not normally available from manual techniques, and only otherwise available from more complex GIS software systems.

MacNetwork is a working tool intended to improve the understanding of urban and regional systems. By providing ready access to these tools we might hope to help raise the general appreciation of the complexity of regional modelling to a range of planners and students who may not otherwise have appreciated these tools.

References

Arentze, T.A., Borgers, A.W.J. and Timmerman, H.J.P. (1996), 'Integrating GIS into the planning process', in Fischer, M.M., Scholten, H.J. and Unwin, D. (eds), *Spatial Analytical Perspectives on GIS*, Taylor & Francis, London, pp. 187–198.

Batty, M. (1993), 'Using geographic information systems in urban and regional planning and policy-making', in Fischer, M.M. and Nijkamp, P. (eds), *Geographic Information Systems, Spatial Modelling and Policy Evaluation*, Springer-Verlag, Berlin, pp. 51–69.

Câmara, A., Ferreira, F. and Castro, P. (1996), 'Spatial simulation modelling', in Fischer, M.M., Scholten, H.J. and Unwin, D. (eds), *Spatial Analytical Perspectives on GIS*, Taylor & Francis, London, pp. 201–212.

Newton, P., Taylor, M. and Sharpe, R. (eds) (1990), *Desktop Planning: Advanced Microprocessor Applications for Physical and Social Infrastructure Planning*, Hargren Publishing, Melbourne.

Openshaw, S. (1993), 'Some suggestions concerning the development of artificial intelligence tools for spatial modelling and analysis in GIS', in Fischer, M.M. and Nijkamp, P. (eds), *Geographic Information Systems, Spatial Modelling and Policy Evaluation*, Springer-Verlag, Berlin, pp. 17–33.

Roy, G.G. and Snickars, F. (1993), 'Computer-aided regional planning: Applications for the Perth and Helsinki regions', in Fischer, M.M. and Nijkamp, P. (eds), *Geographic Information Systems, Spatial Modelling and Policy Evaluation*, Springer-Verlag, Berlin, pp. 181–197.

Roy, G.G. and Snickars, F. (1996), 'CityLife: A cellular automata model of urban dynamics', in Fischer, M.M., Scholten, H.J. and Unwin, D. (eds), *Spatial Analytical Perspectives on GIS*, Taylor & Francis, London, pp. 213–228.

Roy, G.G., Snickars, F. and Zaitseva, G. (1997), 'Simulation modelling of location choice for buyers and sellers in urban systems', Working Paper, Department of Infrastructure and Planning, Royal Institute of Technology, Stockholm.

Sanders, L. (1996), 'Dynamic modelling of urban systems', in Fischer, M.M., Scholten, H.J. and Unwin, D. (eds), *Spatial Analytical Perspectives on GIS*, Taylor & Francis, London, pp. 229–244.

Scholten, H.J. (1988), 'The role of geographical information systems in physical planning', in Giaoutzi, M. and Nijkamp, P. (eds), *Informatics and Regional Development*, Avebury, Aldershot, pp. 292–307.

Xie, Y. and Batty, M. (1997), 'Automata-based exploration of urban form', *Geographical Systems*, vol. 4, pp. 83–102.

21 A Combined Economic Activity and Transportation Model: A Solution Procedure and Application to Multi-Regional Planning

Paul F. Hanley
Tschangho John Kim
Department of Urban and Regional Planning
University of Illinois at Urbana-Champaign
111 Temple Buell Hall, 611 E. Lorado Taft Dr.
Champaign, IL 61820, USA

21.1 Introduction

In this paper, an integrated economic activity and transportation equilibrium model is developed in which the interaction between the public and private sectors are explicitly represented by simultaneously minimizing transportation costs, production costs, and import and export costs. The assumed role of the public sector is to develop highways that connect import and export ports zones to the locations where the private sector produces most economically. In the proposed model, the optimum amounts and locations of production, interzonal traffic flows, and port volumes are determined.

The first objective is to model regional economic activity by integrating transportation with input-output analysis. To achieve the objective, an economic activity model is developed based on an integrated transportation and land-use model formulated by Azis and Kim (1992) and Hanley (1994). The model contains non-linear equations, is static, and provides a heuristic solution. The solution represents the minimum aggregate cost of economic activities nation-wide.

The second objective is to use commercially available software packages to solve the proposed model. Therefore, data transferring interfaces are developed for the dissimilar software packages. The interfaces allow modeling scenarios to be generated by the user without specific knowledge of the data format requirements of the individual software packages.

The planning problem is to identify the optimum transport investment options available that facilitate economic growth, meet export targets, and at the same time enhance the welfare of a nation. A few large scale non-linear transportation and economic activity models have been solved, but only with a super-computer. Since the use of a super-computer is not a viable option for most planners there is a need to develop solution methods not tied to one. Therefore, a solution procedure using a PC or a workstation is desirable. Also, a desirable feature of a solution procedure is user friendliness because many potential users are not familiar with all the subject areas contained in the model such as input-output analysis, linear and non-linear programming solution algorithms, transportation network user-equilibrium analysis, and the integrated nature of transportation and economic activities.

21.2 An Integrated Economic Activity and Transportation Model

Combined trip assignment and distribution model formulations, such as the model formulated by Beckman et al. (1956) as reported by Boyce et al. (1988) require the identification of paths between origin and destination pairs (incident matrix) and the identification of all the links contained in these paths. This information is then stored in a matrix and later used to assign vehicle trips along the paths. Constructing such a matrix where the cell entry equals 1 if the link is an element of a path between origin i and destination j is not difficult for a small-scale network. However, for a large-scale network the matrix becomes prohibitive because of the numerous possible paths between each origin destination pair in an actual highway network.

To overcome the requirement for this matrix, algorithms that rely on minimum paths between origin and destination pairs have been proposed by Evans (1976), Florian (1977), Florian and Nguyen (1978), Kim (1989), and Azis and Kim (1992). As an example, the algorithm proposed by Eash et al. (1979) has a five step process that is a weighted combination of a sequence of all-or-nothing assignments. This procedure requires an iterative process that calculates and recalculates the minimum paths based on updated congestion measurements.

Opting to use such an iterative equilibrium trip assignment algorithm leads to the reformulation of the integrated transportation and land-use model formulated by Azis and Kim (1992) and Hanley (1994). A hybrid approach is proposed here because there is no commercially available software package that solves this type of iterative non-linear mathematical programming. This approach splits the model into distinct mathematical programming components. Each component is solved sequentially with the results of the previous component supplied as the input to the current solution. The process continues until the difference between the current iteration and the previous iteration of objective function in the main program is within a given tolerance.

21.3 The Formulation of the Model

The solution method proposed in this paper, requires the model be developed in two distinct mathematical components, a linear main program and a non-linear sub-program. The main program has an overall objective to minimize the total inter-zonal commodity shipping costs, production costs, and port operating costs using constant cost coefficients. The main program is constrained to meet national and zonal final demand, existing regional production, existing import and export volumes by port, maximum port capacities, and national foreign import and export goals.

The first component, the main program, is a linear program that calculates total national commodity output, production location, commodity flows, and the port of entry and exit of the foreign import and export volumes. The origin and destination of each commodity can be traced and provides the spatial data required by the second component. In the first component, production costs and port operating costs are modeled as functions of the commodity flow. To obtain production costs, the regional commodity production amount is multiplied by a constant production cost. The port operation costs are obtained in a similar manner. Unlike the transportation costs, these cost coefficients are assumed to be independent of volume.

The second component, the sub-program, is non-linear and calculates the transportation link congestion cost based on the origin and destination of the production amounts determined by the main program. The link congestion cost is calculated using an iterative equilibrium trip assignment model that seeks to minimize the users' transportation cost. Calculating the transportation costs outside the main program eliminates the requirement to enumerate all possible paths between origins and destinations. Once the solution to the equilibrium trip assignment is obtained the link costs are passed back to the main program. The main program is again solved using these new transportation costs. The process continues until the difference in the objective function of the main program from the previous iteration and the current iteration is within a given tolerance.

21.4 The Model

Component 1

$$\min \quad \Sigma_i \Sigma_j (c_{ij} \Sigma_r x_{ijr}) + \Sigma_{i \in p} \Sigma_j (c_{ij} \Sigma_r m_{ijr}) + \Sigma_i \Sigma_r p_{ir} pd_{ir} + \Sigma_i \Sigma_r d_{ir}(e_{ir} + m_{ir}) \qquad (1)$$

s.t.

$$\Sigma_i pd_{ir} = \Sigma_q b_{rq}(f_q + e_q) \qquad \text{for all r} \qquad (2)$$

$$\Sigma_j x_{jir} + pd_{ir} + \Sigma_j m_{jir} \geq \Sigma_j x_{ijr} + \Sigma_q a_{rq} pd_{iq} + f_{ir} \qquad \text{for all inland zone i and r} \qquad (3)$$

$$\Sigma_j x_{jir} + pd_{ir} + \Sigma_j m_{jir} \geq \Sigma_j x_{ijr} + \Sigma_q a_{rq} pd_{iq} + f_{ir} + e_{ir} \qquad \text{for all port zone i and r} \qquad (4)$$

$$\Sigma_i e_{ir} = e_r \qquad \text{for all r} \qquad (5)$$

$$e_{ir} \geq ex_{ir} \qquad \text{for all port zone i and r} \qquad (6)$$

$$\Sigma_j m_{ijr} \geq mx_{ir} \qquad \text{for all i and r} \qquad (7)$$

$$\Sigma_i \Sigma_j m_{ijr} = m_r \qquad \text{for all r} \qquad (8)$$

$$\Sigma_j m_{jir} \leq \Sigma_q a_{rq} pd_{iq} \qquad \text{for all i and r} \qquad (9)$$

$pd_{ir} \geq pdx_{ir}$ for all i and r (10)

$e_{ir} + m_{ir} \leq pc_{ir}$ for all port zone i, r (11)

Component 2

$$\min \sum_a \int_0^{v_a} c_a(x)dx \tag{12}$$

$$\sum_p x_{ijrp} = x_{ijr} \tag{13}$$

$$c_a = co_a(1 + 0.15(f_a/h_a)^4) \tag{14}$$

$$v_a = \sum_r g_r \sum_i \sum_j \sum_p x_{ijrp} \delta_{ijpa} \tag{15}$$

δ_{ijpa} = 1, if link $a \in$ path p between i and j
= 0, otherwise

Exogenous Variables

a_{rq}: input-output coefficients.

b_{rq}: elements of the Leontief inverse $(I-A)^{-1}$.

co_a: free flow transport cost at link a.

d_{ir}: unit port handling cost for commodity r at port i.

e_r: total export of commodity r from Jawa.

ex_{ir}: existing total export (1989) for commodity r from port i.

f_r: total household demand for consumption for commodity r in Jawa.

g_r: coefficients converting monetary units to passenger car units.

h_a: capacity of link a.

m_r: total import of commodity r for Jawa.

mx_{ir}: existing import (1989) of commodity r through port i.

p_{ir}: unit production cost for producing commodity r at zone i; r1 = agricultural products, r2 = manufacturing products, and r3 = service products; i = 1 to 82, covering all kabupatens in Jawa.

pdx_{ir}: existing production amount (1989) for commodity r at zone i.

Endogenous Variables

c_a: congested transport cost on link a.

c_{ij}: transport cost between zone i and zone j (based on congested transport cost on links from component 2).

e_{ir}: total amount to be exported for commodity r at port i.

f_{ir}: household demand for commodity r at zone i.

m_{ir}: foreign (including inter-island) import flow of commodity r through port i.

m_{ijr}: foreign (including inter-island) import flow of commodity r through port i going to zone j.

pd_{ir}: total amount to be produced for commodity r at zone i.

x_{ijr}: flow of commodity r from zone i to zone j.

x_{ijrp}: assigned commodity r flow on path p from zone i to zone j.

v_a: equilibrium traffic volume at link a.

The principle objective represented in Equation (1) is to minimize the transportation costs for the finished goods x_{ijr} from zone i to zone j, where i and j are all inland and port zones; plus the transportation costs for the intermediate goods that are foreign imports m_{ijr} from zone i to zone j, where i are port zones ($i \in p$) and j contain all inland and port zones; plus the production costs for producing good pd_{ir} within zone i, where i is all inland and port zones; plus the port handling costs for importing intermediate goods m_r and exporting finished goods e_r from and to foreign markets.

Equation (2) is an identity that requires the sum of regional production to equal the national total output for the finished good r. The national total output is generated by multiplying the Leontief inverse of the technical coefficient matrix by the sum of the household final demand and the foreign export volume. By including international imports in the Leontief matrix, equation (2) computes the volumes of domestic production. The Leontief inverse matrix, household final demand and the foreign export volumes are required inputs to the model.

Equation (3) characterizes a conservation of flow for an inland zone. It requires that the sum of finished goods imported into the region minus the finished goods exported out of the region plus the amount of finished goods produced in the region is greater than or equal to the amount of regional household final demand plus the amount of inter-industrial consumption minus the amount of foreign imports that directly substitute for inter-industrial consumption.

If no production occurs in the region no inter-industrial consumption occurs, but if one industry produces within the region it requires inputs from other industries. These inputs can either be satisfied by having those industries all produce within the region, or the region can import the required amount, or the region can substitute foreign imports directly for the inter-industrial consumption. The substitution of foreign imports require the imported goods be shipped form port zones increasing overall transportation costs. The inter-industrial consumption is calculated by multiplying the industrial technical coefficients for that sector by the level of production for that sector within the region.

Equation (4) is similar to equation (3) except it is a conservation of flow for port zones rather than inland zones. The reason port zones are treated differently is that port zones have to export goods to foreign markets, whereas the inland zones do not. So, a foreign export term is added to the right-hand-side of equation (3).

Equation (5) is an identity that requires the sum of the contribution to foreign export of each port zone equal the national total foreign export.

Equation (6) states that the assigned foreign export volume to a port zone must be greater than or equal to a minimal economically feasible amount.

Equation (7) states that the total foreign import volume to a port zone must be greater than or equal to an initially assigned volume.

Equation (8) is an identity that requires the sum of the portion of foreign imports of each port zone is equal to the national total foreign import.

Equation (9) is a constraint on the amount of foreign imports that can be substituted for inter-industrial consumption within producing regions. The equation limits the amount of substitution to be less than or equal to the amount of inter-industrial consumption required by the level of production within the region.

Equation (10) requires the assigned production volume in every zone is greater than or equal to an initially assigned volumes.

Equation (11) is an upper bound on the handling capacity of the port zones. The equation requires the sum of the foreign export plus the foreign imports be less than or equal to the upper limit of the handling capacity of each port zone.

Equations (12 - 15) comprise the second component of the model that calculates the transportation link costs based on network congestion. This program minimizes the user equilibrium costs equivalent. The cost function used is the standard Bureau of Public Roads capacity constraint formula.

21.5 Solution Approach

The second objective of the paper is to demonstrate the feasibility of solving the above model using existing software that is commercially available. Also, a goal is to demonstrate that the large-scale transportation and land-use model can be solved using a computer platform other than a super-computer. Therefore, the approach is to solve the model by integrating existing software into one package. The software selected to do this is LOTUS, a spread sheet program, GAMS, a package that consists of general purpose mathematical program solution algorithms and TRANPLAN, a transportation planning analysis program.

A LOTUS spread sheet is used to prepare the input data into a suitable format for GAMS. Household final demand and the foreign export volumes are exogenous to the model. These values and all other input to GAMS can be changed using the spread sheet and are easily integrated into the model. This allows for quick and accurate data modifications when different policy scenarios are run.

GAMS is a front end to various mathematical programming solution algorithms distributed by the Scientific Press. GAMS was developed by Alexander Meeraus and Anthony Brook of the World Bank as a high level language for formulating models with concise algebraic statements that are easily read by modelers and computers alike, easily modified, and easily moved from one computer environment to another – it is independent of the solution algorithms of specific solvers Brooke, Kendrick, and Meeraus (1992). GAMS is used to solve the first component of the model.

The final commercial software package used was TRANPLAN which is a transportation modeling software package distributed by the Urban Analysis Group. The equilibrium trip assignment is used to calculate the transportation network link costs.

21.5.1 Integrating Components 1 and 2

The integration of the individual components begins with the generation of the input data. A LOTUS spread sheet is used to create the input data read by GAMS. Also, a transportation network needs to be created using TRANPLAN. In this study the base year transportation network was used for all time periods.

A steering file first calls GAMS to begin solving component 1. With the solution obtained by the GAMS\MINOS solver, a fortran program which was developed for this particular job reads the solution and translates it into the required input format of TRANPLAN. It is noted that component 1 is solved using data in monetary units as required by the input-output system so, the commodity flows are converted from monetary units to passenger car unit equivalents (PCU) as shown in equation (15). Excluded from the commodity flows is the flow of service because services are not transferred using the highway network. Also, the initial solution to component 1 does not contain the effects of transportation costs because at this time they are unknown. After the completion of the FORTRAN program, TRANPLAN is called and calculates the transportation cost based on the commodity flows in PCUs generated by component 1. The transportation costs can reflect both passenger and freight movement if passenger trips are initially included in the transportation network model. Once the transportation costs are obtained, a second FORTRAN based program reads the output from TRANPLAN and translates it into GAMS input. GAMS is again called and solves component 1 accounting for transportation costs. The iterative process of solving the component and interchanging their results repeats itself until the stopping criterion is satisfied.

After every GAMS-to-TRANPLAN iteration, a third FORTRAN program is called to check the change in the objective function value of component 1. The steering file controls the loop by checking if the results of the stopping criterion is satisfied. It checks if the percent absolute difference between the current and the previous objective function value is greater than the tolerance begin another GAMS-to-TRANPLAN iteration, if it is less than or equal to the tolerance stop.

When the model was developed a detailed test case was used to evaluate its behavior characteristics. The test case used a four-zone three-economic-sector nation. The zones selected were the first two inland and the first two port zones that are included in the larger data set for Jawa, Indonesia. In addition, sensitivity analysis has been performed based on the test case. The results for the test case can be obtained from the authors.

The convergence of the model was studied using the test case. The results showed that the change in the objective function consistently decreased to the tolerance level selected. Although there is no way of guaranteeing this is an optimum solution it does show the stability of the model. The results of the convergence tests can be obtained from the authors.

21.5.2 Limitations of the Solution Approach

A limitation of this approach is the selection of an appropriate stopping tolerance. If the tolerance is too large the solution stops without achieving a satisfactory result; a tolerance too small could mean an unacceptably long time for a solution to be obtained, specially on a PC.

The overall limitations of the approach focus around the data requirements and modeling an economic system using mostly linear relationships. The industrial technical coefficients table required by the input-output analysis is the main driving force for the economic model and is difficult to obtain. Additionally, a critical assumption is required regarding the availability and suitability of land and labor. Currently, the model implicitly assumes that all regions have the sufficient quantity and quality of land and labor to support the forecasted development. This is derived from assuming that the given import and export targets have been constrained by the quantity and quality of land and labor in Jawa.

An alternative to an integrated model could be to run each model separately. Another alternative is to create a dynamic model that considers change over time. To avoid the linear relationships in input-output models, an econometric model with non-linear functions could be considered. However, the objective that the final version run on a PC limits the practical alternatives.

21.6 Application to Jawa, Indonesia

This section is included to demonstrate that the proposed model and solution procedure has been applied successfully to a large-scale problem as reported in Azis and Kim (1992).

The data used in the application to Jawa, Indonesia, were either obtained through official publications or independently estimated and forecasted by a research team from the University of Indonesia.

Data were collected for 82 zones corresponding to political boundaries called kabupatens in Jawa and for 10 economic sectors, but later converted to 3 sectors. The three sectors used to produce the results are agriculture, manufacturing, and services. The conversion from 10 to 3 sectors was necessary to solve the model on a PC, the 10 sector version can be solved using a Workstation platform.

21.6.1 Goods Production Results

The results are shown in Tables 21.1 to 21.8. Table 21.1 to 21.3 contain the results for the year 2005 and in Tables 21.4 to 21.8 for the year 2020. These tables selectively list forecasted zones in rank order of percentage growth for the periods of 1989-2005 (Tables 21.1 to 21.3) and 1989-2020 (Tables 21.4 to 21.6). An interesting result is (refer to Tables 21.1–21.5) that these high growth zones for both agricultural (Tables 21.1 and 21.4) and manufacturing (Tables 21.2 and 21.5) sectors are those which have been lagging traditionally in agricultural and manufacturing productions. Pandeglang, Indramayu, Cilacap, Subang, and Karawang, are the exception in which agricultural production activities have been high and still the model suggested a high activity level. Unlike in the agriculture and manufacturing sectors, high growth in service sector is suggested in the zones where traditionally service production activities currently prevail.

Perhaps the most surprising result in some regions is that the increase of production in the agricultural and manufacturing sectors is not accompanied by any increase of production in the service sector. This seems to contradict production activity based on input-output theory. But the contradiction is not true, the behavior of the service sector is exactly the same as that of the other sectors: it will seek regions where total production and transportation costs are lowest. Actually, that is what has happened in many developing countries: production in agricultural and manufacturing occurs in less developed regions but the supporting activities take place in well developed regions. This includes the primate city where production cost for the service sector is relatively inexpensive, creating agglomeration of the service sector in big cities.

In general, the model suggests some economic activity relocation in order to minimize total cost of production, transportation and port operation. This conclusion comes from the fact that rapid growth of agricultural and manufacturing production occurs in what are referred to as the poor regions. In addition, some zones where existing levels of agricultural and manufacturing productions are relatively high do not remain so according to the forecast. Consequently, the simultaneous minimizing of transportation, production, and port operation costs has indicated that the role of the poor regions in distributing goods tends to become more important, while that of the traditionally high growth regions tends to become less important. The prevailing factor behind such a forecast is that, in the developed areas, transportation costs due to congestion outweigh the advantages of having lower production costs. The significant contribution of the model is that congestion costs are endogenously determined, and at the same time, transportation costs and production costs are simultaneously considered in determining the amounts and location of productive activities.

Some of the results shown in Tables 21.1 to 21.8 may seem, at first, inconsistent and, in some cases, contradictory over time. This is particularly evident since relocation costs are not included in this model. By not including relocation costs, establishments can move without penalty to other locations in the next time period. At the same time, this may well be the impact of the congestion that causes higher transportation costs which, in turn, make that region less attractive in the next time period.

21.6.2 Commodity Flow Results

As a consequence of the high increase of production activities in the traditionally poor regions, some transportation corridors will be heavily used. Tables 21.7 and 21.8 show corridors where heavy interzonal traffic flows were forecasted by the model. The heaviest traffic is forecasted in the Sampang-Gresik and Sampang-Jakarta corridors for the year 2005. For the year 2020, Pandeglang-Kudus and Pandeglang-Jakarta corridors are forecasted to be the heaviest ones.

21.6.3 Planning Policy Implications

The model suggests that there is a limit to the growth of regions in Jawa. Beyond certain limits, any increase in production located in highly developed regions would cause intolerable overall activity congestion. Thus, the most efficient spatial development pattern in Jawa is to develop the traditionally poor regions, while considering both transportation congestion and production costs. Most of these regions are in the traditionally non productive poor regions.

21.7 Conclusion

The paper documents the development of an economic activity and transportation model based on input-output analysis and a system that merges existing software packages into one modeling package. It has been demonstrated that the model can be used for evaluating alternative regional development policy options. Moreover, the model has been implemented on a PC. The major contribution of this paper is that any office that has a 486-PC can use the model for policy evaluation.

Should further work continue, variable production and port operation costs will be included. Using non-linear econometric equations, it is felt, would provide a description of the interrelation between transportation and industrial production at a less aggregated level providing greater insight on policy decision impacts. Also, it is suggested that further software interfaces be developed and to include an expert system to assist the user in altering the data, running the model and analyzing the results.

Table 21.1. Sample model results: Agricultural production amounts (billion Rp)

		1989	2005	1989–2005	Percent/year
SAMPANG	AGR	107	89575.82	89468.82	52.29
PEMALANG	AGR	253.79	85956.5	85702.71	43.92
TUBAN	AGR	165	50373.75	50208.75	42.99
PACITAN	AGR	89	12171.54	12082.54	35.99
WONOSOBO	AGR	123	11896.62	11773.62	33.07
BLITAR	AGR	592.53	51522.11	50929.58	32.19
CIAMIS	AGR	485	19933.95	19448.95	26.14
KLATEN	AGR	251	9590.58	9339.58	25.57
CIANJUR	AGR	693	23491.59	22798.59	24.63
PANDEGLAN	AGR	8265.4	187016.55	178751.15	21.52
KENDAL	AGR	231	3118.46	2887.46	17.66
PONOROG	AGR	157	1808.96	1651.96	16.51
KARAWANG	AGR	1087.73	9768.08	8680.35	14.70
TEMANGGUN	AGR	340.98	2491.74	2150.76	13.24
CILACAP	AGR	1253.67	8831.58	7577.91	12.98
LUMAJANG	AGR	229.41	913.54	684.13	9.02
LEBAK	AGR	182.47	583.22	400.75	7.53
JEPARA	AGR	145	421.08	276.08	6.89
BANYUWANG	AGR	343	953.97	610.97	6.60
INDRAMAYU	AGR	3581.65	8519.79	4938.14	5.57
KUNINGAN	AGR	260	520.56	260.56	4.43
BOYOLALI	AGR	146	247.42	101.42	3.35
PURWOREJO	AGR	194.12	324.73	130.61	3.27
TRENGGALE	AGR	108.41	148.89	40.48	2.00
BREBES	AGR	209	232.47	23.47	0.67
SUBANG	AGR	1950.87	2029.49	78.62	0.25

Table 21.2. Sample model results: Manufacturing production amounts (billion Rp)

		1989	2005	1989–2005	Percent/year
SAMPANG	SECT3	1	305881.25	305880.25	120.22
WONOSOBO	SECT3	2	40939.79	40937.79	85.97
BLITAR	SECT3	8	13805.07	13797.07	59.33
SUBANG	SECT3	10	6407.62	6397.62	49.77
LEBAK	SECT3	31.86	1864.3	1832.44	28.96
PANDEGLAN	SECT3	16724.5	596009.43	579284.93	25.02
PURWAKART	SECT3	843	1103.4	260.4	1.70

Table 21.3. Sample model results: Service production amounts (billion Rp)

		1989	2005	1989–2005	Percent/year
SEMARANG	SERV	622	183885.74	183263.74	42.70
SUKOHARJO	SERV	195	10920.87	10725.87	28.61
GRESIK	SERV	8005.64	315746.65	307741.01	25.82
TANGGERAN	SERV	186	5699.58	5513.58	23.85
BEKASI	SERV	1133	30304.02	29171.02	22.80
SIDOARJO	SERV	1806	17906.06	16100.06	15.42
YOGYAKART	SERV	1598	8621.45	7023.45	11.11
JAKARTA	SERV	24438.23	70127.25	45689.02	6.81

Table 21.4. Sample model results: Agricultural production amounts (billion Rp)

		1989	2020	1989–2020	Percent/year
PEMALANG	AGR	253.79	369087.37	368833.58	26.48
TRENGGALE	AGR	108.41	85041.63	84933.22	23.99
BLITAR	AGR	592.53	332974.9	332382.37	22.66
PANDEGLAN	AGR	8265.4	4057583.17	4049317.77	22.13
JEPARA	AGR	145	45230.04	45085.04	20.35
PONOROG	AGR	157	29452.18	29295.18	18.39
WONOGIRI	AGR	161	21469.32	21308.32	17.10
CIANJUR	AGR	693	87766.56	87073.56	16.90
BATANG	AGR	134	15047.05	14913.05	16.45
GARUT	AGR	642	69322.98	68680.98	16.30
KARAWANG	AGR	1087.73	110828.9	109741.17	16.09
BOYOLALI	AGR	146	12083.51	11937.51	15.31
REMBANG	AGR	121	9738.68	9617.68	15.21
KEBUMEN	AGR	213	14082.17	13869.17	14.48
WONOSOBO	AGR	123	7992.26	7869.26	14.41
PEKALONGA	AGR	246	15623.2	15377.2	14.33
DEMAK	AGR	161	10202.81	10041.81	14.32
BLORA	AGR	150	9502.34	9352.34	14.32
BANYUMAS	AGR	244	15454.66	15210.66	14.32
KLATEN	AGR	251	15883.22	15632.22	14.32
PATI	AGR	236	14920.56	14684.56	14.31
KENDAL	AGR	231	14598.41	14367.41	14.31
BANJARNEG	AGR	147	9228.7	9081.7	14.29
LEBAK	AGR	182.47	11279.79	11097.32	14.23
CIAMIS	AGR	485	29968.61	29483.61	14.23
LUMAJANG	AGR	229.41	14161.36	13931.95	14.22
KUNINGAN	AGR	260	15486.01	15226.01	14.09
SUMEDANG	AGR	430	25594.94	25164.94	14.09
SUKABUMI	AGR	612	36417.43	35805.43	14.09
MAJALENGK	AGR	396	23537.34	23141.34	14.08
CILACAP	AGR	1253.67	73291.78	72038.11	14.02
MAGELANG	AGR	305	15497.57	15192.57	13.51
PURWOREJO	AGR	194.12	9504.86	9310.74	13.37
BANYUWANG	AGR	343	12670.91	12327.91	12.35
BOJONEGO	AGR	148	5398.79	5250.79	12.30
INDRAMAYU	AGR	3581.65	116053.05	112471.4	11.87
BONDOWOS	AGR	202.67	5771.24	5568.57	11.41
TEMANGGUN	AGR	340.98	9295.87	8954.89	11.25
SAMPANG	AGR	107	2873.65	2766.65	11.20
SRAGEN	AGR	434.42	10372.08	9937.66	10.78
PASURUAN	AGR	357	7401.97	7044.97	10.27
MADIUN	AGR	178	3656.1	3478.1	10.24
MAGETAN	AGR	165	3385.26	3220.26	10.24
SUMENEP	AGR	163	3339.59	3176.59	10.23
BREBES	AGR	209	4278.12	4069.12	10.23
PACITAN	AGR	89	1817.85	1728.85	10.22
BANGKALAN	AGR	131	2674.36	2543.36	10.22
TUBAN	AGR	165	3365.27	3200.27	10.22
NGAWI	AGR	158	3221.76	3063.76	10.21
SUBANG	AGR	1950.87	39428.26	37477.39	10.18
YOGYAKART	AGR	607	11301.5	10694.5	9.89

		1989	2020	1989–2020	Percent/year
GROBOGAN	AGR	1594.37	15031.21	13436.84	7.51
NGANJUK	AGR	1051.98	3538.59	2486.61	3.99
PAMEKAS	AGR	88	146.6	58.6	1.66

Table 21.5. Sample model results: Manufacturing production amounts (billion Rp)

		1989	2020	1989–2020	percent/year
BLITAR	SECT3	8	887353.52	887345.52	45.46
SAMPANG	SECT3	1	2374.33	2373.33	28.50
PANDEGLAN	SECT3	16724.5	13703995.35	13687270.85	24.16
SUBANG	SECT3	10	7753.41	7743.41	23.94
GARUT	SECT3	20	14452.49	14432.49	23.66
CIAMIS	SECT3	9	3897.13	3888.13	21.63
WONOSOBO	SECT3	2	665	663	20.60
BATANG	SECT3	72	22412	22340	20.34
REMBANG	SECT3	24	7028.05	7004.05	20.11
BANJARNEG	SECT3	1	262	261	19.68
KEBUMEN	SECT3	11	2164	2153	18.58
BOJONEGO	SECT3	47	8064.33	8017.33	18.05
BANYUWANG	SECT3	123	16528.55	16405.55	17.13
LEBAK	SECT3	31.86	1567.01	1535.15	13.39
CILACAP	SECT3	1344	13232	11888	7.66

Table 21.6. Sample model results: Service production amounts (billion Rp)

		1989	2020	1989-2020	percent/year
SEMARANG	SERV	622	2620901.4	2620279.4	30.90
MALANG	SERV	2198	780018.42	777820.42	20.85
SURAKARTA	SERV	230	69019.43	68789.43	20.20
TANGGERAN	SERV	186	51685.13	51499.13	19.90
BANDUNG	SERV	1787	414942.94	413155.94	19.21
BEKASI	SERV	1133	228074.71	226941.71	18.66
CIREBON	SERV	3978	624611.19	620633.19	17.72
PACITAN	SERV	14	1839.47	1825.47	17.04
BANYUWANG	SERV	349	25387.18	25038.18	14.83
SIDOARJO	SERV	1806	116434.59	114628.59	14.38
BOGOR	SERV	1121	69379.02	68258.02	14.23
JAKARTA	SERV	24438.23	1158490.13	1134051.9	13.26
GRESIK	SERV	8005.64	315717.82	307712.18	12.59
YOGYAKART	SERV	1598	49177.3	47579.3	11.69

Table 21.7. Heavy traffic corridors forecasted (year 2005)

FROM	TO	PCU's per year
SAMPANG	GRESIK	4905086987
SAMPANG	JAKARTA	2782023835
TUBAN	KUDUS	1640056746
PEMALANG	SERANG	1003434735
BLITAR	KEDIRI	894847690
CIAMIS	BANDUNG	655813642

FROM	TO	PCU's per year
PEMALANG	TANGGERAN	626213368
SAMPANG	SEMARANG	608216684
BLITAR	MALANG	564827670
PEMALANG	GRESIK	538193563
SAMPANG	BEKASI	508532458
CIANJUR	BOGOR	481657929

Table 21.8. Heavy traffic corridors forecasted (year 2020)

FROM	TO	PCU's per year
PANDEGLAN	KUDUS	43991313919
PANDEGLAN	JAKARTA	43209832377
PANDEGLAN	GRESIK	30352797399
PANDEGLAN	SERANG	29783961755
PANDEGLAN	BANDUNG	23668754436
PANDEGLAN	TANGGERAN	19009822495
BLITAR	KEDIRI	18179806026
BLITAR	MALANG	17528868816
PANDEGLAN	SEMARANG	17376562652
PANDEGLAN	BOGOR	15434811889
PANDEGLAN	PURWAKART	9176838607

References

Azis, I.J. and Kim, T.J. (1992), 'Trade and transport: Economic evaluation of road transport in Jawa for promoting exports – Final report', Inter-University Center for Economics, University of Indonesia, Jakarta.

Beckmann, M., McGuire, C.B. and Winsten, C.B. (1956), *Studies in the Economics of Transportation,* Yale University Press, New Haven, CT.

Brooke, A., Kendrick, D. and Meeraus, A. (1992), *Release 2.25: GAMS – A User's Guide*, The Scientific Press, CA.

Boyce, D.E., LeBlanc, L.J. and Chon, K.S. (1988), 'Network equilibrium models of urban location and travel choices: A retrospective survey', *Journal of Regional Science*, vol. 28, pp. 159–183.

Eash, R., Janson, B. and Boyce, D.E. (1979), 'Equilibrium Trip Assignment: Advantage and Implications for Practice', *Transportation Research Record*, vol. 728, pp. 1–8.

Evans, S.P. (1976), 'Derivation and analysis of some models for combining trip distribution and assignment', *Transportation Research*, vol. 10, pp. 37–57.

Florian, M. (1977), 'A traffic equilibrium model of travel by car and public transit modes', *Transportation Science*, vol. 11, pp. 166–179.

Florian, M. and Nguyen, S. (1978), 'A combined trip distribution modal split and trip assignment model', *Transportation Research*, vol. 12, pp. 241–246.

Hanley, P.F. (1994), 'A combined economic activity and transportation equilibrium model: A solution procedure and application to Jawa, Indonesia', Masters Thesis, Department of Urban and Regional Planning, University of Illinois at Urbana-Champaign, IL.

Kim, T.J. (1989), *Integrated Urban Systems Modeling: Theory and Applications*, Kluwer Academic Publishers, Boston, MA.

22 Road Infrastructure and Corridor Development

Sytze A. Rienstra
Piet Rietveld
Maarten T.H. Hilferink
Frank R. Bruinsma
Department of Spatial Economics, Free University,
De Boelelaan 1105
1081 HV Amsterdam, The Netherlands

22.1 Introduction

The impacts of network infrastructure are usually studied for spatial units such as cities or regions. *Cities* can be conceived as nodes in a network, the development of which is influenced by the quality of links (roads, railways, canals) and of transport nodes (airports, seaports). Thus, one can study the effect of changes in transport links (e.g. high speed rail) on the relative position of cities (Bruinsma and Rietveld, 1993).

Impacts of transport infrastructure change on *regions* can be analyzed when a transport network is connected to a spatial system in terms of zones or regions. By connecting the centroid of each region to a point in the network, changes in the network can be translated into changes in accessibility of regions. Usually, regions are defined in administrative terms.

An opportunity, which has received less attention in research, is to focus on *corridors*. Corridors can be defined as areas near to links in networks. The difference between corridors and regions is that regions are defined a priori (usually in an administrative way), and corridors can be defined in terms of the shape of a network. During the last 10–15 years the availability of data with a very high spatial detail (individual, postal code) has increased substantially. As a consequence, opportunities to analyze spatial processes by means of spatial units that can be modified according to the features of the topic studied have increased substantially. This has also led to an improvement of possibilities to study the effects of network development on corridors. This is a welcome development since by using more appropriate spatial units of analysis, research can be carried out in a more satisfactory way. A focus on corridors is also important in the eyes of policy makers. Corridors seem to be relevant elements of mental maps of countries and have an intuitive appeal. In physical planning and regional economic policies, corridors regularly appear as relevant units (RPD, 1991).

A closer look at the concept of a corridor reveals that corridors usually relate to road infrastructure. This is because road infrastructure has a relatively high number of entry and exit points, so that it can be expected that land use is affected in the whole area surrounding a road or a highway. The lower the number of entry or exit points, the less relevant the notion of a corridor becomes. For most railway lines, for example, distances between subsequent stops are so far that the corridor concept no longer applies. An exception may occur in the case of rail systems within metropolitan areas where stops are near to each other.

This chapter presents results of a broader study on the impact of highway development on corridors in the Netherlands (see also Rietveld and Bruinsma, 1998). Two spatial approaches have been used with respect to corridors. One is to use given statistical regions (Corop-regions[1]). These regions are broader than the corridors, so that they give rise to the disadvantages mentioned above. The advantage is that they allow a rather long time series from 1970 onwards. The other approach is to use postal code areas and to combine them to form corridors. This leads to an analysis of an appropriately defined spatial unit. A disadvantage is that data are available for a much shorter period.

In this paper we present mainly results on the basis of Corop-regions. Results of the more refined spatial data are reported in another paper (Bruinsma et al., 1997). A more detailed account of the results of this paper can be found in Bruinsma et al. (1994).

22.2 Theory on Economic Impacts of Transport Infrastructure

22.2.1 General Introduction

As indicated in Table 22.1, transport infrastructure investments have both temporary and permanent effects on the economy. A major temporary effect concerns the stimulation of employment and income during the construction phase via the demand side.

Table 22.1. Temporary and permanent effects of transport infrastructure investments

	Demand side	Supply side
Temporary effect	construction effects crowding out	–
Permanent effect	operations and maintenance	effect on productivity and locations of activities

This effect can be studied by means of input-output analysis and there is little controversy about it, except for the fact that one should pay attention to the question how the infrastructure is financed. A tax increase or an increase in interest rates due to government borrowing on the capital market would have a negative impact on consumption or investments, which would counter the initial demand stimulating effect of government spending. Such crowding-out effects are often ignored in regional or urban studies where the infrastructure project is considered to be 'small' compared with the size of the national economy. However, if local

[1] Corop-regions are statistical areas. There are 40 such areas in the Netherlands. The average population of a Corop-region is about 400.000 inhabitants (see Fig. 22.4).

projects are financed by means of local financial resources (local taxes), one should of course take into account the impacts of these taxes on investment behaviour of firms in the area concerned.

Another demand related effect of infrastructure occurs in the field of operations and maintenance; this effect is non-temporary. Although maintenance is not an activity which strikes the imagination of the general public as much as new infrastructure projects may, it is nevertheless an important activity (OECD, 1986), with high rates of return on investments.

In this paper, we will focus on the programme effects, i.e. the permanent effects at the supply side. Fig. 22.1 shows a number of possible effects that are relevant for an economic analysis of transport infrastructure investments: generalized transport costs, GDP, employment, environment, welfare and equity.

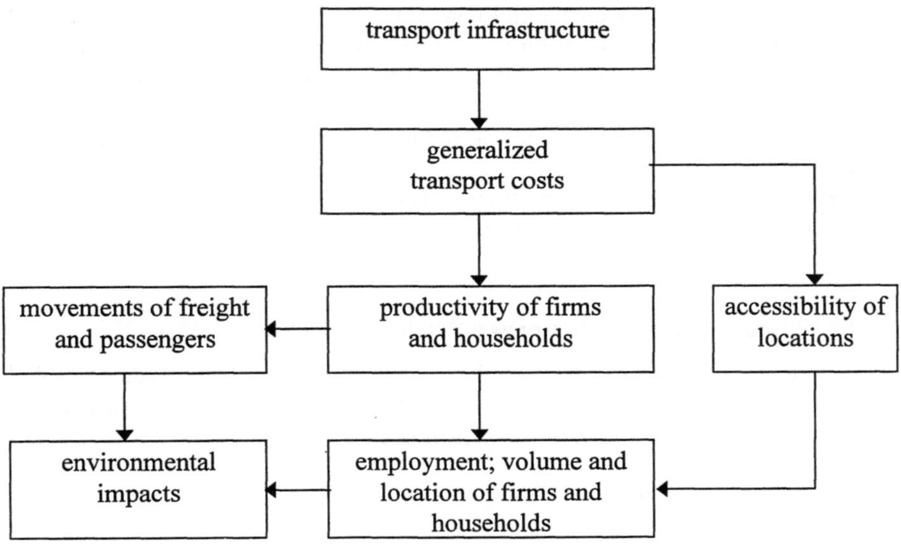

Fig. 22.1. Relationships between transport infrastructure and spatial development

Transport infrastructure investments lead to changes in *generalized transport costs* via shorter distances or higher speeds which give rise to reductions in fuel, capital and labour costs. Such changes will have impacts in the transport system in the form of mode choice, choice of time of day (in the case of congested networks), and the generation and attraction of trips per zone.

The reduction in generalized transport costs in combination with the changes in transport flows of firms lead to an increase in *productivity* in the firms concerned. This increase in productivity will manifest itself in the form of a change in value added which will, in turn, lead to a growth of gross domestic product (GDP) in the region or country concerned.

There will be several types of *environmental* impacts of road infrastructure. Development of a major highway will make road transport more competitive compared to rail and inland navigation. The resulting modal shift is disadvantageous from the perspective of transport-related emissions. In addition, it induces more transport in the form of longer trips. Also, highway construction may affect natural areas and it will lead to visual intrusion. Highway construction also has some positive environmental effects, however. It leads to changes in route choice so that usually local environmental effects are reduced in urban areas. In addition, highways may imply more continuous traffic flows having favourable effects on energy use. The balance of these effects depends strongly on the local circumstances. It falls outside the scope of this paper to discuss these environmental effects in more detail.

Effects on *employment* of infrastructure investment, among others, take place via substitution and complementarity relationships between labour, private capital and infrastructure. They also occur via differences in growth rates of economies in regions due to the differences in advantages they receive from changes in infrastructure networks.

Next, we will focus the analysis on the issue of employment impacts in more detail by discussing:
- transport infrastructure as a production factor;
- transport infrastructure and interregional/international trade.

22.2.2 Transport Infrastructure as a Production Factor

Transport infrastructure can be considered as a stock of a certain type of capital available to a region or a country. A general formulation of a production function for sector *i* in region *r*, with various types of infrastructure is:

$$Q_{ir} = f_{ir}(L_{ir}, K_{ir}, IA_r ... IM_r) \tag{1}$$

where:

Q_{ir}	value added in sector *i*, region *r*
L_{ir}	employment in sector *i*, region *r*
K_{ir}	private capital in sector *i*, region *r*
$IA_r ... IM_r$	infrastructure of various types in region *r*

After the contribution of Aschauer (1989) to the estimation of production functions, a large stream of literature has appeared on this theme. A review can be found in Munnell (1993) and Gramlich (1994).

Among the types of infrastructure distinguished are transportation, communication, energy supply, water supply, education, health services, etc. As far as transport infrastructure is concerned, it is not easy to take into account its *network* properties in the production function approach. One possibility is to distinguish various types of transport infrastructure according to their spatial range: intraregional, interregional and possibly international.

A related problem with infrastructure is that its impact may transcend the boundaries of regions. A certain region may benefit from a university or airport, even though these facilities are not located in the region itself. This problem of spatial spill-overs may be solved by using the concept of *accessibility* to certain types of infrastructure in the production function (see e.g., Johansson, 1992). An alternative approach to study the role of infrastructure in production processes is to use cost functions (Seitz, 1993).

It should be noted that accessibility is a concept that can be operationalized in various ways (cf Bruinsma and Rietveld, 1998). It indicates the potential for interactions of a certain place or region taking into account generalized costs. Since the generalized costs depend on the time of the day when congestion is relevant, accessibility itself is also time dependent. For example, the accessibility in highly urbanized regions is higher during the off-peak period compared to the peak period. In non-congested regions such a difference does not play a role.

What are the services offered by transport infrastructure that increase productivity? First, the improvement of transport infrastructure enables a reduction of the costs for the collection of inputs and the distribution of outputs. Second, improvement of transport infrastructure leads to better functioning of labour markets, which in turn may lead to a higher labour productivity. Also, productivity of private capital may increase.

Production functions of type (1) can be used to derive demand functions for labour and private capital. With respect to labour demand, two effects can be distinguished. First, an increase in the availability of infrastructure leads to a shift in the optimum allocation of labour L and private capital K at a given level of production. This shift may be such that the total level of both L and K needed to produce the same volume of output is decreased (see Fig. 22.2(a)). But it may also occur that demand for L increases and demand for K decreases (see Fig. 22.2(b)), or vice versa. In all cases, the total costs of private production factors will decrease.

In other words, the direction of the first effect of infrastructure improvement on labour demand is not certain. A second effect is that, due to the decrease in the costs of production, output can be expanded. This will under the usual assumptions lead to an increase in the demand for private production factors.

Thus, we arrive at the conclusion that when output remains constant, improvement of infrastructure may lead to a decrease in labour demand. On the other hand, a decrease in production costs can induce a higher output, which has a positive effect on employment. These results hold true when we ignore the impacts of transport infrastructure on interregional or international trade. A discussion of the latter is given in the next section.

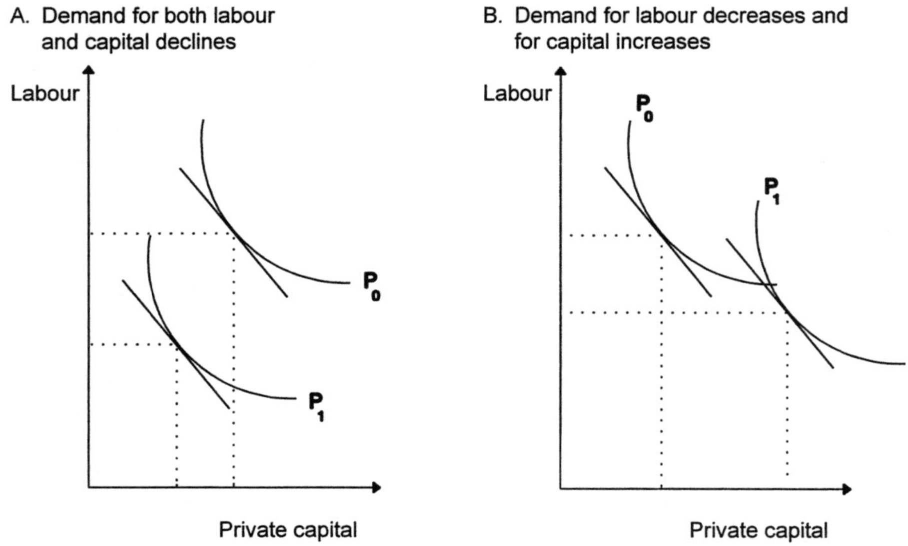

Fig. 22.2. Allocation of labour and capital before and after improvement of infrastructure

22.2.3 Transport Infrastructure and Interregional Competition

The standard model of interregional trade is illustrated in Fig. 22.3. Export takes place from region 1 to region 2 when transportation cost is less than the difference in equilibrium price for a certain good in both regions. Compared to the situation without trade an additional surplus is created consisting of areas A (accruing to producers in region 1) and B (accruing to producers in region 2). Thus, both regions benefit from trade according to the model.

Improvement of infrastructure leads to a decrease in transportation costs and hence to an increase in transportation volumes. The equilibrium price in region 1 will increase and the price in region 2 will decrease. Thus, in region 2, consumers benefit from the improvement in infrastructure, whereas producers are negatively affected. In region 1, it is the other way around. In employment terms, region 1 benefits, but region 2 is adversely affected by the improvement of transportation infrastructure.

An important aspect of the above observations is that improved transport infrastructure enables firms to make use of economies of scale in production. This leads to specialization tendencies of regional economies with positive impacts on certain sectors and negative impacts on others.

In terms of the types of infrastructure considered, it can be noted that in the productivity approach the intraregional aspects of the services of infrastructure are emphasized, whereas in the trade approach the interregional aspects come to the fore. Examples of empirical studies based on the interregional competition approach can be found in Botham (1983), Evers et al. (1987), Mills and Carlino (1989) and Plassard (1994).

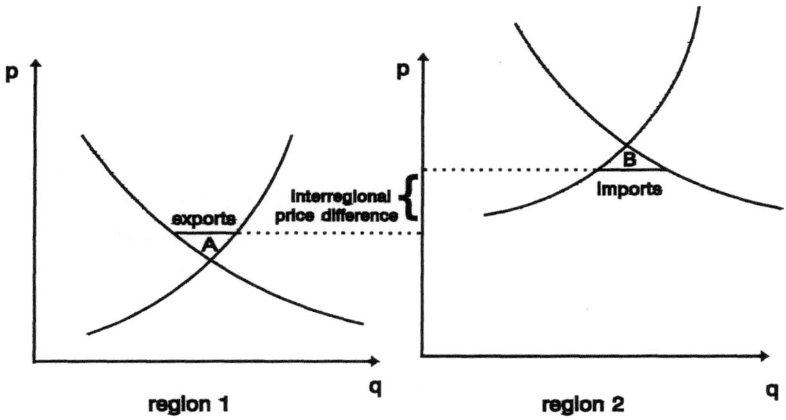

Fig. 22.3. Supply and demand in two regions

22.3 The Development of Accessibility in Different Zones in the Netherlands

22.3.1 The Development of Regional Employment in the Netherlands

Many European countries have witnessed an 'urban-rural manufacturing shift' during the past decades (Keeble et al., 1983). To analyze the urban-rural manufacturing shift in the Netherlands it is useful to divide the country into three zones: the Randstad, the intermediary zone and the peripheral zone (see Fig. 22.4).

The Randstad is a highly urbanized area, which was characterized by the highest economic and employment growth before 1970. The four largest Dutch cities are located in this region (Amsterdam, Rotterdam, The Hague and Utrecht). After 1970, a shift took place from the Randstad to the intermediary zone. This shift did not only relate to an out-migration of population from the Randstad, but also to a low rate of employment growth in this area; employment growth in the Randstad was below the national average. The intermediary zone benefited from this trend; this area showed growth rates of employment, which are above the national average (see Table 22.2).

Table 22.2. Development of employment in the Netherlands (index: 1970 = 100)

	1970	1975	1980	1985	1990
Netherlands	100	101	105	100	115
Randstad	100	98	102	97	109
Intermediary zone	100	104	109	104	123
Peripheral zone	100	99	105	100	112

Source: Own calculations based on: CBS, REJ.

Fig. 22.4. The research area

The peripheral zone was characterized by a relatively strong emphasis on agriculture in the past. Economic development in this zone stayed behind the other zones before the 1970s. In the period 1970–1990, employment growth was only slightly lower than that of the national average; in most periods, its growth was even higher than that of the Randstad.

A shift-share analysis reveals the extent to which these observations are caused by differences in the sectoral structure between the regions. The results of such an analysis based on 12 sectors are shown in Table 22.3.

The shifts in the Randstad are negative in all five-year periods. This means that the negative development of the Randstad is not caused by its sectoral composition, but by other factors. The shifts are clearly positive in the intermediary zone; here, the locational profile appears to be clearly favourable compared to the average.

Table 22.3. Relative shifts in the different zones in the Netherlands

	'70–'75	'75–'80	'80–'85	'85–'90
Randstad	-0.04	-0.02	-0.02	-0.04
Intermediary zone	0.06	0.01	0.02	0.04
Peripheral zone	1.00	0.01	0.00	-0.02

Source: Own calculations from: CBS, REJ.

22.3.2 The Development of the Accessibility in the Different Zones

Highway infrastructure in the Netherlands was mainly constructed in the period 1960–1975. After 1975, the investments in road infrastructure decreased, which was partly caused by the growing deficits of the government, but also because of the growing attention to the negative external effects of road transport (see Table 22.4).

Table 22.4. The development of length of roads with a dual carriage way[2] outside residential areas (index: 1966 = 100)

	1966	1970	1975	1980	1985	1990
Netherlands	100	168	250	296	315	372
Randstad	100	150	189	196	196	241
Intermediary zone	100	161	252	301	313	346
Peripheral zone	100	233	474	654	791	960

Source: Own calculations based on: CBS, Statistiek van de wegen.

It is interesting to analyze in which zones infrastructure construction took place most intensively. It appears that the growth rate of kilometres of highway was the highest in the peripheral zone, whereas the network in the Randstad was not extended that much. A reason for this may be that the construction of highways started earlier in the densely populated Randstad. The construction in the intermediary zone was about the national average. By comparing Table 22.4 with the preceding tables, it may be concluded that the strong emphasis on construction activities in the peripheral zone has not led to a clear improvement of the economic position of this area.

---

[2] The highway system consists exclusively of dual carriage ways.

22.3.3 Using an Accessibility Index

The above data relate to infrastructure as an input. The importance of infrastructure is that it provides services to users, however. One way to analyze this is by using an accessibility index. Such an index can be defined as (see e.g., Bruinsma and Rietveld, 1993):

$$B_i = a \sum_j M_j / c_{ij} \qquad (2)$$

where:
B_i = accessibility index for region i
a = constant
M_j = mass of zone j (mass = employment)
c_{ij} = travel time from zone i to j

The travel time is calculated by means of detailed data on speeds on many links (the Basnet system). Travel times between Corop-regions have been computed by means of a shortest route algorithm. From the definition of the accessibility measure, it is clear that there are two major reasons for changes in accessibility: changes in mass M, and changes in the network itself. In the present context, we are only interested in the changes in the network. Therefore, we computed changes in accessibility while holding the masses constant.

Table 22.5. Growth of the accessibility index in different zones (1970 = 100)

	1970	1975	1980	1985	1990
Netherlands	100	113	120	125	130
Randstad	100	112	115	118	122
Intermediary zone	100	115	122	126	132
Peripheral zone	100	115	125	133	140

From Table 22.5, it appears that accessibility in the peripheral zone grew most strongly, while that of the Randstad grew least. The differences are smaller than of the road densities as shown in Table 22.4, because the construction of roads in a zone does not only lead to an increase in its internal accessibility, but also of its external accessibility, so that other regions will also benefit. In addition, it should be noted that the accessibility figures given here are based partly on the underlying road network, the development of which has been less dynamic than the highway system. As a result, the increase in accessibility is less than the increase in the length of the highways. Again we observe that, although accessibility in the peripheral zone grew most rapidly, its relative economic position did not improve.

It should be noted that in this formulation of the accessibility index no attention is paid to the role of congestion (see Section 22.2.2). If we take into account the fact that congestion has increased during the period between 1985 and 1990 in the Randstad area, it is clear that Table 22.5 may give an overly favourable picture of the development of accessibility in this part of the country.

22.3.4 Correlations in the Accessibility Growth

We consider the question whether the growth in accessibility in the different regions is correlated during various periods. The correlation coefficients between the growth in the above-mentioned accessibility index in the different periods are given in Table 22.6.

A positive correlation coefficient for subsequent periods means that accessibility increases do not take place in a random way, but that there is a pattern whereby some regions experience above average accessibility growth and others have below average accessibility growth during successive periods.

It appears that between the periods 1975–1980 and 1980–1985 there is a positive and significant correlation between the growth in the accessibility indices. In the other periods, the correlations are not significant and have both positive and negative signs. Thus, at the level of Corop-regions, the indications are not strong that infrastructure supply leads to a systematic bias for certain regions.

Table 22.6. Correlations in the growth of the accessibility index

	'70–'75	'75–'80	'80–'85	'85–'90
'70–'75	1.00	-0.08	-0.24	0.22
'75–'80	-0.08	1.00	0.58*	0.21
'80–'85	-0.24	0.58*	1.00	0.13
'85–'90	0.22	0.21	0.13	1.00

Note: * significant correlation

22.4 Impact of the A1 Highway Construction; a Reference Region Approach

22.4.1 Introduction to the A1 Highway and the A1-Regions

The A1 is the highway that runs from Amsterdam, via Amersfoort-Apeldoorn-Almelo to the German border. In Germany, this highway (E8) goes on to Berlin and Poland. In the central parts of the Netherlands it was mainly completed in 1972 and 1973; in the eastern part (Twente) several parts were opened between 1978 and 1992. The road runs through several Corop-regions: Larger-Amsterdam, Utrecht, the Veluwe, Southwest-Overijssel and Twente (Fig. 22.4).

In order to study the impact of the A1-construction on the regional economy the reference region approach will be used in this section. In this approach we try to determine the impact of A1-construction by comparing the employment growth in the A1-regions with that of reference regions. These reference regions are comparable with the A1-regions in terms of economic structure and location, but in these regions little or no construction of infrastructure took place in the period(s) considered.

The study is carried out for three A1-regions, which are situated in the intermediary zone: the Veluwe, Southwest-Overijssel and Twente. In the observed period, the A1-regions all had different economic developments. In Table 22.7 the index shows, for example, the growth of employment.

Table 22.7. Employment growth in the Netherlands and the A1-regions (1970 = 100)

	1970	1975	1980	1985	1990
Netherlands	100	101	105	100	115
Veluwe	100	111	118	117	136
SW-Overijssel	100	100	106	98	120
Twente	100	97	99	90	104

Source: Own calculations based on: CBS, REJ.

The *Veluwe* is one of the biggest natural areas in the Netherlands, it is a popular location region for people as well as companies; the regional economy is characterized by a large share of defence employment and by a relatively high growth of the service sectors leading to a positive shift in most periods (see Table 22.8). The A1 in this region was constructed in 1972 and this highway is one of the main transport corridors in this region.

Southwest-Overijssel is a small, mainly rural region, traditionally focused on the industry sector. The growth rate of the employment lagged behind the national average until 1985, the shift was positive, however, between 1970 and 1980. After 1985, the employment growth as well as the relative shift was relatively high. The A1 was constructed in 1972 and 1973, and is the only highway in this region.

Finally, *Twente* was traditionally characterized by a strong textile-industry sector, which declined to a large extent in the period under analysis. Until 1985 this region experienced a negative employment growth, after 1985 the employment grew rapidly, however. When the growth is corrected for the sectoral structure, the region had a positive shift in three of the four five-year periods. The A1 was mainly opened between 1978 and 1992, this highway is the only one in this region.

Next, we will carry out the reference region approach in order to investigate employment impacts of the A1-construction.

22.4.2 The A1-regions Compared with the Remainder of the Intermediary Zone

In Table 22.8 a comparison is made between the relative shift in the three A1-regions and the remainder of the intermediary zone.

From Table 22.8, it appears that in the *Veluwe* the shift in the first period is much higher than in the intermediary zone. At first sight this seems to be the result of the A1-construction. When the sectoral structure is investigated, however, it appears that the high relative shift is mainly caused by growth of the defence sector, in the form of a growth of defence employment. It is not likely that this employment is attracted by the construction of the A1.

Table 22.8. Relative shifts in the intermediary zone and A1-regions

	'70–'75	'75–'80	'80–'85	'85–'90
Intermediary zone	0.05	0.01	0.02	0.04
Veluwe	**0.10**	-0.00	0.03	0.05
SW-Overijssel	**0.01**	0.02	-0.02	0.08
Twente	**0.03**	**0.02**	**-0.02**	**0.02**

Note: The period in which the A1 is opened is printed bold.
Source: Own calculations based on: CBS, REJ.

In the other periods during which construction took place, the relative shift in *Twente* is somewhat higher in the period 1975–1980. In the other relevant periods the shift is lower than in the intermediary zone, however. Also, an investigation of lagged effects does not lead to clear results on an impact of highway construction on regional employment growth. We conclude that a positive influence of A1-construction on the shift cannot be proven by using the remainder of the intermediary zone as reference region.

Although it is certainly relevant to compare the development of the A1-regions with that in the intermediary zone, it is clear this zone is not an entirely satisfactory reference region. This is because also other highway construction projects took place in the intermediary zone. Therefore, we carried out a more detailed analysis with reference regions that did not experience extensive highway construction projects in the periods concerned.

22.4.3 The A1-Regions Compared to Reference Regions

In Table 22.9, the relative shifts of the A1-regions and those of several reference regions are presented (see also Fig. 22.4 for their location).

The Achterhoek is, like the *Veluwe*, a rural region that is considered to be an attractive living and location region, in which little main road infrastructure has been constructed in the observed period. When we compare the relative shift of both regions in the period of opening of the A1 (1970–1975), we see a higher shift in the Veluwe; as already mentioned above however, this is mainly caused by a growth of defence employment. In the Achterhoek, the relative shift was also high during this period, but here it was caused by market sectors, so a clear influence of A1-construction on the employment in the Veluwe is not proven during the 1970–1975 period. The same holds for a lagged effect.

Middle-Limburg and Northeast-North-Brabant are chosen as reference regions for *Southwest-Overijssel* because they are situated in the intermediary zone and also have an economic structure with an emphasis on industry. When we comparethe relative shifts, it appears that the shift in both reference regions is much higher in the period of opening of the A1, so it follows clearly that no effect of the opening can be proven. The same holds for a lagged influence.

Table 22.9. Relative shifts in the A1- and reference regions (%)

	'70–'75	'75–'80	'80–'85	'85–'90
Veluwe	**0.10**	-0.00	0.02	0.05
ref. region: Achterhoek	*0.05*	*0.05*	*-0.04*	*0.07*
Southwest-Overijssel	**0.01**	0.02	-0.02	0.08
ref. region: NE-North-Brabant	*0.10*	*0.03*	*0.05*	*0.10*
Middle-Limburg	*0.11*	*0.09*	*0.02*	*0.06*
Twente	**0.03**	**0.02**	**-0.02**	**0.02**
ref. region: Aggl- Arnhem/Nijmegen	-0.00	-0.02	-0.03	0.01
SE-North-Brabant	0.03	-0.00	0.02	0.06

Note: The period in which the A1 is opened is printed bold.
Source: Own calculations based on: CBS, REJ.

Arnhem/Nijmegen and Southeast-North-Brabant are chosen as reference regions for *Twente*, because they are characterized by medium-sized cities surrounded by a rural area. In addition, Arnhem/Nijmegen was characterized by a restructuring of the industry, while Southeast-North-Brabant has a large emphasis on the manufacturing sector. Both regions are well supplied with main road infrastructure, in Twente relatively more infrastructure was constructed. When the relative shifts of the regions are compared, we see that the relative shifts in Twente are in every period higher than in Arnhem/Nijmegen, whereas they are in most periods lower than in Southeast-North-Brabant. Therefore, no clear conclusion can be drawn.

We conclude that no clear impact of the construction of the A1 is found by carrying out the reference region approach. Only in some cases are the relative shifts higher in the A1 regions, but for most regions this is not the case.

22.4.4 Remarks on the Reference Region Approach

It is clear that the results of the reference region approach are dependent on the choice of the reference region; but with other choices the conclusion appears to remain the same. Another problem is that this approach is rather crude, because no other specific features of the regions are considered. Clearly, in a relatively small country with only 40 statistical regions, one cannot select reference regions in such a way that they satisfy all requirements for a quasi-experimental method (Isserman et al., 1989). To meet this objection we will next estimate a model, where several region-specific factors are taken into account. Finally, it should be mentioned that since the spatial scale level of Corop-regions is rather high, it is certainly possible that an influence can be proven on a lower spatial level. This would mean that highways induce a relocation of employment *within* regions towards points of access leading to corridor effects, but that regional totals as such are not influenced.

22.5 A Regional Labour Market Approach

22.5.1 The Regional Labour Market: An Introduction

Given the limitations of the reference region approach, it is useful to estimate a model in which the various regional features are included in an explicit way. Let us take the next production function as a starting point:

$$Q_r = f(L_r, K_r, B_r, A_r) \tag{3}$$

where:

Q_r volume of production in region *r*
K_r capital in region *r*
L_r labour in region *r*
B_r accessibility of region *r*
A_r other factors in region *r*, for example, other types of infrastructure and urbanization (dis-)economies.

Transport infrastructure is represented by means of the factor B, thus it enters the production function via the services it provides as measured by means of accessibility (Forslund and Johansson, 1995). This formulation of the role of transport infrastructure in the production function allows one to take into account the spatial spill-over effects associated with the supply of infrastructure.

If we may assume that resources are allocated in such a way that regional profits are maximized given the price of the output at the world market, we arrive at a labour demand function with the following arguments:

$$L_r = h(s_r, w_r, p, B_r, A_r) \tag{4}$$

where:

s_r price of capital in region *r*
w_r price of labour in region *r*
p given price of the output

In our application we operationalize these arguments as follows:

Price of capital (s_r). The price of capital is equal for all regions with one exception: firms may receive subsidies on investments. Therefore, we define a regional policy variable to take these subsidies into account. The variable assumes the value 1 in regions receiving subsidies and the value 0 in the other regions. The sign of impact of investment subsidies on employment is uncertain. If substitution effects dominate, it will be negative, but if output effects dominate, it will be positive.

Price of labour (w_r). Because of the strong national component in the annual wage negotiations and the uniform national minimum wage policy, interregional wage differentials are very small in the Netherlands. This is one of the causes of frictions on regional labour markets. Therefore, we use the percentage of the labour force with an unemployment duration of less than one year as a proxy for the difficulty to recruit labour. Long-term unemployed are not taken into account in this measure because the probability of their recruitment is often low.

A second labour market variable taken into account is the level of education of the regional labour force.

Price of output (p). The impact of the price cannot be estimated with this model, because p is assumed to be the same for all regions.

Accessibility (B_r). This concept has already been defined in Section 22.3.3. We will use two accessibility indicators: one for domestic destinations and one for international destinations in neighbour countries to test whether the process of European economic integration has an impact on regional development. In the given formulation we may expect a positive impact of accessibility on employment. However, if we take into account the theoretical considerations given in Section 22.2, we note that substitution effects may lead to a negative impact on employment. In addition, from an interregional trade perspective, an increase in accessibility may make regions more vulnerable to competition by producers from elsewhere. In order to test the latter consideration, one may investigate whether importing sectors in a region are influenced in a different way by a change in accessibility compared with exporting sectors. Whether a sector is importing or exporting can be operationalized by means of the locational quotient.

Other regional factors (A_r). The degree of urbanization is used as an explanatory variable to account for urbanization (dis-)economies.

For a detailed definition of the variables we refer to Bruinsma et al. (1994). We estimated the model for total regional employment as well as for the separate sectors.

22.5.2 The Results of the Model for Total Regional Employment

An empirical application of the theoretical model allows one to use a formulation in terms of changes or in terms of absolute values. We have chosen the first specification. This is because with a specification in terms of absolute values one may have serious causality problems. The supply behaviour of the public sector is usually such that densely populated areas have high infrastructure densities. A positive relationship between infrastructure supply and employment does not say much therefore when one wants to investigate the contribution of infrastructure improvement to regional growth.

In our estimation (see Table 22.10) the relative shift in employment of a region during a five-year period is explained by:

acc the relative change in domestic accessibility
int the relative change in international accessibility
unemp the relative change in short term unemployment
educ the relative change in the level of education
urb the level of urbanization
pol the regional policy variable (a dummy)

It appears that the change in domestic accessibility had a significant, negative impact on the employment in the period 1970–1975, while in the period 1985–1990 this impact was positive. In the other periods no significant impact is proven. International accessibility does not have a significant impact on regional employment. So a clear impact of a change in accessibility on the shift in regional employment cannot be proven.

Of the other variables the change in unemployment rate has the expected significant sign in two five-year periods. The level of education does not have the expected impact. The level of urbanization does have a significant, negative sign in three periods, so the urban-rural manufacturing shift (see Section 22.3) seems to have more influence than agglomeration effects. The regional policy has a negative, significant sign in three periods, which may mean that the substitution effect is larger than the output effect. An alternative interpretation is that the regions receiving government support are regions with structural weaknesses, not captured by the other variables. Government policies are not strong enough to overcome these structural weakness and as a consequence, a negative sign is found for the policy variable.

Table 22.10. Estimation results of the model for total regional employment (relative shift)

	'70–'75		'75–'80		'80–'85		'85–'90	
	Coeff	*t*	Coeff	*t*	Coeff	*t*	Coeff	*t*
acc	-0.94	-2,07*	0.74	0.81	0.22	0.54	2.12	2.11*
int	0.08	0.51	0.20	0.28	1.47	1.26	-0.67	-1.08
unemp	-0.01	-1.10	0.22	4.86*	0.05	2.21*	0.15	1.18
educ	-0.54	-3.36*	-0.09	-0.66	0.03	0.30	-0.07	-0.73
urb	-0.65	-5.04*	-0.15	-1.15	-0.19	-2.96*	-0.28	-2.14*
pol	-0.10	-1.92*	-0.04	-0.91	-0.06	-2.55*	-0.07	-1.75*
Cst	0.51	5.36*	-0.02	-0.23	0.02	0.33	0.00	0.05
R^2	0.52		0.59		0.50		0.26	

Note: * significant at the 5% level

22.5.3 Results for Separate Sectors: The Transport and Communication Sector

One sector where a significant impact of a change in accessibility on employment growth may be expected is the transport and communication sector. We estimated the model with the above-mentioned variables. The results are given in Table 22.11.

Domestic accessibility has in the three periods between 1975 and 1990 a significant, positive influence on employment in this sector. The same holds for international accessibility in the periods 1970–1975 and 1985–1990. We conclude that the change in accessibility has a positive impact on the employment in this sector.

Table 22.11. Results of the model for the transport and communication sector (relative employment growth)

	'70–'75		'75–'80		'80–'85		'85–'90	
	Coeff	*t*	Coeff	*t*	Coeff	*t*	Coeff	*t*
acc	-0.44	-0.54	2.45	0.89	3.45	2.12*	5.54	2.37*
int	0.54	1.95*	-1.76	-0,80	10.29	2.26*	-1.81	-1.25
unemp	0.08	0.29	-0.22	-0.54	0.26	0.78	-0.18	-0.74
educ	-0.00	-0.36	0.31	2.23*	0.19	1.99*	0.74	2.51*
urb	-0.11	-1.27	0.06	0.41	0.21	-2.13*	-0.24	-2.48*
pol	-0.04	-0.17	-0.21	-0.54	-0.34	-1.39	-0.90	-2.97*
Cst	0.04	0.27	0.16	0.74	-0.04	-0.15	0.37	1.72*
R^2	0.11		0.26		0.55		0.39	

Note: * significant at the 5% level

The other variables give about the same results as the model for total employment: the change in unemployment rate has a positive influence on employment growth in this sector; the regional policy has twice a significant, negative sign and the level of urbanization has once a negative, significant sign. The education level does not have a significant influence on the employment of this sector.

The estimates for the other sectors are given in Bruinsma et al. (1994). In general it can be said that with individual sectors the explained part of total variance (R^2) is lower than for total regional employment. Domestic as well as international accessibility have only incidentally a significant influence on employment growth in the other sectors.

In an alternative specification we tested whether the accessibility variable has a different impact for importing and exporting sectors, but this did not yield better results for the individual sector estimates. We conclude that the transport and communication sector is the only one where an influence can be proven on the Corop-level.

22.6 Conclusions

In this contribution we have investigated the link between highway construction and the development of regional employment in the Netherlands during the past decades. In terms of employment growth, the intermediate zone has been most successful during the period considered. However, the increase in accessibility has been largest in the peripheral zone and this indicates that there is no simple mono-causal relationship between the development of accessibility and employment.

In the reference region approach we compared A1-corridor regions with reference regions, which had similar locational features and sectoral structure, but had less highway construction during the construction of the A1. On a comparison of the relevant shift in employment no clear impact of highway construction on regional employment could be proven, the shift in the A1-regions only incidentally was higher than in the control region.

Finally, a regional labour market model was estimated with regional features as well as domestic and international accessibility as variables. For total employment no unambiguous impact could be proven; the impact of the domestic accessibility was significantly negative in one period and significantly positive in another period. Only the transport and communication sector showed a clear positive impact.

From this empirical research, we conclude that there is no evidence that the construction of main road infrastructure and the resulting change in accessibility has a clear impact on overall employment in regions. This conclusion holds for the spatial level of Corop-regions. This is an important result for regional politicians who often propose investments in road infrastructure because they believe that this will boost regional employment growth. It is certainly possible, however, that a positive impact on the regional employment may be found in a corridor analysis at a lower spatial level. This is the subject of another paper (Bruinsma et al., 1997).

It is important to realize that our research scope has been rather narrow in this paper. By focusing on employment effects we have ignored the productivity and transport cost effects which receive ample attention in the cost-benefit analyses usually carried out for large infrastructure projects. The result that we did not observe substantial employment effects does, of course, not imply that the other effects will be small as well. Note also that a cost-benefit analysis would entail the inclusion of environmental effects which have been ignored in our study.

A point that deserves particular attention in subsequent research is the impact of traffic congestion on accessibility. Since travel time is an important component of generalized costs, accessibility becomes dependent on the time of the day (peak versus off-peak). In addition, the notion of generalized costs also includes the costs related to uncertainty of travel time. Transport infrastructure investments in congested areas do not only imply lower travel times but also lower uncertainty costs. Inclusion of these notions in accessibility measures would be most welcome.

Acknowledgment

The research on which this paper is based has been funded by the project bureau IVVS. The authors thank Peter Jorritsma and Gijsbert Borgman for their useful advice during the course of the project.

References

Aschauer, D.A. (1989), 'Is public expenditure productive', *Journal of Monetary Economics*, vol. 23, pp. 177–200.

Botham, R. (1983), 'The road programme and regional development: The problem of the counterfactual' in Button, K. and Gillingwater, D. (eds), *Transport Location and Spatial Policy*, Gower, Aldershot, pp. 23–56.

Bruinsma, F.R. and Rietveld, P. (1993), 'Urban agglomerations in European infrastructure networks', *Urban Studies*, vol. 30, pp. 919–934.

Bruinsma, F.R. and Rietveld, P. (1998), 'The accessibility of European cities', *Environment and Planning A*, vol. 30, pp. 499–521.

Bruinsma, F.R., Rienstra, S.A and Rietveld, P. (1997), 'Economic impacts of the construction of a transport corridor; A multi-level and multi-approach case study for the construction of the A1 highway in the Netherlands', *Regional Studies,* vol. 31, pp. 391–402.

Bruinsma, F.R., Rienstra, S.A., Rietveld, P. and Hilferink, M.T.H. (1994), *De Structurerende Effekten van Infrastructuur op Interregionaal Niveau langs Verbindingsassen; Een Analyse op Corop-niveau*, RE-VU study for Pb IVVS, Amsterdam.

Centraal Bureau voor de Statistiek (CBS) (1970–1992), *Regionaal Economische Jaarcijfers (REJ)* Voorburg/Heerlen.

CBS (1966–1992), *Statistiek van de Wegen*, Voorburg/Heerlen.

Evers, G.H.M., Meer, P.H. van der, Oosterhaven J. and Polak, J.B. (1987), 'Regional impacts of new transport infrastructure: A multisectoral potentials approach', *Transportation*, vol. 14, pp. 113–126.

Forslund, U.M. and Johansson, B. (1995), 'Assessing road investments: Accessibility changes, cost benefit and production effects', *The Annals of Regional Science*, vol. 29, pp. 155–174.

Gramlich E.M. (1994), 'Infrastructure investment: A review essay', *Journal of Economic Literature*, vol. 32, pp. 1176–1196.

Isserman, A.M., Rephann, T. and Sorenson, D.J. (1989), 'Highways and rural economic development: Results from a quasi-experimental approaches', Seminar on Transportation Networks and Regional Development, Leningrad.

Johansson, B. (1992), *Infrastructure, Accessibility and Economic Growth*, Department of Regional Planning, Royal Institute of Technology, Stockholm.

Keeble, D., Owens, P.L. and Thompson, C. (1983), 'The urban-rural manufacturing shift in the European Community', *Urban Studies*, vol. 20, pp. 405–418.

Mills, E.S. and Carlino, G. (1989), 'Dynamics of county growth', in Andersson, Å.E., Batten, D., Johansson, B. and Nijkamp, P. (eds), *Advances in Spatial Theory and Dynamics*, North-Holland, Amsterdam, pp. 195–205.

Munnell, A.H. (1993), 'An assessment of trends in and economic impacts of infrastructure investment', in *Infrastructure Policies for the 1990s*, OECD, Paris.

OECD (1986), *Infrastructure Investments and their Contribution to Employment and Economic Growth*, Paris.

Plassard F. (1994), 'High speed transport and regional development', Paper presented to Round Table no. 94 on 'Regional Policy, Transport Networks and Communications', ECMT, OECD, Paris.

Rietveld, P. and Bruinsma, F.R. (1998), *Is Transport Infrastructure Effective? Transport Infrastructure, Accessibility and the Space Economy*, Springer-Verlag, Berlin.

Rijksplanologische Dienst (RPD) (1991), *Perspectieven in Europa*, The Hague.

Seitz, H. (1993), 'A dual economic analysis of the benefits of the public road network', *The Annals of Regional Science*, vol. 27, pp. 223–239.

MIX
Papier aus verantwortungsvollen Quellen
Paper from responsible sources
FSC® C105338

If you have any concerns about our products,
you can contact us on
ProductSafety@springernature.com

In case Publisher is established outside the EU,
the EU authorized representative is:
Springer Nature Customer Service Center GmbH
Europaplatz 3, 69115 Heidelberg, Germany

Printed by Libri Plureos GmbH
in Hamburg, Germany